Advances in Atom and Single Molecule Machines

More information about this series at http://www.springer.com/series/10425

Dimas G. de Oteyza · Celia Rogero
Editors

On-Surface Synthesis II

Proceedings of the International Workshop On-Surface Synthesis, San Sebastián, 27–30 June 2016

Springer

Editors
Dimas G. de Oteyza
Donostia International Physics Center
San Sebastián
Spain

and

Materials Physics Center
Centro de Física de Materiales CSIC-UPV/EHU
San Sebastián
Spain

and

Ikerbasque, Basque Foundation for Science
Bilbao
Spain

Celia Rogero
Materials Physics Center
Centro de Física de Materiales CSIC-UPV/EHU
San Sebastián
Spain

ISSN 2193-9691 ISSN 2193-9705 (electronic)
Advances in Atom and Single Molecule Machines
ISBN 978-3-030-09331-0 ISBN 978-3-319-75810-7 (eBook)
https://doi.org/10.1007/978-3-319-75810-7

Printed on acid-free paper

This Springer imprint is published by the registered company Springer International Publishing AG part of Springer Nature
The registered company address is: Gewerbestrasse 11, 6330 Cham, Switzerland

Preface

Due to the rise of new technologies building on ever smaller structures, there is an increasing need of control in the construction of architectures with atomic precision. Molecular materials are very promising candidates, for which the ultimate goal is first being able to design them with pre-defined properties according to our needs, and then, synthesizing them with atomic precision to allow their use in the ultimate device implementation. In this context, on-surface synthesis is appearing as a revolutionary production method at the nanoscale.

Zero-, one-, or two-dimensional materials, such as quantum dots, nanoribbons, or covalent organic frameworks, respectively, are nowadays routinely synthesized in solution or with different top-down methods. However, the fabrication of molecular electronic or spintronic devices requires their controllable arrangement, as well as well-defined connections between the molecular part and the contacts or substrates. Profiting from the clean environment, ultra-high vacuum and surface science in general stand as ideal platforms for the atomically precise creation and characterization of such functional structures. Thus, the use of molecular building blocks that link covalently into functional materials, driven by self-assembly, is an extremely powerful bottom-up strategy toward the controllable construction of nanoarchitectures.

This very attractive and relatively new concept is drawing the interest of an increasing number of researchers from different fields. Organic chemistry, condensed matter physics, or electronic engineering are only some examples of the disciplines that interact within this emerging field. Important efforts are being devoted to augment the still scarcely equipped on-surface synthesis toolbox, to improve the understanding of the chemical reaction mechanisms on the surfaces, which may in turn feed back into the design and synthesis of new optimized precursors, as well as to the characterization of molecular architectures of potential functionality in diverse applications. In this context, this book comprises contributions from expert researchers that have participated in the second edition of the

"On-Surface Synthesis" international workshop celebrated in San Sebastian, Spain, in June 2016, providing a state-of-the-art description of the current activities and research directions being followed in this blooming field.

San Sebastián, Spain

Dimas G. de Oteyza
Celia Rogero

Contents

Mechanistic Insights into Surface-Supported Chemical Reactions

Alexander Riss

Abstract Its excellent spatial resolution makes scanning probe microscopy a capable method to investigate chemical reactions at the single-molecule level and obtain fascinating and unprecedented insights into the mechanisms of chemical transformations. Particularly exciting are recent advances in atomic force microscopy that allow bond-resolved imaging and thus make the chemical identification of organic molecular reaction intermediates and products possible. In this chapter we will give an overview about recent fundamental research on reaction mechanisms and kinetics of surface-supported reactions by scanning probe microscopy. Particular emphasis will be placed on the stabilization and statistical analysis of intermediates, which provides fundamental understanding of the microscopic driving forces of complex chemical transformations of organic molecules.

1 Introduction

One of the greatest challenges in fundamental as well as technological research in chemistry is the determination of mechanistic aspects of reactions of organic molecules. Design of novel synthesis protocols and thus controlled fabrication of new materials critically rely on the understanding of reaction mechanisms. Information about the reaction mechanisms can be obtained from measurements of the time-dependence of the concentration of reactants and products of a chemical reaction. But such analysis of chemical kinetics cannot uniquely determine the precise sequence of transformation steps, because certain details of the mechanisms, such as the occurrence of intermediate steps, might be overlooked [1, 2].

Therefore—particularly for complex organic reactions—it is crucial to identify reaction intermediates and their respective structures. In chemistry, structure identification is commonly based on spectroscopic techniques, such as nuclear

A. Riss (✉)
Physics Department E20, Technical University of Munich,
James-Franck-Str.1, 85748 Garching, Germany
e-mail: a.riss@tum.de

D. G. de Oteyza and C. Rogero (eds.), *On-Surface Synthesis II*,
Advances in Atom and Single Molecule Machines,
https://doi.org/10.1007/978-3-319-75810-7_1

magnetic resonance, infrared spectroscopy, and mass spectrometry. These techniques provide a plethora of information about the chemical (and electronic) structure and in most cases yield unique fingerprints for different species. However, chemical reactions of organic molecules commonly yield reaction mixtures containing different chemical species, which hampers the use of such ensemble-averaging techniques. The respective reaction intermediates are difficult to separate from the reaction mixture and often are not available in sufficiently high concentrations. Also, for unknown intermediate species (which in many cases are highly reactive), reference spectra are not available and thus structure determination can be difficult.

Recently, progress has been made in the visualization of intermolecular reactions by aberration corrected transmission electron microscopy in real space [3, 5–9]. Reactions can be induced via the electron beam and followed in situ. The rather high energy of the electron beam imposes requirements on the stability of the chemical species to be investigated, as well as on the support. In a recent study, hetero-atoms (such as Cl and S) have been introduced for enhanced contrast for single-molecule imaging and higher stability under the electron beam [3] (Fig. 1a, b).

Scanning probe techniques, scanning tunneling microcopy (STM) and atomic force microscopy (AFM), provide superior spatial resolution for single-molecule investigations of chemical transformations [10–16] (Fig. 1c–e). Recent advances in AFM with functionalized tips allow to image the chemical structure of organic molecules [17–20] making this technique particularly suitable for investigations of on-surface chemistry. One of the main drawbacks of this technique, however, is the relatively long acquisition time that is needed to record images with chemical-structure resolution. In principle it is possible to control the reaction rate of a chemical transformation by variation of the reaction temperature, which opens up the possibility to investigate reactions in situ. In practice, however, it is substantially more difficult to overcome experimental difficulties associated with tip functionalization and sample drift when scanning at non-cryogenic temperatures. Another possibility to follow reactions in situ and even observe intermediates is to trigger reactions with light [21–25], by a change of the chemical potential of the surface by gating [26, 27], or via the scanning probe tip [11, 28–33] (different mechanisms have been reported, such as the creation of electronically and/or vibrationally excited states in the molecules, as well as force-induced chemical changes [28, 33]). However, so far this has only been demonstrated for relatively simple chemical transformations. Thus, a major research focus is put on thermally induced chemical transformations that allow exploration of a wide variety of chemical synthesis routes.

To investigate reaction mechanisms of thermally induced chemical transformations, it is necessary to stabilize potential reaction intermediates. As will be discussed below, whether and which intermediates are stabilized depends on a variety of parameters: the potential-energy landscape, energy flow at the microscopic scale, as well as entropic effects. One of the main parameters that can be controlled is the reaction temperature, the adjustment of which can be used to slow down reaction

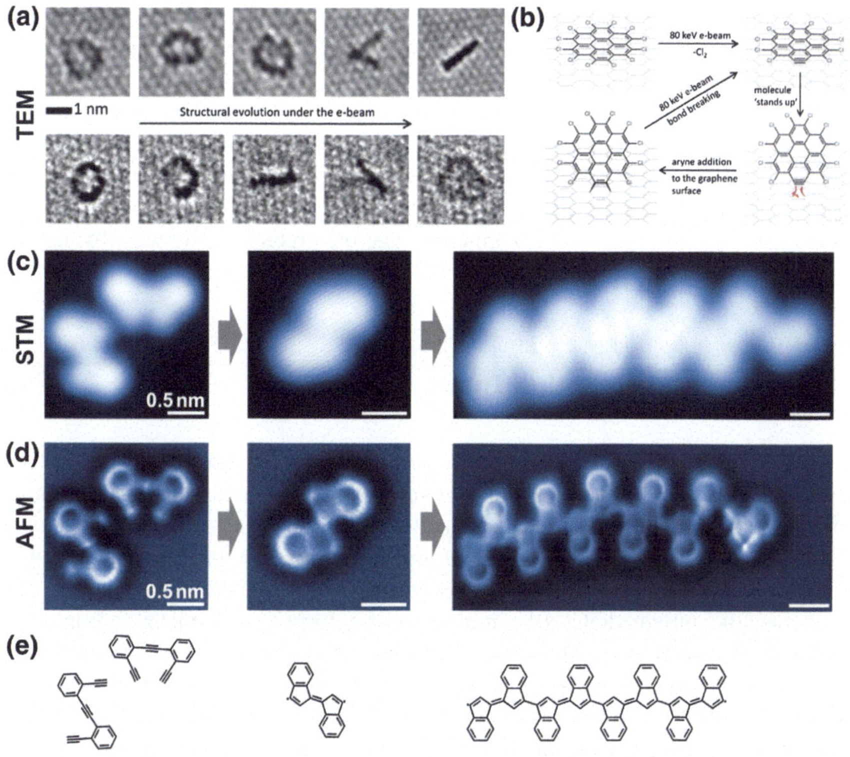

Fig. 1 Real-space imaging of chemical reactions at the single-molecule level by TEM, STM, and AFM. **a** Aberration-corrected high-resolution transmission electron microscopy (AC-HRTEM) images showing the transition of perchlorocoronene from a face-on orientation to an edge-on orientation on graphene. This reaction is triggered by the 80 keV electron beam and has been recorded in situ [3]. **b** Reaction scheme derived from the experimental data and additional DFT calculations shows cleavage of C–Cl bonds by the electron beam and subsequent Diels-Alder cycloaddition of the formed aryne to the graphene substrate [3]. **c** STM and **d** AFM images along with **e** the derived chemical structure models show the thermally induced cyclization of 1,2-bis (2-ethynylphenyl)ethyne on Au(111) towards an (*E*)-1,1′-bi(indenylidene) diradical, which then reacts towards *poly*-(*E*)-1,1′-bi-(indenylidene) chains via radical step growth [4]. **a**, **b** were included with permission from Ref. [3]. Copyright 2017 American Chemical Society; original figure is available under the terms of the ACS AuthorChoice license

rates and in certain cases lead to "freezing" of the reaction at different steps (see for example [34–45]). Furthermore, it is possible to chemically quench a reaction at a certain step, for instance by use of additional reactants that are dosed to passivate (reversibly or irreversibly) specific reactive centers of intermediates along the reaction pathway.

In the following, a few examples will be shown demonstrating what level of understanding can be achieved based on scanning probe investigation of surface-supported reactions. In most of these examples, the experimental studies are corroborated by theoretical calculations [46]. Based on data gained through experiments, calculations can unveil chemical and physical details of molecular behavior at length and time scales that are not accessible experimentally. In the context of the investigation of reaction mechanisms and intermediate species it is particularly important to provide compelling evidence that the theoretical simulations are indeed resulting in (or based on) a relevant reaction path, i.e. a reaction path that is observed in experiment. Such crucial connections between theory and experiment can be established indirectly (for instance by comparing simulated and experimentally observed reaction kinetics) and/or, preferably, directly by identifying reaction intermediates.

2 Imaging and Counting Intermediates

In a recent study the reaction pathway of Ullmann-type coupling between bromotriphenylene molecules on a Cu(111) surface was elucidated using AFM measurements supported by theoretical calculations [37]. The exact reaction mechanism of the Ullmann reaction, i.e. coupling of aryl halides, is still under debate [10, 47–51]. Of particular interest is the occurrence and nature of radicals and organometallic intermediates (the molecules can form carbon-metal-carbon moieties with surface atoms), which can shed light on the catalytic role of substrate.

Intermediates of the transformation and coupling of bromotriphenylene molecules can be observed when subjecting the sample to a series of annealing steps to temperatures between 100 and 660 K (Fig. 2). After each annealing step, the sample was cooled down to cryogenic temperatures for AFM experiments (high-resolution AFM measurements are commonly performed at 4–6 K with CO modified tips [17, 18, 37, 52]). When the sample temperature is kept below 100 K, the precursor molecule bromotriphenylene can be detected (labeled "Br–TP" in Fig. 2). Increase of the sample temperature triggers cleavage of the bromine atom to form triphenylene molecules (labeled "TP" in Fig. 2) that—as the asymmetric distortions and local reduction of the brightness in the AFM images suggest—exhibit radical character. In the next step, at temperatures closer to room temperature, where TP molecules can more freely diffuse across the surface, intermolecular coupling leads to organometallic intermediates (labeled "TP–Cu–TP" in Fig. 2). These exist in two configurations: *cis* and *trans*. In both cases, the experimental data in conjunction with theoretical calculations shows that the molecular units are connected by Cu adatoms (as opposed to surface atoms). Finally, when the sample is heated above 500 K, bistriphenylene ("TP–TP") are formed via transformation of the C–Cu–C bond into a covalent C–C bond.

Statistical analysis of the observed chemical species as a function of annealing temperature is shown in Fig. 2c. The precursor molecule "Br–TP" can only be observed when the sample temperature is kept below 300 K. The intermediates "TP" and "TP–Cu–TP" start to form at sample temperatures of 100 and 200 K, respectively. The reaction product "TP–TP" starts to form at 500 K. At 600 K the majority of the molecules have been transformed into "TP–TP".

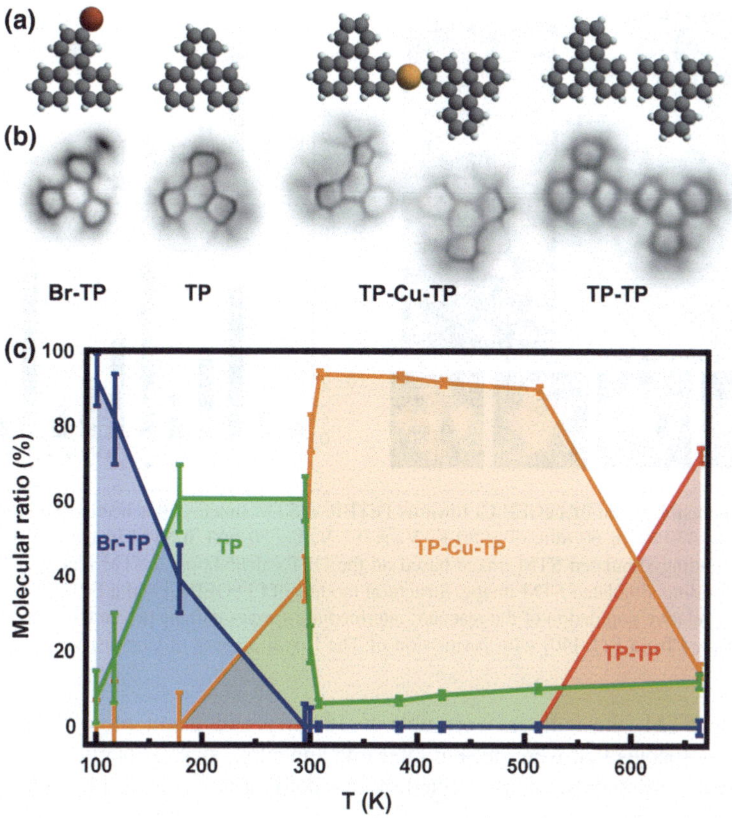

Fig. 2 Ullmann-type coupling reaction of Br–TP via TP and TP–Cu–TP towards TP–TP. **a** Chemical structure models and **b** corresponding AFM images show the stepwise transformation that includes an intermediate with radical character ("TP") and an organometallic intermediate ("TP–Cu–TP"). The *trans* configurations of TP–TP and TP–Cu–TP are shown, *cis* configurations were observed as well. **c** The relative counts of the respective chemical species were determined as a function of annealing temperature. Adapted with permission from Ref. [37]. Copyright 2017 American Chemical Society

A recent scanning tunneling microscopy study investigated the kinetics of the transformation of 2,3,7,8,12,13,17,18-octaethylporphyrin Fe(III) chloride (FeOEP-Cl) towards iron-II-tetra-benzo-porphyrin (FeTBP) on a Cu(111) surface [40, 53]. The reaction proceeds via a dechlorination step and dehydrogenation of the molecule's eight terminal ethyl groups. Dehydrogenation of an ethyl group leads to the formation of a vinyl group. When neighboring ethyl group are dehydrogenated, benzene rings are formed. Even though only these few types of reaction steps occur, 22 intermediate states are found depending on the sequence of these steps (Fig. 3). These transformations give rise to characteristic changes of molecular features, based on which the different species can be identified by STM measurements (with support of theoretical simulations).

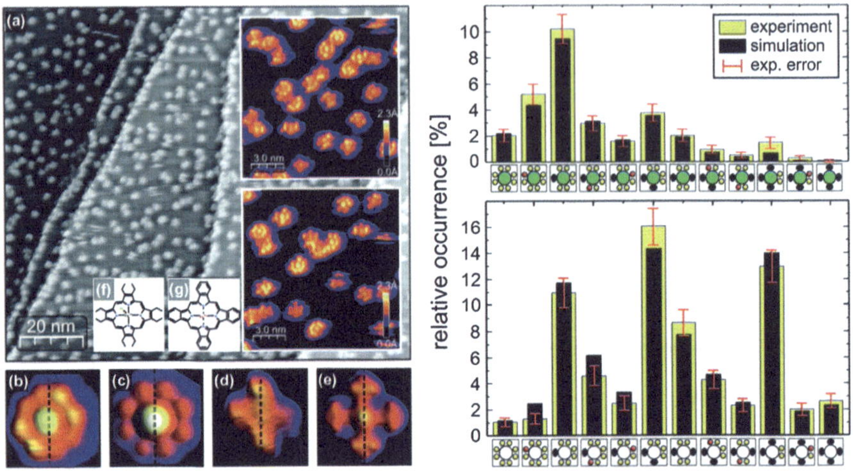

Fig. 3 Transformation of FeOEP-Cl towards FeTBP. **a** STM images after heating FeOEP-Cl on Cu(111) to 430 K for 60 min. (T = 80 K, V_S = 0.2 V, I = 50 pA). **b** STM image of FeOEP-Cl, **c** corresponding simulated STM image based on the DFT calculations. **d** STM image of FeTBP, **e** corresponding simulated STM image. Structural models of **f** FeOEP-Cl and **g** FeTBP. The graph shows the relative population of the reactant, intermediate species and the product after the heating step. Adapted from Ref. [40] with permission of The Royal Society of Chemistry

Using STM, the authors were able to identify and count more than 10,000 molecular species that were present after one annealing step. This gives a snapshot of the reaction kinetics, which is fitted to a model that is based on a few reasonable assumptions about the reaction sequence and its energetics (these assumptions are needed to reduce the number of free fitting parameters). Based on this fit, differences in activation energies of the respective types of reaction steps (vinyl group formation of chlorinated molecules, benzene ring formation of chlorinated molecules, dechlorination, vinyl group formation of dechlorinated molecules, benzene ring formation of dechlorinated molecules) can be estimated.

The prowess of single molecule bond-resolved AFM measurements is shown in recent studies on enediyne cyclization reactions [4, 29, 35, 36, 54]. These reactions are more complex than the reactions presented above, as multiple chemical bonds are affected in such transformations. Furthermore, different reaction pathways can occur leading to a complex reaction mixture that needs to be analyzed molecule by molecule [4, 36, 54, 55].

On a Ag(100) surface, 1,2-bis(2-ethynyl phenyl)ethyne (**1**) can undergo intermolecular coupling and different types of intramolecular cyclization reactions upon thermal annealing. These transformations yield a plethora of different chemical species that were analyzed using bond-resolved AFM using CO-functionalized tips [36]. The major focus of this study was put on the dimeric species, which are formed through monomer coupling by formation of covalent bonds between the terminal alkyne groups of two precursor molecules (**1**). Different chemical coupling

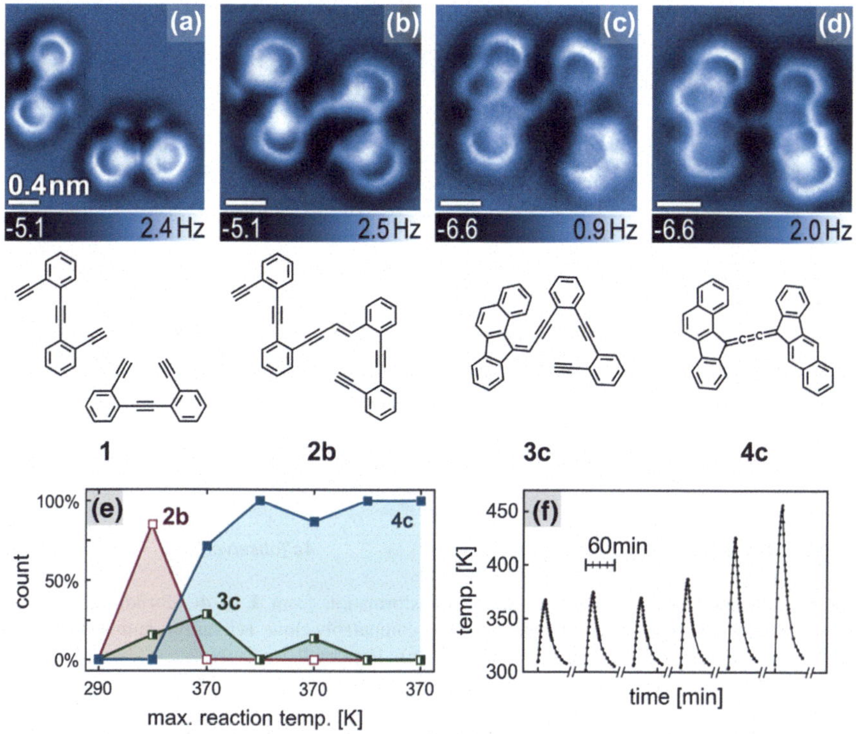

Fig. 4 Thermally induced intermolecular coupling and intramolecular cyclization of enediyne molecules. **a–d** AFM images show the reaction pathway from the precursor molecules **1** via the intermediate species **2b** and **3c** towards the product **4c**. **e** The relative abundance of the experimentally observed species after each annealing step was determined by single-molecule counting. **f** Sample temperature measured for each annealing cycle. Adapted by permission from Macmillan Publishers Ltd: Nat. Chem. 8, 678 (2016), copyright 2016

of the precursors (**1**) and different types of cyclization reactions can occur giving rise to a complex reaction mixture consisting of a wide variety of chemically distinct dimers. However, the dimers can be grouped based on their cyclization grade: uncyclized dimers (dimers that consist of two uncyclized subunits), half-cyclized dimers (dimers that consist of one cyclized subunit and one uncyclized monomer subunit) and fully cyclized dimers (dimers that consist of two cyclized monomer subunits). With each annealing step the relative ratio of these different forms of dimers gradually shifts towards increasing cyclization grades.

Figure 4 shows molecular species representing one of many competitive reaction pathways, identified by high-resolution AFM. An intermolecular C–C bond is formed between two precursor molecules (**1**) leading to the eneyne intermediate **2b**. Via C1–C6 and C1–C5 cyclizations and hydrogen transfers within **2b**, the benzo[a] fluoren intermediate **3c** is formed. Another cyclization and hydrogen transfer

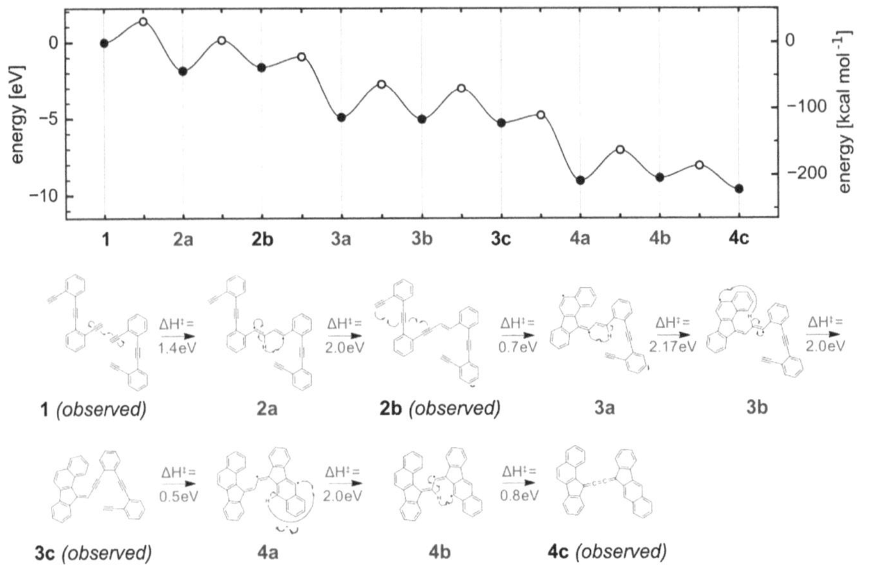

Fig. 5 Calculated energy landscape of the transformation from **1** to **4c**. Strikingly, the two experimentally observed intermediates exhibit comparably low activation enthalpy barriers ($\Delta H^{\ddagger}$ = 0.7 eV for **2b** and $\Delta H^{\ddagger}$ = 0.5 eV for **3c**). Reprinted by permission from Macmillan Publishers Ltd: Nat. Chem. 8, 678 (2016), copyright 2016

sequence leads to formation of the cumulene **4c**, the final product of this reaction sequence.

The relative abundance of the species associated with this reaction pathway was determined by identification of individual species in the complex reaction mixture via AFM and single-molecule counting. As can be seen in Fig. 4e, sequential annealing of the sample initially causes an increase of the relative abundance of the intermediate species (**2b** and **3c**) and eventually leads to the prevalence of the reaction product **4c** (Fig. 4e, f).

Theoretical calculations reveal the reaction barriers associated with every transformation step in the reaction pathway from **1** to **4c** (Fig. 5). Surprisingly, the experimentally observed intermediates **2b** and **3c** exhibit the lowest transformation enthalpy barriers: $\Delta H^{\ddagger}$ = 0.7 eV for **2b** and $\Delta H^{\ddagger}$ = 0.5 eV for **3c**. This can at first seem like a contradiction to the experiment. However, the next section will give an in-depth look at why intermediates are stabilized and which other factors need to be considered to assess reaction kinetics of surface-supported reactions.

3 Why Are Intermediates Stabilized?

The concentration c_I of a particular reaction intermediate I is increased by reaction steps leading towards intermediate I, and at the same time decreased by any possible reaction steps, in which the intermediate I is transformed into other intermediates or a reaction product. For simplicity we will restrict our considerations to reactions consisting of a sequence of unimolecular transformations of first order without side-reactions (the main insights obtained for this simple case are also relevant for more complex reaction types). In this case the change of the concentration of the reaction intermediate I with time (t) can be written as:

$$\frac{dc_I}{dt} = k_{I-1} \cdot c_{I-1} - k_I \cdot c_I \quad (1)$$

where c_I and c_{I-1} are the concentrations of the intermediate I, as well as the preceding intermediate $I-1$, while k_I and k_{I-1} are the respective rate constants [2]. The temperature-dependence of the rate constants can be expressed by the Arrhenius equation [56–59]:

$$k = A \cdot \exp\left(-\frac{\Delta E_A}{k_B T}\right) \quad (2)$$

Here, A is the so-called preexponantial factor, ΔE_A is the activation energy for the respective transformation step, k_B is the Boltzmann constant and T is the temperature. Using the Eqs. (1) and (2), one can establish a system of coupled differential equations that—even for the simple case of a sequence of unimolecular transformations of first order without side-reactions—can give rise to complex reaction kinetics, where stabilization of potential intermediates is strongly influenced by the applied reaction temperature.

However, this apparently simple description of reaction kinetics hides certain intricacies. In particular, the local temperature T can deviate from the "macroscopic" sample temperature due to energy release associated with chemical transformations in previous reactions steps. Particularly for surface-supported reactions, this can lead to complex dynamics demanding further refinements of the kinetic modelling. Energy dissipation to the surface can occur at the same timescale as subsequent chemical transformations. The efficiency of the energy exchange between substrate and adsorbed molecular species can strongly vary dependent on the chemical structure and adsorption configuration of the respective intermediates [36].

Furthermore, entropy changes along the reaction pathway need to be taken into consideration. These can stem from vibrational entropy changes of the respective species, as well as from changes in rotational and or/translational entropy of the adsorbates. Similarly, desorption of atomic or molecular species can be associated with a substantial entropy change. Transition state theory provides a way to include the effect of entropy changes on the reaction kinetics [59–62]. According to the Eyring equation, the rate constant k can be expressed as [61]:

$$k = k_B T/h \cdot \exp\left(\frac{\Delta S^{\ddagger}}{k_B}\right) \cdot \exp\left(-\frac{\Delta H^{\ddagger}}{k_B T}\right) \tag{3}$$

where h is Planck's constant, $\Delta S^{\ddagger}$ is the activation entropy and $\Delta H^{\ddagger}$ is the activation enthalpy for the respective reaction step.

4 Bringing It All Together

The influence of microscopic energy dissipation and entropy changes along the reaction pathway was investigated in detail for the case of the enediyne coupling and cyclization cascade from **1** to **4c** [36]. According to the reaction pathway shown in Fig. 5, a system of temperature-dependent kinetic rate equations can be established [see Eqs. (1) and (3)]. The temperature-dependent concentrations of the reactants, intermediates and product species are obtained by numerically solving this system of differential equations (Fig. 6). Different models can be considered:

- In an *adiabatic approximation* (Fig. 6a) it is assumed that the chemical energy released in previous reaction steps is not dissipated to the substrate, but instead remains within the molecule. That means that extra energy is available to trigger the subsequent reactions steps. The simulations for this case show that no intermediates are stabilized (contradicting experimental observations). This can directly be rationalized based on the energy landscape shown in Fig. 5: once the first transition state is reached, sufficient energy is available within the molecule to overcome all subsequent transformation barriers. The reaction kinetics are determined by the first reaction barrier.
- *Instant thermalization* (Fig. 6b): This approximation, which is often used in heterogeneous catalysis [63–66], assumes that the chemical energy released in previous reaction steps is dissipated to the surface before the next reaction step takes place. In this case the intermediates with the highest reaction barriers are stabilized, i.e. **2a** and **3a**—again contradictory to experimental observations.
- *Selective dissipation* (Fig. 6c): In this approximation the dissipation of the released chemical energy is determined for each reaction step by theoretical simulations. Energy dissipation is strongly dependent on the phonon coupling between vibrational modes of the respective adsorbate and the substrate. Thus the energy dissipation can strongly vary for different intermediates. However, this model predicts stabilization of the intermediate **3b** and thus also contradicts the experimental observations.
- *Selective dissipation and entropy* (Fig. 6d): In addition to selective dissipation, entropy changes along the reaction pathway are taken into account [see Eq. (3)].

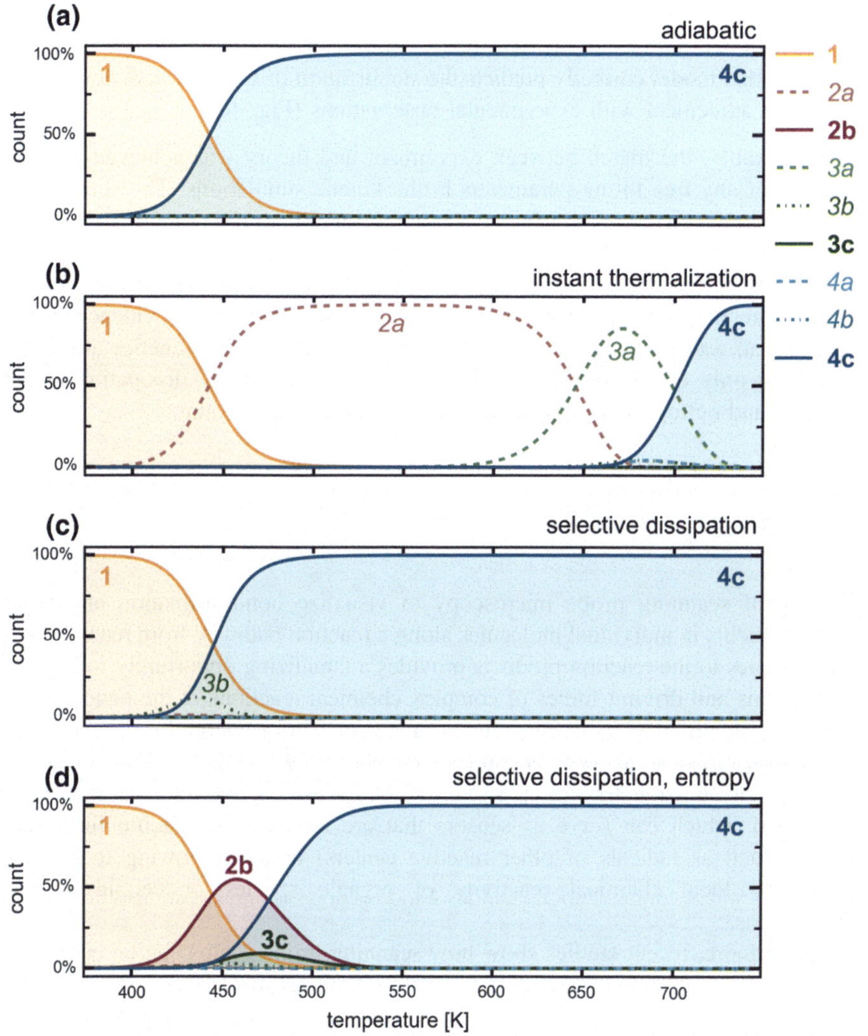

Fig. 6 Simulation of reaction kinetics for the reaction pathway from **1** to **4c**. **a** In the adiabatic model only the reactant **1** and the product **4c** are stabilized. **b** Instant thermalization stabilizes intermediates with the highest transformation barriers (**2a** and **3a**). **c** Taking into account selective dissipation lowers the formation temperature of **4c** and leads to the stabilization of intermediate **3b**. **d** A model taking into account selective dissipation and entropy predicts stabilization of the intermediates **2b** and **3c** in agreement with experimental observations. Solid (dashed) lines represent the relative concentrations of species that are (are not) experimentally observed. Adapted by permission from Macmillan Publishers Ltd: Nat. Chem. 8, 678 (2016), copyright 2016

In this case entropy differences are to a large extent determined by the roto-translational mobility of the adsorbates (vibrational contributions play a smaller role). This model correctly predicts the stabilization of the intermediates **2b** and **3c**—in agreement with experimental observations (Fig. 4e).

Remarkably, the match between experiment and theory was achieved without the need of any free fitting parameters in the kinetic simulations. The simulations are based only on the theoretical calculations of the energy landscape, energy dissipation, and entropy. Thus, the imaging and identification of the chemical structure of intermediate species together with the agreement between simulated and experimentally observed reactions kinetics provides conclusive evidence for the reaction pathway presented in Fig. 5. Importantly, the reaction kinetics are determined not only by the potential-energy landscape, but energy dissipation to the substrate and entropy changes also need to be taken into account.

5 Outlook

The use of scanning probe microscopy to visualize bond formation and bond rearrangements in individual molecules along a reaction pathway from reactants via intermediates to the reaction products provides a tantalizing opportunity to explore mechanisms and driving forces of complex chemical reactions at the nanoscale.

It is even possible to extract three-dimensional information by atomic force microscopy [67–70], as well as information about the chemical nature of atoms [71–74]. Tip-functionalization [18, 52, 75–77] with organic molecules can be envisioned, which can serve as sensors that are sensitive to specific functional groups (such as radicals or other reactive centers) thereby allowing to directly assess the local chemical reactivity of organic species or certain surface terminations.

Furthermore, recent studies show how scanning probe setups can be combined with optical pump-probe techniques facilitating time-resolved measurements down to the femtosecond regime [78, 79]. Such new developments bring the realm of real-time observation of molecular movement and chemical transformations at the single-molecule level within reach.

There are exciting times ahead.

Acknowledgements I want to thank everyone. Support from the European Research Council Consolidator Grant NanoSurfs (No. 615233) is acknowledged. A.R. acknowledges fellowship support by the Austrian Science Fund (FWF): J3026-N16.

References

1. Wintterlin, J., Volkening, S., Janssens, T.V.W., Zambelli, T., Ertl, G.: Atomic and macroscopic reaction rates of a surface-catalyzed reaction. Science **278**, 1931–1934 (1997)
2. Laidler, K.J.: Chemical kinetics. Harper & Row, New York (1987)
3. Chamberlain, T.W., Biskupek, J., Skowron, S.T., Markevich, A.V., Kurasch, S., Reimer, O., Walker, K.E., Rance, G.A., Feng, X., Müllen, K., Turchanin, A., Lebedeva, M.A., Majouga, A.G., Nenajdenko, V.G., Kaiser, U., Besley, E., Khlobystov, A.N.: Stop-frame filming and discovery of reactions at the single-molecule level by transmission electron microscopy. ACS Nano **11**, 2509–2520 (2017)
4. Riss, A., Wickenburg, S., Gorman, P., Tan, L.Z., Tsai, H.-Z., de Oteyza, D.G., Chen, Y.-C., Bradley, A.J., Ugeda, M.M., Etkin, G., Louie, S.G., Fischer, F.R., Crommie, M.F.: Local electronic and chemical structure of oligo-acetylene derivatives formed through radical cyclizations at a surface. Nano Lett. **14**, 2251–2255 (2014)
5. Koshino, M., Niimi, Y., Nakamura, E., Kataura, H., Okazaki, T., Suenaga, K., Iijima, S.: Analysis of the reactivity and selectivity of fullerene dimerization reactions at the atomic level. Nat Chem. **2**, 117–124 (2010)
6. Harano, K., Takenaga, S., Okada, S., Niimi, Y., Yoshikai, N., Isobe, H., Suenaga, K., Kataura, H., Koshino, M., Nakamura, E.: Conformational analysis of single perfluoroalkyl chains by single-molecule real-time transmission electron microscopic imaging. J. Am. Chem. Soc. **136**, 466–473 (2014)
7. Khlobystov, A.N., Porfyrakis, K., Kanai, M., Britz, D.A., Ardavan, A., Shinohara, H., Dennis, T.J.S., Briggs, G.A.D.: Molecular motion of endohedral fullerenes in single-walled carbon nanotubes. Angew. Chemie Int. Ed. **43**, 1386–1389 (2004)
8. Ke, X., Bittencourt, C., van Tendeloo, G.: Possibilities and limitations of advanced transmission electron microscopy for carbon-based nanomaterials. Beilstein J. Nanotechnol. **6**, 1541–1557 (2015)
9. Sloan, J., Matthewman, G., Dyer-Smith, C., Sung, A.-Y., Liu, Z., Suenaga, K., Kirkland, A.I., Flahaut, E.: Direct imaging of the structure, relaxation, and sterically constrained motion of encapsulated tungsten polyoxometalate Lindqvist ions within carbon nanotubes. ACS Nano **2**, 966–976 (2008)
10. Grill, L., Dyer, M., Lafferentz, L., Persson, M., Peters, M.V., Hecht, S.: Nano-architectures by covalent assembly of molecular building blocks. Nat Nano. **2**, 687–691 (2007)
11. Hla, S.-W., Bartels, L., Meyer, G., Rieder, K.-H.: Inducing all steps of a chemical reaction with the scanning tunneling microscope tip: towards single molecule engineering. Phys. Rev. Lett. **85**, 2777–2780 (2000)
12. Lindner, R., Kühnle, A.: On-surface reactions. ChemPhysChem **16**, 1582–1592 (2015)
13. Mali, K.S., Pearce, N., De Feyter, S., Champness, N.R.: Frontiers of supramolecular chemistry at solid surfaces. Chem. Soc. Rev. **46**, 2520–2542 (2017)
14. Shen, Q., Gao, H.-Y., Fuchs, H.: Frontiers of on-surface synthesis: from principles to applications. Nano Today. **13**, 77–96 (2017)
15. Held, P.A., Fuchs, H., Studer, A.: Covalent-bond formation via on-surface chemistry. Chem. A Eur. J. 1–20 (2017)
16. Klappenberger, F., Zhang, Y.Q., Björk, J., Klyatskaya, S., Ruben, M., Barth, J.V.: On-surface synthesis of carbon-based scaffolds and nanomaterials using terminal alkynes. Acc. Chem. Res. **48**, 2140–2150 (2015)
17. Gross, L., Mohn, F., Moll, N., Liljeroth, P., Meyer, G.: The chemical structure of a molecule resolved by atomic force microscopy. Science **325**, 1110–1114 (2009)
18. Jarvis, S.P.: Resolving intra- and inter-molecular structure with non-contact atomic force microscopy. Int. J. Mol. Sci. **16**, 19936–19959 (2015)
19. Iwata, K., Yamazaki, S., Mutombo, P., Hapala, P., Ondrácek, M., Jelínek, P., Sugimoto, Y.: Chemical structure imaging of a single molecule by atomic force microscopy at room temperature. Nat. Commun. **6**, 7766 (2015)

20. Huber, F., Matencio, S., Weymouth, A.J., Ocal, C., Barrena, E., Giessibl, F.J.: Intramolecular force contrast and dynamic current-distance measurements at room temperature. Phys. Rev. Lett. **115**, 66101 (2015)
21. Palma, C.A., Diller, K., Berger, R., Welle, A., Björk, J., Cabellos, J.L., Mowbray, D.J., Papageorgiou, A.C., Ivleva, N.P., Matich, S., Margapoti, E., Niessner, R., Menges, B., Reichert, J., Feng, X., Räder, H.J., Klappenberger, F., Rubio, A., Müllen, K., Barth, J.V.: Photoinduced C-C reactions on insulators toward photolithography of graphene nanoarchitectures. J. Am. Chem. Soc. **136**, 4651–4658 (2014)
22. Colazzo, L., Sedona, F., Moretto, A., Casarin, M., Sambi, M.: Metal-free on-surface photochemical homocoupling of terminal alkynes. J. Am. Chem. Soc. **138**, 10151–10156 (2016)
23. Basagni, A., Colazzo, L., Sedona, F., Marino, M.Di, Carofiglio, T., Lubian, E., Forrer, D., Vittadini, A., Casarin, M., Verdini, A., Cossaro, A., Floreano, L., Sambi, M.: Stereoselective photopolymerization of tetraphenylporphyrin derivatives on Ag(110) at the sub-monolayer level. Chem. A Eur. J. **20**, 14296–14304 (2014)
24. Basagni, A., Ferrighi, L., Cattelan, M., Nicolas, L., Handrup, K., Vaghi, L., Papagni, A., Sedona, F., Valentin, C.Di, Agnoli, S., Sambi, M.: On-surface photo-dissociation of C-Br bonds: towards room temperature Ullmann coupling. Chem. Commun. **51**, 12593–12596 (2015)
25. Gao, H.-Y., Zhong, D., Mönig, H., Wagner, H., Held, P.-A., Timmer, A., Studer, A., Fuchs, H.: Photochemical Glaser coupling at metal surfaces. J. Phys. Chem. C **118**, 6272–6277 (2014)
26. Riss, A., Wickenburg, S., Tan, L.Z., Tsai, H.-Z., Kim, Y., Lu, J., Bradley, A.J., Ugeda, M.M., Meaker, K.L., Watanabe, K., Taniguchi, T., Zettl, A., Fischer, F.R., Louie, S.G., Crommie, M.F.: Imaging and tuning molecular levels at the surface of a gated graphene device. ACS Nano **8**, 5395–5401 (2014)
27. Wickenburg, S., Lu, J., Lischner, J., Tsai, H.-Z., Omrani, A.A., Riss, A., Karrasch, C., Bradley, A., Jung, H.S., Khajeh, R., Wong, D., Watanabe, K., Taniguchi, T., Zettl, A., Neto, A.H.C., Louie, S.G., Crommie, M.F.: Tuning charge and correlation effects for a single molecule on a graphene device. Nat. Commun. **7**, 13553 (2016)
28. Morgenstern, K.: Switching individual molecules by light and electrons: From isomerisation to chirality flip. Prog. Surf. Sci. **86**, 115–161 (2011)
29. Schuler, B., Fatayer, S., Mohn, F., Moll, N., Pavliček, N., Meyer, G., Peña, D., Gross, L.: Reversible Bergman cyclization by atomic manipulation. Nat. Chem. **8**, 220–224 (2016)
30. Pavlicek, N., Schuler, B., Collazos, S., Moll, N., Pérez, D., Guitián, E., Meyer, G., Peña, D., Gross, L.: On-surface generation and imaging of arynes by atomic force microscopy. Nat. Chem. **7**, 623–628 (2015)
31. Mohn, F., Repp, J., Gross, L., Meyer, G., Dyer, M.S., Persson, M.: Reversible bond formation in a gold-atom-organic-molecule complex as a molecular switch. Phys. Rev. Lett. **105**, 266102 (2010)
32. Albrecht, F., Neu, M., Quest, C., Swart, I., Repp, J.: Formation and characterization of a molecule-metal-molecule bridge in real space. J. Am. Chem. Soc. **135**, 9200–9203 (2013)
33. Ladenthin, J.N., Frederiksen, T., Persson, M., Sharp, J.C., Gawinkowski, S., Waluk, J., Kumagai, T.: Force-induced tautomerization in a single molecule. Nat. Chem. **8**, 1–6 (2016)
34. Lafferentz, L., Eberhardt, V., Dri, C., Africh, C., Comelli, G., Esch, F., Hecht, S., Grill, L.: Controlling on-surface polymerization by hierarchical and substrate-directed growth. Nat. Chem. **4**, 215–220 (2012)
35. Kawai, S., Haapasilta, V., Lindner, B.D., Tahara, K., Spijker, P., Buitendijk, J.A., Pawlak, R., Meier, T., Tobe, Y., Foster, A.S., Meyer, E.: Thermal control of sequential on-surface transformation of a hydrocarbon molecule on a copper surface. Nat. Commun. **7**, 12711 (2016)
36. Riss, A., Paz, A.P., Wickenburg, S., Tsai, H.-Z., De Oteyza, D.G., Bradley, A.J., Ugeda, M. M., Gorman, P., Jung, H.S., Crommie, M.F., Rubio, A., Fischer, F.R.: Imaging

single-molecule reaction intermediates stabilized by surface dissipation and entropy. Nat. Chem. **8**, 678–683 (2016)
37. Zint, S., Ebeling, D., Schlöder, T., Ahles, S., Mollenhauer, D., Wegner, H.A., Schirmeisen, A.: Imaging successive intermediate states of the on-surface Ullmann reaction on Cu(111): role of the metal coordination. ACS Nano. **11**, 4183–4190 (2017)
38. Papageorgiou, A.C., Fischer, S., Oh, S.C., Sağlam, Ö., Reichert, J., Wiengarten, A., Seufert, K., Vijayaraghavan, S., Écija, D., Auwärter, W., Allegretti, F., Acres, R.G., Prince, K.C., Diller, K., Klappenberger, F., Barth, J.V.: Self-terminating protocol for an interfacial complexation reaction in vacuo by metal-organic chemical vapor deposition. ACS Nano **7**, 4520–4526 (2013)
39. Wiengarten, A., Lloyd, J.A., Seufert, K., Reichert, J., Auwärter, W., Han, R., Duncan, D.A., Allegretti, F., Fischer, S., Oh, S.C., Saglam, Ö., Jiang, L., Vijayaraghavan, S., Écija, D., Papageorgiou, A.C., Barth, J.V.: Surface-assisted cyclodehydrogenation; break the symmetry, enhance the selectivity. Chem. A Eur. J. **21**, 12285–12290 (2015)
40. van Vörden, D., Wortmann, B., Schmidt, N., Lange, M., Robles, R., Brendel, L., Bobisch, C. A., Möller, R.: Following the steps of a reaction by direct imaging of many individual molecules. Chem. Commun. **52**, 1–4 (2016)
41. Blake, M.M., Nanayakkara, S.U., Claridge, S.A., Fernández-Torres, L.C., Sykes, E.C.H., Weiss, P.S.: Identifying reactive intermediates in the Ullmann coupling reaction by scanning tunneling microscopy and spectroscopy. J. Phys. Chem. A. **113**, 13167–13172 (2009)
42. Wang, W., Shi, X., Wang, S., Van Hove, M.A., Lin, N.: Single-molecule resolution of an organometallic intermediate in a surface-supported ullmann coupling reaction. J. Am. Chem. Soc. **133**, 13264–13267 (2011)
43. Treier, M., Pignedoli, C.A., Laino, T., Rieger, R., Müllen, K., Passerone, D., Fasel, R.: Surface-assisted cyclodehydrogenation provides a synthetic route towards easily processable and chemically tailored nanographenes. Nat Chem. **3**, 61–67 (2011)
44. Onishi, H.: Reaction intermediates on TiO2(110) identified by time-lapse scanning tunneling microscopy. Catal. Surv. Jpn **6**, 1–8 (2002)
45. Matthiesen, J., Wendt, S., Hansen, J., Madsen, G., Lira, E., Galliker, P., Vestergaard, E., Schaub, R., Lægsgaard, E., Hammer, B., Besenbacher, F.: Observation of all the intermediate steps of a chemical reaction on an oxide surface by scanning tunneling microscopy. ACS Nano **3**, 517–526 (2009)
46. Björk, J.: Reaction mechanisms for on-surface synthesis of covalent nanostructures. J. Phys.: Condens. Matter **28**, 83002 (2016)
47. Xi, M., Bent, B.E.: Iodobenzene on Cu(111): formation and coupling of adsorbed phenyl groups. Surf. Sci. **278**, 19–32 (1992)
48. Xi, M., Bent, B.E.: Mechanisms of the Ullmann coupling reaction in adsorbed monolayers. J. Am. Chem. Soc. **115**, 7426–7433 (1993)
49. Weiss, P.S., Kamna, M.M., Graham, T.M., Stranick, S.J.: Imaging benzene molecules and phenyl radicals on Cu(111). Langmuir **14**, 1284–1289 (1998)
50. Dong, L., Liu, P.N., Lin, N.: Surface-activated coupling reactions confined on a surface. Acc. Chem. Res. **48**, 2765–2774 (2015)
51. Fan, Q., Gottfried, J.M., Zhu, J.: Surface-catalyzed C-C covalent coupling strategies toward the synthesis of low-dimensional carbon-based nanostructures. Acc. Chem. Res. **48**, 2484–2494 (2015)
52. Bartels, L., Meyer, G., Rieder, K.-H.: Controlled vertical manipulation of single CO molecules with the scanning tunneling microscope: a route to chemical contrast. Appl. Phys. Lett. **71**, 213 (1997)
53. Heinrich, B.W., Ahmadi, G., Müller, V.L., Braun, L., Pascual, J.I., Franke, K.J.: Change of the magnetic coupling of a metal-organic complex with the substrate by a stepwise ligand reaction. Nano Lett. **13**, 4840–4843 (2013)
54. de Oteyza, D.G., Gorman, P., Chen, Y.-C., Wickenburg, S., Riss, A., Mowbray, D.J., Etkin, G., Pedramrazi, Z., Tsai, H.-Z., Rubio, A., Crommie, M.F., Fischer, F.R.: Direct imaging of

covalent bond structure in single-molecule chemical reactions. Science **340**, 1434–1437 (2013)
55. de Oteyza, D.G., Pérez Paz, A., Chen, Y.-C., Pedramrazi, Z., Riss, A., Wickenburg, S., Tsai, H.-Z., Fischer, F.R., Crommie, M.F., Rubio, A.: Noncovalent dimerization after enediyne cyclization on Au(111). J. Am. Chem. Soc. **138**, 10963–10967 (2016)
56. Arrhenius, S.: Über die Reaktionsgeschwindigkeit bei der Inversion von Rohrzucker durch Säuren. Z. Phys. Chem. **4**, 226–248 (1889)
57. Winzor, D., Jackson, C.: Interpretation of the temperature dependence of equilibrium and rate constants. J. Mol. Recognit. **19**, 389–407 (2006)
58. Pollak, E., Talkner, P.: Reaction rate theory: what it was, where is it today, and where is it going? Chaos. **15**, 26116 (2005)
59. Marbach, H., Steinrück, H.-P.: Studying the dynamic behaviour of porphyrins as prototype functional molecules by scanning tunnelling microscopy close to room temperature. Chem. Commun. **50**, 9034 (2014)
60. Ditze, S., Stark, M., Buchner, F., Aichert, A., Jux, N., Luckas, N., Görling, A., Hieringer, W., Hornegger, J., Steinrück, H.P., Marbach, H.: On the energetics of conformational switching of molecules at and close to room temperature. J. Am. Chem. Soc. **136**, 1609–1616 (2014)
61. Eyring, H.: The activated complex in chemical reactions. J. Chem. Phys. **3**, 107–115 (1935)
62. Laidler, K.J., King, M.C.: Development of transition-state theory. J. Phys. Chem. **87**, 2657–2664 (1983)
63. Reuter, K., Scheffler, M.: First-principles kinetic Monte Carlo simulations for heterogeneous catalysis: application to the CO oxidation at RuO2(110). Phys. Rev. B. **73**, 45433 (2006)
64. Stoltze, P., Nørskov, J.K.: Theoretical modelling of catalytic reactions. In: Handbook of Heterogeneous Catalysis. Wiley-VCH Verlag GmbH & Co. KGaA (2008)
65. Meyer, J., Reuter, K.: Modeling heat dissipation at the nanoscale: an embedding approach for chemical reaction dynamics on metal surfaces. Angew. Chemie Int. Ed. **53**, 4721–4724 (2014)
66. Neurock, M.: The microkinetics of heterogeneous catalysis. By J. A. Dumesic, D. F. Rudd, L. M. Aparicio, J. E. Rekoske, and A. A. Treviño, ACS Professional Reference Book, American Chemical Society, Washington, DC, 1993, 315p (1994)
67. Schuler, B., Liu, W., Tkatchenko, A., Moll, N., Meyer, G., Mistry, A., Fox, D., Gross, L.: Adsorption geometry determination of single molecules by atomic force microscopy. Phys. Rev. Lett. **111**, 106103 (2013)
68. Moreno, C., Stetsovych, O., Shimizu, T.K., Custance, O.: Imaging three-dimensional surface objects with submolecular resolution by atomic force microscopy. Nano Lett. **15**, 2257–2262 (2015)
69. Albrecht, F., Pavliček, N., Herranz-Lancho, C., Ruben, M., Repp, J.: Characterization of a surface reaction by means of atomic force microscopy. J. Am. Chem. Soc. **137**, 7424–7428 (2015)
70. Albrecht, F., Bischoff, F., Auwärter, W., Barth, J.V., Repp, J.: Direct identification and determination of conformational response in adsorbed individual nonplanar molecular species using noncontact atomic force microscopy. Nano Lett. **16**, 7703–7709 (2016)
71. Schmid, M., Stadler, H., Varga, P.: Direct observation of surface chemical order by scanning tunneling microscopy. Phys. Rev. Lett. **70**, 1441–1444 (1993)
72. Sugimoto, Y., Pou, P., Abe, M., Jelinek, P., Pérez, R., Morita, S., Custance, O.: Chemical identification of individual surface atoms by atomic force microscopy. Nature **446**, 64 (2007)
73. van der Heijden, N.J., Hapala, P., Rombouts, J.A., van der Lit, J., Smith, D., Mutombo, P., Švec, M., Jelinek, P., Swart, I.: Characteristic contrast in Δfmin maps of organic molecules using atomic force microscopy. ACS Nano **10**, 8517–8525 (2016)
74. Onoda, J., Ondráček, M., Jelínek, P., Sugimoto, Y.: Electronegativity determination of individual surface atoms by atomic force microscopy. Nat. Commun. **8**, 15155 (2017)
75. Mohn, F., Schuler, B., Gross, L., Meyer, G.: Different tips for high-resolution atomic force microscopy and scanning tunneling microscopy of single molecules. Appl. Phys. Lett. **102**, 73109 (2013)

76. Okabayashi, N., Gustafsson, A., Peronio, A., Paulsson, M., Arai, T., Giessibl, F.J.: Influence of atomic tip structure on the intensity of inelastic tunneling spectroscopy data analyzed by combined scanning tunneling spectroscopy, force microscopy, and density functional theory. Phys. Rev. B Condens. Matter Mater. Phys. **93**, 1–6 (2016)
77. Xin, X., Gan, L.-Y., Van Hove, M.A., Ren, X., Wang, H., Guo, C.-S., Zhao, Y.: Exploring molecules beyond CO as tip functionalizations in high-resolution noncontact atomic force microscopy: a first principles approach. ACS Omega. **1**, 1004–1009 (2016)
78. Cocker, T.L., Peller, D., Yu, P., Repp, J., Huber, R.: Tracking the ultrafast motion of a single molecule by femtosecond orbital imaging. Nature **539**, 263–267 (2016)
79. Peplow, M.: The next big hit in molecule Hollywood. Nature **544**, 408–410 (2017)

Kinetic and Thermodynamic Considerations in On-Surface Synthesis

Jonas Björk

Abstract In this chapter it will be explained how kinetic and thermodynamic aspects of on-surface reactions may be accounted for by electronic structure theory together with transition state theory. The focus of the chapter is to discuss what free energy contributions, particularly in terms of entropy, we need to account for to properly describe chemical reactions on surfaces. For example, dehydrogenation reactions are often endothermic and their occurrence on surfaces can be explained by the entropy gain of associatively desorbing hydrogen, which make them thermodynamically favorable. In another example, experimentally observed intermediate structures of a bimolecular enediyne coupling were concluded to be stabilized by differences in surface dissipation of excess energy and translational entropy, requiring a quite complex free energy description to understand the reaction. Calculating reaction pathways is becoming a frequent practice within on-surface synthesis, but are often considered at 0 K to reduce computational efforts. The recent advances in describing in particular entropic contributions of on-surface reactions provide important guidelines for how calculations can be refined, and for what kind of scenarios we may expect the necessity for more sophisticated descriptions of kinetics and thermodynamics of reactions.

1 Introduction

Kinetics and thermodynamics are well-used concepts to describe chemical reactions and we often refer to reactions as being either kinetically or thermodynamically driven. In a thermodynamically governed reaction the system is allowed to reach its (global) free energy minimum, while for a kinetically driven reaction thermodynamically less favorable products are formed due to smaller activation energies. The concept is demonstrated for a simplified reaction in Fig. 1, where the reactant can

J. Björk (✉)
Department of Physics, Chemistry and Biology, IFM,
Linköping University, Linköping, Sweden
e-mail: jonas.bjork@liu.se

D. G. de Oteyza and C. Rogero (eds.), *On-Surface Synthesis II*,
Advances in Atom and Single Molecule Machines,
https://doi.org/10.1007/978-3-319-75810-7_2

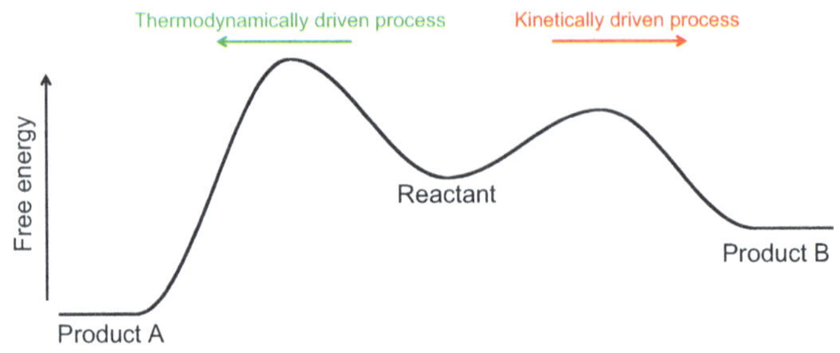

Fig. 1 Comparison between a thermodynamically driven process, leading from reactant to product A, and a kinetically driven process, ending up in product B

end up in either of two products. The free energies of both product A and B are smaller than that of the reactant, and product A represents the global free energy minimum of the system. On the contrary, the activation energy to form product A is larger than the one yielding product B. A reaction leading to product A is thus known as a thermodynamically driven process, while a reaction leading to product B we call a kinetically driven process. Reality is of course more complicated. In particular, it is difficult to anticipate what is the global free energy minimum of a system, and we need to define this within some boundaries.

Independent of whether a reaction should be considered being a thermodynamically or kinetically driven process, it is of great interest being able to investigate kinetic and thermodynamic aspects within on-surface synthesis. To do so we need methods that can calculate the energy landscapes of reactions. The highest level of theory that could feasibly achieve this is density functional theory (DFT), which has been used in several examples for studying reaction mechanisms [1], while also the computationally less demanding density functional-based tight binding (DFTB) has been used [2]. One may even anticipate the use of reactive force fields, but which has not yet found its way into the on-surface synthesis community.

The approach for studying kinetics and thermodynamics of reactions is the same independent of the underlying theoretical machinery we prefer. We need to calculate the Gibbs free energy of reactant, product and eventual intermediates (local minima), and the transition state(s) separating these. For a reaction with only one transition state (i.e. without intermediate states) the reaction energy is defined as the energy difference between product and reactant, and the activation energy as the energy difference between transition state and reactant. Knowing these factors, the rate v going between two states at a temperature T is given by the Eyring equation

$$v = \frac{k_B T}{h} e^{\Delta S^{\ddagger}/k_B} e^{-\Delta H^{\ddagger}/k_B T} = \frac{k_B T}{h} e^{-\Delta G^{\ddagger}/k_B T} \tag{1}$$

where k_B is Boltzmann's constant and h is Planck's constant. $\Delta G^{\ddagger} = \Delta H^{\ddagger} - T\Delta S^{\ddagger}$ is Gibb's free energy of activation, where $\Delta H^{\ddagger}$ and $\Delta S^{\ddagger}$ are the activation enthalpy and entropy, respectively. This is the essence of transition state theory (TST), which assumes that a reaction step leading from one local minimum to another is associated with a transition state—a saddle point on the free energy surface—separating the two states. The preferred pathway, from a kinetic point of view, may be found by considering the activation energies, where, as a rule of thumb, the path with lowest activation energy will be preferred. At least for a kinetically driven process, whereas a thermodynamically driven process prefers the pathway with the most negative reaction energy.

To calculate activation and reaction energies we need to evaluate the free energies of local minima and transition states, for which we need the enthalpy H and entropy S. The enthalpy is given by

$$H = H_{\text{elec}} + H_{\text{vib}} + H_{\text{rot}} + H_{\text{trans}}, \tag{2}$$

where H_{elec} is the electronic enthalpy, H_{vib} is the vibrational enthalpy, H_{rot} is the rotational enthalpy and H_{trans} is the translational enthalpy. The two first terms are quite straightforward to obtain. H_{elec} is simply the total electronic energy of the system while H_{vib} can be obtained from the vibrational frequencies

$$H_{\text{vib}} = k_{\text{B}} \sum \left(\frac{h\nu_i}{2k_{\text{B}}} + \frac{h\nu_i}{k_{\text{B}}} \frac{1}{e^{h\nu_i/k_{\text{B}}T} - 1} \right), \tag{3}$$

where ν_i are the vibrational frequencies of the system. The final two terms $H_{\text{rot}} + H_{\text{trans}}$ are easily evaluated for molecules in an ideal gas, but are more complicated for molecules on surfaces, in particular if the molecules are chemically interacting with the substrate. For example, chemical bond(s) may compromise some, or all, rotational degrees of freedom, and also, to an even larger extent, translational degrees of freedom. Within harmonic transition state theory, however, these terms are ignored and we only consider the contributions from H_{elec} and H_{vib} to the enthalpy.

Similarly, the total entropy is given by

$$S = S_{\text{elec}} + S_{\text{vib}} + S_{\text{rot}} + S_{\text{trans}}, \tag{4}$$

where S_{elec} is the electronic entropy, S_{vib} is the vibrational entropy, S_{rot} is the rotational entropy and S_{trans} is the translational entropy. The electronic entropy is typically ignored, as it differs insignificantly between different chemical states of the same system. As for the enthalpy, the rotational and translational contributions are non-trivial to evaluate for molecules on surfaces, while the vibrational entropy can be obtained from the vibrational frequencies

$$S_{\text{vib}} = k_{\text{B}} \sum \left[\frac{h\nu_i}{k_{\text{B}}T} \frac{1}{e^{h\nu_i/k_{\text{B}}T} - 1} - \ln\left(1 - e^{-h\nu_i/k_{\text{B}}T}\right) \right]. \tag{5}$$

It should be noted that although translational and rotational contributions of the free energy are often neglected, they might play important roles. In particular, if the strength of the molecule-surface interactions is strongly affected throughout a reaction, as in the final example of this chapter. Notably, as including vibrational contributions is already quite demanding from a computational point of view, the majority of theoretical studies take only the electronic enthalpy into consideration, neglecting any effect of temperature.

2 Dehydrogenation Reactions

Dehydrogenation processes—the removal of hydrogen atoms from molecules—are important in several on-surface synthesis protocols. The cyclodehydrogenation of linear polymers results in graphene nanoribbons [3, 4], the homo-coupling of terminal alkynes [5–8] gives graphdiyne structures, and a range of other reactions are associated with removal of hydrogen [9–11].

Dehydrogenation reactions on surfaces are often endothermic. For example, the dehydrogenation of quaterphenyl molecules on Cu(111) has a reaction (electronic) enthalpy of around 1.5 eV [9], and the reaction enthalpy has shown to be positive for a series of model dehydrogenation reactions on the (111) facets of the coinage metals, with a few exceptions [12]. Recently, an electrocyclic ring closure reaction was reported of an octaethylporphyrin into a tetrabenzoporphyrin [13] and later a similar type of reaction was reported for the transformation of a tetraazaporphyrin into a freestanding phthalocyanine [14]. The reaction mechanism in the latter case, was studied for the model reaction shown in Fig. 2, with an overall reaction (electronic) enthalpy of 2.32 eV [15].

Evidently, many dehydrogenation reactions are endothermic. But still, they take place on surface, and their occurrence is often explained—handwavingly—by the entropy gain by the associative desorption of H_2 into the vacuum. The question is then still remaining; how large is the entropy of the desorbing H_2?

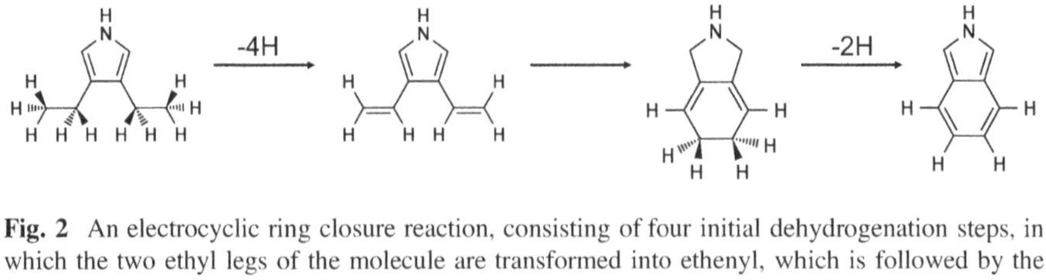

Fig. 2 An electrocyclic ring closure reaction, consisting of four initial dehydrogenation steps, in which the two ethyl legs of the molecule are transformed into ethenyl, which is followed by the actual ring closure and finalized by two dehydrogenation steps. On the Au(111), the overall reaction is highly endothermic, with a reaction energy of 2.23 eV [15]. Reprinted with permission from [15]. Copyright 2016 American Chemical Society

3 Entropy of Desorbed Hydrogen Molecules

To estimate the entropy of H_2 in the vacuum, all we need is the entropy of an H_2 gas at the temperature T of our system and the pressure p of the ultrahigh vacuum. This is a quite straightforward procedure. Especially, if knowing the entropy at a standard pressure p_0, it may be calculated for any pressure p with the equation [12]

$$S_{H_2}(T,p) = S_{H_2}(T,p_0) - k_B ln\left(\frac{p}{p_0}\right). \tag{6}$$

This assumes an ideal gas, i.e. that the vibrational and rotational components of the entropy are not dependent on pressure and we only need to consider changes in translational entropy, leading directly to this expression. The entropy at a standard pressure p_0 can be found tabulated for a range of temperatures [16].

Svane and Hammer calculated the entropy of a hydrogen gas as a function of temperature and pressure [12], as shown in Fig. 3. This gives only a rough estimate for the entropy-gain per split-off hydrogen, as we need to remember that a hydrogen atom bonded to a molecule has some vibrational entropy, and it does not cover the full temperature range at which dehydrogenation reactions within on-surface synthesis are relevant. Nevertheless, it shows how greatly the experimental conditions influence the entropy of the desorbed hydrogen, and to what extent split-off hydrogen contributes to the free energy.

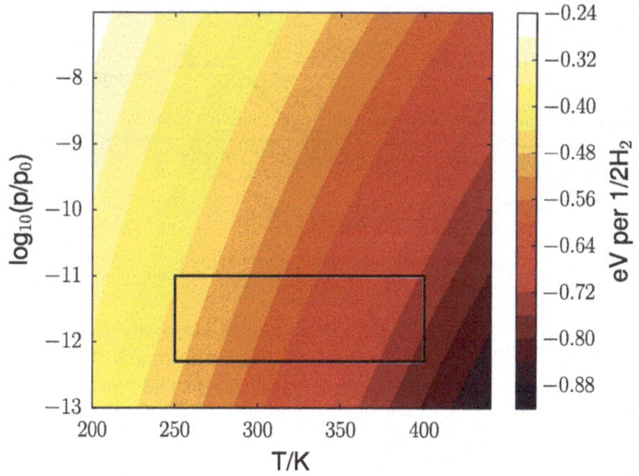

Fig. 3 Entropic contribution to the free energy of a hydrogen gas $-\frac{1}{2}TS_{H_2}(T,p)$ as a function of temperature and pressure, with the pressure dependence given in Eq. (6). Typical experimental conditions are indicated by the rectangle, although higher temperatures are commonly used in on-surface synthesis. The reference pressure p_0 is 1 bar. Reprinted with permission from [15]. Copyright 2014 American Institute of Physics

4 Electrocyclic Ring-Closure Reaction

Recently, I studied both the thermodynamics and the kinetics of the electrocyclic ring-closure reaction on Au(111) previously mentioned, for the model compound shown in Fig. 2. This reaction is initiated by four dehydrogenation steps, followed by the actual ring-closure, and finalized by splitting-off two hydrogen atoms. Thus, the overall reaction is associated with six dehydrogenation steps. Given that the reaction occurs at 300 °C under an ultrahigh vacuum well below 10^{-10} bar the entropic contribution from the split-off hydrogen atoms can be expected to be quite significant [15].

Figure 4 shows the reaction pathway for the first four dehydrogenation steps of the reaction, comparing the electronic enthalpy at 0 K to the free energy obtained at a temperature of 300 °C and a pressure of 10^{-10} bar. These four dehydrogenation steps can be summarized by the transformation of the two ethyl groups into ethenyl. First, a

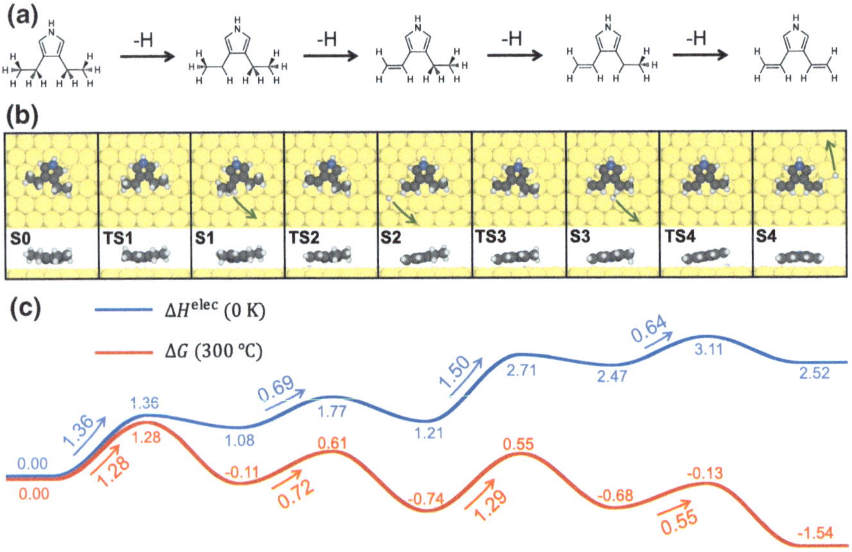

Fig. 4 Reaction mechanism of the four dehydrogenation steps initiating the electrocyclic ring closure reaction, with **a** valence bond structures of local minima, **b** top and side views of the structures of local minima (**S0–S4**) and transition states (**TS1–TS4**) on Au(111), and **c** electronic enthalpy profile in blue and the free-energy profile (calculated at T = 300 °C and p = 10^{-10} bar) in red. Units in eV. Reprinted with permission from [15]. Copyright 2016 American Chemical Society

methylene bridge is dehydrogenated followed by dehydrogenation of the associated methyl group, and in an equivalent process for the other ethyl group. In these processes, the dehydrogenation steps of the methylene bridges have the highest barriers. Importantly, initiating the reaction by abstracting hydrogen from a terminal methyl was shown to have an even larger barrier. Notably, the four initial dehydrogenation steps in Fig. 4 are highly endothermic, considering the electronic enthalpy [15].

The free energy differences in Fig. 4 were calculated as

$$\Delta G(T,p) = \Delta H^{\text{elec}} + \Delta H^{\text{nuclei}}(T) - T\Delta S^{\text{vib}}(T) - \frac{1}{2}n_H T S_{H_2}(T,p). \tag{7}$$

ΔH^{elec} gives the electronic enthalpy difference, ΔS^{vib} the vibrational entropy difference and the entropy S_{H_2} of the H_2 gas is defined by Eq. (6). The nuclei enthalpy difference ΔH^{nuclei} contains the vibrational enthalpy difference, and in addition the translational and rotational enthalpy of the H_2 gas. Since S_{H_2} contains all entropy contributions of the H_2 gas it is appropriate that also the enthalpy does, since the entropy and enthalpy to some extent cancel out each other. The integer n_H is the number of split-off hydrogen atoms.

When comparing the free energy to the electronic enthalpy curve in Fig. 4 it is clear that the barriers are only weakly affected by temperature. However, the overall free energy is lowered after the abstraction of each H atom. In fact, following each dehydrogenation step, the free energy path is lowered by $\sim$1 eV compared to the electronic enthalpy path, which will be explained below.

The initial dehydrogenation steps are followed by the actual ring closure and two additional dehydrogenation reactions. These final three steps of the overall reaction are shown in Fig. 5. The ring closure has a relatively large barrier, considering all other steps of the reaction, while the final two dehydrogenations are considered having small barriers. Again, the barriers do not differ significantly between the free energy and electronic enthalpy path, while for each split-off hydrogen the free energy path is shifted in a negative direction of $\sim$1 eV with respect to the electronic enthalpy path [15].

In the overall reaction, the free energy of the final state (**S7**) is 6.06 eV lower than the corresponding electronic enthalpy. This is in agreement with what we observed previously, that each split-off hydrogen lowers the free energy by $\sim$1 eV compared to the electronic enthalpy. Thus, it is quite tempting to say that the difference between the two paths is due to the abstracted hydrogen atoms. And it appears this is the case. Notably, the contribution $-\frac{1}{2}n_H T S_{H_2}$ has a value of -6.08 eV for $n_H = 6$, almost perfectly coinciding with the overall difference between the two curves. Thus, we can say that for each split-off hydrogen we gain

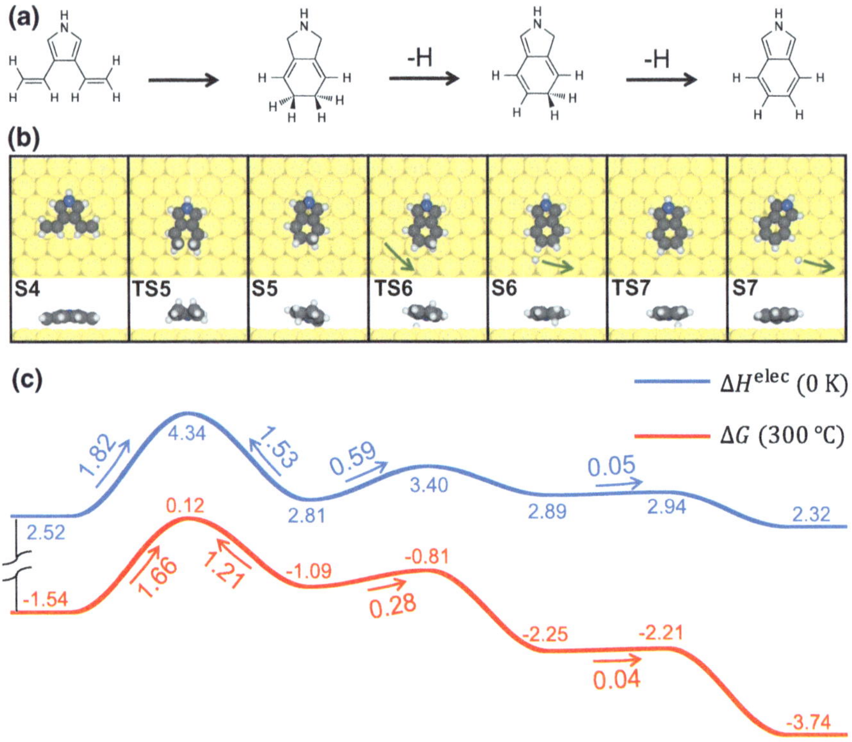

Fig. 5 Reaction mechanism of the steps finalizing the electrocyclic ring closure reaction, including the actual ring closure and two dehydrogenation steps, with **a** valence bond structures of local minima, **b** top and side views of the structures of local minima (**S4**–**S7**) and transition states (TS5–TS7) on the Au(111) surface, and **c** electronic enthalpy profile in blue and the free energy profile (calculated at T = 300 °C and p = 10^{-10} bar) in red. Notice that the blue and red curves have been moved closer together to minimize the amount of white space in the figure, which makes the energy difference between the two curves appear too small. Units in eV. Reprinted with permission from [15]. Copyright 2016 American Chemical Society

1 eV due to the entropy of the H_2 gas, perfectly explaining the differences between the two curves. One may make the interpretation that the two terms ΔH^{nuclei} and $-T\Delta S^{\text{vib}}$ are not important, and that we could add temperature effects by simply just adding $-\frac{1}{2}n_H T S_{H_2}(T,p)$. However, both ΔH^{nuclei} and $-T\Delta S^{\text{vib}}$ are quite significant, but cancel out each other almost perfectly by chance at these specific conditions. Thus, although the main difference between the free energy and electronic enthalpy curves comes from the entropy of the split-off hydrogen, ΔH^{nuclei} and $-T\Delta S^{\text{vib}}$ could still give important contributions [15]. For future studies, it would be interesting to investigate whether there is a general rule under which conditions these two terms cancel each other and how such a rule would differ between types of reactions. With prevailing knowledge, however, it is impossible to make such an assessment.

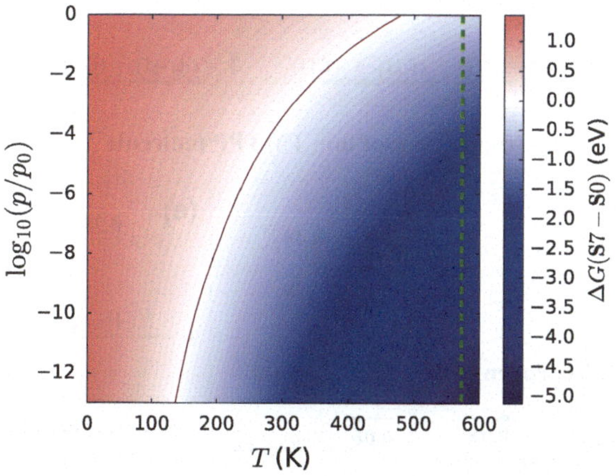

Fig. 6 Free-energy difference between the product (**S7**) and reactant (**S0**), of the electrocyclic ring closure reaction shown in Figs. 4 and 5, as a function of temperature and pressure of the hydrogen gas. The solid black line shows where the free-energy difference is zero, and the dashed green line indicates T = 300 °C, at which the reaction was observed experimentally. The reference pressure p_0 is 1 bar. Reprinted with permission from [15]. Copyright 2016 American Chemical Society

To understand how the thermodynamics of the electrocyclic ring closure is governed by the temperature and pressure, the reaction free energy of the overall reaction (thus the free energy difference between product and reactant) was calculated, as shown in Fig. 6. At pressures of 10^{-12}–10^{-10} bar, typical for UHV experiments, the reaction is thermodynamically favorable (free energy difference is negative) even at temperatures below 200 K. As the reaction occurs experimentally around 300 °C, kinetics rather than thermodynamics restricts the reaction from occurring at lower temperatures. Furthermore, it is worthwhile noticing that the reaction is thermodynamically favorable at relatively high pressures at 300 °C, indicating the possibility of inducing the reaction outside the UHV chamber, at least under a pure hydrogen atmosphere [15].

5 C–H Activation and Coupling of Porphyrins

Floris and coworkers investigated the driving forces between the C–H activation and coupling between tetra-trimethyl-phenyl-porphyrin (TMTPP) molecules on the Cu(110) surface [17]. In this reaction, the TMTPP molecules dehydrogenate in a selective fashion, such that only specific C–H bonds are dissociated. This is followed by the intermolecular coupling between activated C atoms. The dehydrogenation is endothermic, while the coupling is slightly exothermic, making the overall reaction endothermic with a reaction electronic enthalpy of almost 1 eV. The barriers of the different processes are compared in Fig. 7.

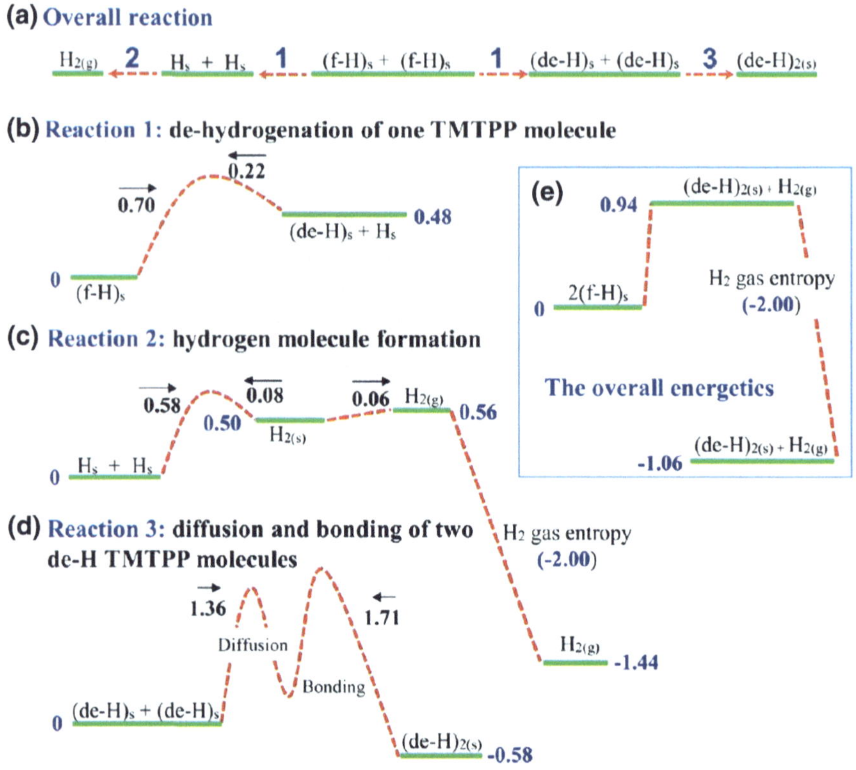

Fig. 7 Energetics for the dehydrogenation and subsequent intermolecular coupling of porphyrin molecules, with the overall reaction shown in (**a**), each of the three reaction steps in (**b–d**) and the overall energetics of the reaction in (**e**). The reaction is initiated by a **b** dehydrogenation step, in which an adsorbed TMTPP molecule (f-H)$_s$ forms a dehydrogenated molecule (de-H)$_s$ and a hydrogen atom H$_s$ adsorbed on the surface. **c** Two adsorbed hydrogen atoms may desorb associatively, in which the entropy of the H_2 gas contributes with −2.00 eV to the free energy. Furthermore, **d** the dehydrogenated TMPTPP molecules can diffuse and couple to each other on the surface, resulting in a dimer. **e** The overall reaction is endothermic with a reaction energy of 0.94 eV, but becomes thermodynamically favorable due to the entropy of the desorbed H_2. Units in eV. Reprinted with permission from [17]. Copyright 2016 American Chemical Society

Similar to the dehydrogenation steps associated with the electrocyclic ring closure above, the C–H activation and concomitant coupling of porphyrins can only be understood when accounting for entropy of the desorbed H_2. This becomes quite evident just by considering the associative desorption of hydrogen, shown in Fig. 7c. The associative desorption is endothermic, with a reaction energy of 0.56 eV. This process is thermodynamically favorable due to the entropic contribution of −2 eV per hydrogen molecule (compare to −1 eV per hydrogen atom above). Adding this free energy gain to the overall reaction (Fig. 7e), it becomes thermodynamically favorable. Again, a lot of insight is provided to a chemical reaction by simply considering the entropy of the desorbing H_2 molecule.

6 Reaction Intermediates Stabilized by Surface Dissipation and Translational Entropy

Typically, within on-surface synthesis we do not take into account the translation and rotation for molecules on surface as it would lead to quite complex calculations. Ignoring these terms is perfectly justified for molecules strongly chemisorbed on surfaces, with large barriers of translation and rotation. However, for weakly adsorbed (physisorbed) molecules these may be important, but are still difficult to account for.

Riss and coworkers suggested a crude, but elegant, way to tackle this problem [2]. They proposed that a molecule interacting weakly with the surface can be considered as an ideal two-dimensional gas and are able to rotate freely around the normal axis to the surface, for which translational and rotational entropies are quite trivial. On the other hand, if the molecule is chemisorbed on the surface, these terms are ignored. Of course, it is not always possible to make the classification if a molecule interacts strongly or not and even though a molecule is chemisorbed to the surface it may be able to rotate around a chemical bond, and the methodology will possibly experience some refinement. But it can still be used to understand fundamental aspects of on-surface reactions.

In particular, it was used to study the surface-catalyzed cross-coupling and sequential cyclization cascade of 1,2-bis(2-ethynyl phenyl)ethyne molecules on Ag (110) [2]. Non-contact atomic force microscopy (nc-AFM) experiments were able to resolve reactant, product as well as two intermediate structures for a dimerization process, as shown in Fig. 8.

To understand the dimerization reaction in Fig. 8, the reaction pathway was calculated at the DFTB level of theory [2]. The resulting pathway is illustrated in Fig. 9. Without going into detail about the specific reaction steps, it is quite interesting to notice that the two intermediate structures observed experimentally (**2b** and **3c**) are followed by relatively small activation enthalpies. Thus, at first sight, it is unclear why these intermediates are observed in the experiments.

To better understand the experimentally observed reaction behavior, kinetic rate equations were solved for different temperatures [2]. First, an adiabatic model was used, in which the energy gained during the reaction is assumed to be kept within the molecular system (Fig. 10a). This model results in that only the reactant and product may be observed, independent of temperature. Second, the instant thermalization model was used (Fig. 10b). In this model, it is assumed that the molecular system has time to release all its excess energy at every intermediate step, before continuing to the next step. Such a model is bound to favor the occurrence of states followed by large energy barriers. This is also the case, as it predicts state **2a** and **3a** to be stable intermediates. Furthermore, the final product is predicted to be formed at significantly higher temperatures than observed experimentally.

As a third option, Riss and coworkers adapted a selective dissipation model, by calculating dissipative energy transfer with molecular dynamics simulations within

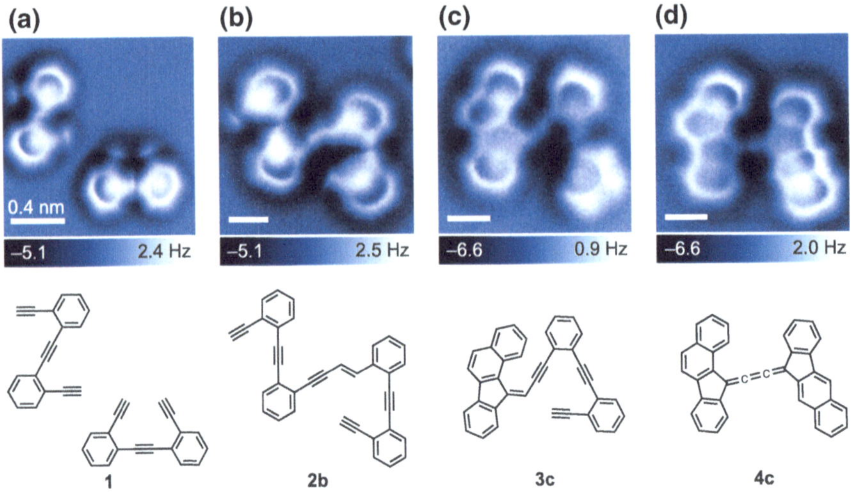

Fig. 8 Experimental observations of intermediates in a stepwise bimolecular enediyne coupling and cyclization cascade. **a–d** nc-AFM images depicting intermediates along the coupling and cyclization cascade of 1,2-bis(2-ethynyl phenyl)ethyne molecules from reactant (**1**) to product (**4c**), with corresponding chemical structures depicted below the nc-AFM images. Reprinted by permission from Macmillan Publishers Ltd: Ref. [2], copyright 2016

the DFTB framework (Fig. 10c). The redistribution of energy from the molecular system to the substrate is a quick process which significantly reduces the available energy to overcome subsequent barriers. Furthermore, it was found that this dissipation of energy varies significantly for each reaction step. By adding this selective dissipation (the details of which are explained in Ref. [2]) it was found that the product can form at a lower temperature compared to the instant thermalization model, but only one (and wrong) intermediate is predicted, and could not explain the experimentally observed reaction behavior.

To explain the observed intermediates, the entropy along the reaction was, at last, included to the selective dissipation model (Fig. 10d). Vibrational, rotational and roto-translational entropies were accounted for, with the observation that intermediates **2b** and **3c** can undergo translational and rotational motion, while other intermediates interact strongly by chemical interactions to the surface, and are not allowed to move. In other words, translational and rotational entropies were only added to intermediates **2b** and **3c**, lowering the free energy of these states significantly compared to other states. The resulting relative concentrations of the different structures are shown in Fig. 10d, in which both **2b** and **3c** are predicted to exist as stable intermediates, in agreement with the experimental reaction behavior.

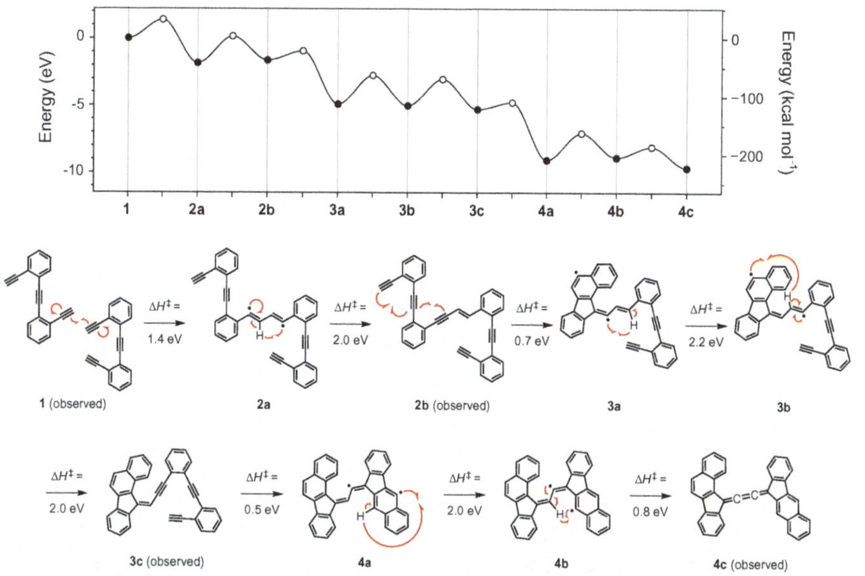

Fig. 9 Calculated pathway for the stepwise enediyne coupling and cyclization cascade. The energy profile shows the enthalpy of intermediates (filled circles) and transition states (open circles) along the reaction pathway from reactant (**1**) to product (**4c**). The chemical structures indicate metastable intermediates along the reaction coordinate (transition-state structures not shown), and calculated activation enthalpies are shown below the reaction arrows. Two out of seven theoretically predicted intermediates were observed experimentally—both structures are followed by comparably low activation enthalpy barriers ($\Delta H^{\ddagger}$ = 0.7 eV for **2b** and $\Delta H^{\ddagger}$ = 0.5 eV for **3c**). Reprinted by permission from Macmillan Publishers Ltd: Ref. [2], copyright 2016

The study by Riss and coworkers [2] has quite a few interesting aspects. First of all, they demonstrated that by explicitly accounting for energy dissipation between the molecule and substrate, instead of assuming instant thermalization which somehow is the norm, the anticipated reaction temperature is lowered significantly. Furthermore, they made an interesting notation that the translational and rotational entropy may differ quite significantly between different intermediate structures. A physisorbed molecule is free to move, and may be treated as an ideal 2D gas, while a molecule chemisorbed to the surface is not allowed to move, leading to significant differences in the free energy. In fact, both these aspects may, for example, explain the relatively low temperature required by the homo-coupling of thermal alkynes [5], despite the quite significant energy barriers leading from a chemisorbed intermediate to the physisorbed product of this reaction [18].

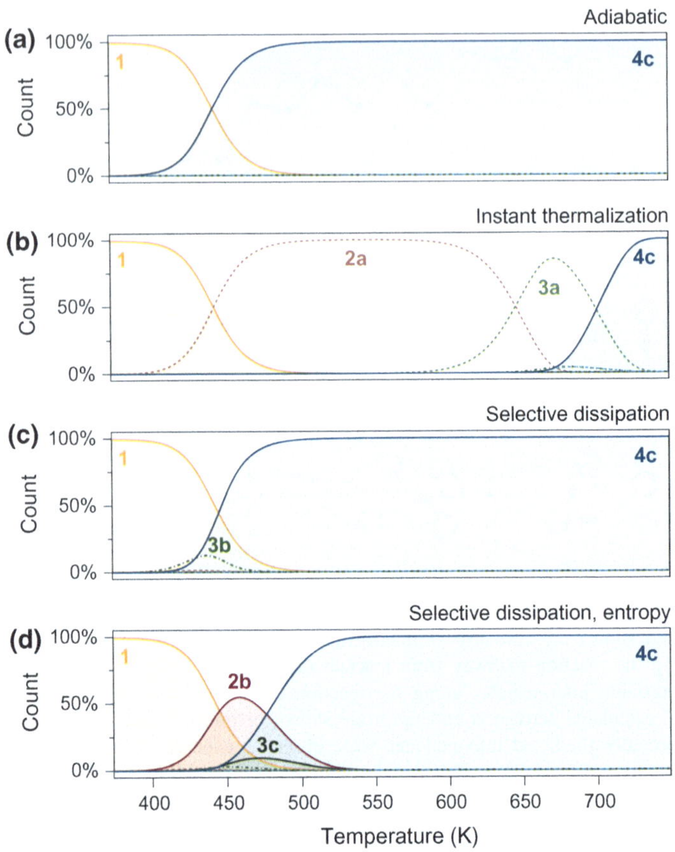

Fig. 10 Calculated temperature-dependent relative concentrations of reactant, intermediates and product determined by solving kinetic rate equations for the reaction between two 1,2-bis (2-ethynyl phenyl)ethyne molecules, shown in Fig. 8. **a** Adiabatic model: only reactant **1** and product **4c** are stabilized. **b** Instant thermalization model: only intermediates followed by the highest activation barriers are stabilized (**2a** and **3a**) and the product **4c** is predicted to be formed only at high temperatures. **c** Selective dissipation effects: the addition of selective dissipation causes **4c** to form at a lower temperature and leads to the stabilization of **3b**. **d** Entropic effects: the addition of entropic effects results in a predominant stabilization of intermediates **2b** and **3c**, in agreement with experimental observations. Reprinted by permission from Macmillan Publishers Ltd: Ref. [2], copyright 2016

7 Summary and Outlook

Studying kinetics and thermodynamics with electronic structure theory and transition state theory can provide important insight about chemical reactions on surfaces. Properly accounting for both enthalpy and entropy contributions of the free energy can be important to understand all aspects of chemical reactions. For example, the entropy of the associative desorption of hydrogen explains the occurrence of endothermic dehydrogenation reactions on surfaces, and the inclusion of entropic terms for molecular adsorbates may explain the stabilization, or destabilization, of particular structures.

It is needless to say that the theoretical description of reactions related to on-surface synthesis has become quite sophisticated, provided by the possibilities of ever-increasing computational resources. However, a complete free energy description of an on-surface reaction is still a costly endeavor, and only a few studies have attempted obtaining this so far. It is important to have this in mind when judging future work. In particular, do we want theoretical studies to provide extremely detailed insight into only a few on-surface reactions, or do we want mechanisms to be calculated for a large number of reactions? The former option would properly minimize the number of theoreticians being able—or willing—to provide input about reaction mechanisms as support to original experimental work. In either case, there are some quite simple measures in order to improve our appreciation of on-surface reactions provided by calculated pathways. For example, to account for the entropy gain from desorbed hydrogen molecules—as well as other leaving-groups—which can be done as a simple correction based on tabulated data, without additional computational resources required.

References

1. Björk, J.: Reaction mechanisms for on-surface synthesis of covalent nanostructures. J. Phys.: Condens. Matter **28**, 083002–083016 (2016)
2. Riss, A., Pérez Paz, A., Wickenburg, S., Tsai, H.-Z., De Oteyza, D.G., Bradley, A.J., Ugeda, M.M., Gorman, P., Jung, H.S., Crommie, M.F., Rubio, A., Fischer, F.R.: Imaging single-molecule reaction intermediates stabilized by surface dissipation and entropy. Nat. Chem. **8**, 678–683 (2016)
3. Cai, J., Ruffieux, P., Jaafar, R., Bieri, M., Braun, T., Blankenburg, S., Muoth, M., Seitsonen, A.P., Saleh, M., Feng, X., Müllen, K., Fasel, R.: Atomically precise bottom-up fabrication of graphene nanoribbons. Nature **466**, 470–473 (2010)
4. Talirz, L., Ruffieux, P., Fasel, R.: On-surface synthesis of atomically precise graphene nanoribbons. Adv. Mater. **28**, 6222–6228 (2016)
5. Zhang, Y.-Q., Kepčija, N., Kleinschrodt, M., Diller, K., Fischer, S., Papageorgiou, A.C., Allegretti, F., Björk, J., Klyatskaya, S., Klappenberger, F., Ruben, M., Barth, J.V.: Homo-coupling of terminal alkynes on a noble metal surface. Nat. Commun. **3**, 1286 (2012)
6. Gao, H.-Y., Wagner, H., Zhong, D., Franke, J.-H., Studer, A., Fuchs, H.: Glaser coupling at metal surfaces. Angew. Chem. Int. Ed. **52**, 4024–4028 (2013)
7. Cirera, B., Zhang, Y.-Q., Björk, J., Klyatskaya, S., Chen, Z., Ruben, M., Barth, J.V., Klappenberger, F.: Synthesis of extended graphdiyne wires by vicinal surface templating. Nano Lett. **14**, 1891–1897 (2014)
8. Klappenberger, F., Zhang, Y.-Q., Björk, J., Klyatskaya, S., Ruben, M., Barth, J.V.: On-surface synthesis of carbon-based scaffolds and nanomaterials using terminal alkynes. Acc. Chem. Res. **48**, 2140–2150 (2015)
9. Sun, Q., Zhang, C., Kong, H., Tan, Q., Xu, W.: On-surface aryl–aryl coupling via selective C–H activation. Chem. Commun. **50**, 11825–11828 (2014)

10. Li, Q., Yang, B., Lin, H., Aghdassi, N., Miao, K., Zhang, J., Zhang, H., Li, Y., Duhm, S., Fan, J., Chi, L.: Surface-controlled mono/diselective ortho C–H bond activation. J. Am. Chem. Soc. **138**, 2809–2814 (2016)
11. Haq, S., Hanke, F., Sharp, J., Persson, M., Amabilino, D.B., Raval, R.: Versatile bottom-up construction of diverse macromolecules on a surface observed by scanning tunneling microscopy. ACS Nano **8**, 8856 (2014)
12. Svane, K.L., Hammer, B.: Thermodynamic aspects of dehydrogenation reactions on noble metal surfaces. J. Chem. Phys. **141**, 174705–174711 (2014)
13. Heinrich, B.W., Ahmadi, G., Müller, V.L., Braun, L., Pascual, J.I., Franke, K.J.: Change of the magnetic coupling of a metal–organic complex with the substrate by a stepwise ligand reaction. Nano Lett. **13**, 4840–4843 (2013)
14. Cirera, B., Giménez-Agulló, N., Björk, J., Martínez-Peña, F., Martin-Jimenez, A., Rodriguez-Fernandez, J., Pizarro, A.M., Otero, R., Gallego, J.M., Ballester, P., Galan-Mascaros, J.R., Ecija, D.: Thermal selectivity of intermolecular versus intramolecular reactions on surfaces. Nat. Commun. **7**, 11002 (2016)
15. Björk, J.: Thermodynamics of an Electrocyclic Ring-Closure Reaction on Au(111). J. Phys. Chem. C **120**, 21716–21721 (2016)
16. Chase, M.W.: NIST-JANAF themochemical tables, 4th edn. J. Phys. Chem. Ref. Data **9**, 1310 (1998)
17. Floris, A., Haq, S., In't Veld, M., Amabilino, D.B., Raval, R., Kantorovich, L.: Driving forces for covalent assembly of porphyrins by selective C–H bond activation and intermolecular coupling on a copper surface. J. Am. Chem. Soc. **138**, 5837–5847 (2016)
18. Björk, J., Zhang, Y.-Q., Klappenberger, F., Barth, J.V., Stafström, S.: Unraveling the mechanism of the covalent coupling between terminal alkynes on a noble metal. J. Phys. Chem. C **118**, 3181–3187 (2014)

Reactivity *on* and *of* Graphene Layers: Scanning Probe Microscopy Reveals

Oleksandr Ivasenko and Steven de Feyter

Abstract In this chapter we give an overview of different chemical transformations that can be done on graphene layers and characterized using scanning tunneling (STM) and atomic force microscopies (AFM). We place particular emphasis on the diversity of reactions, systems and synthetic strategies that are now available to surface scientists working in various fundamental and applied research fields. Using imine formation as the model reaction we discuss common principles of building block design and reaction outcomes specific to interfacial synthesis. Then other reactions are briefly overviewed, including: photo- and electrochemically assisted processes, transformations initiated by STM, and finally, reactions involving the covalent modification of graphene layers.

1 Introduction

Chemistry of and on surfaces is of tremendous importance in material and bio sciences. It will not come as a surprise that surface chemistry takes a special place in nanoscience and nanotechnology as scaling down increases the surface to volume ratio. Among other techniques, scanning probe microscopy tools made it possible to probe surfaces, sometimes at the atomic scale. As these techniques can be used under ambient conditions or even in liquids, they became attractive to chemists and material scientists too. This review aims at giving an overview of molecule-based surface reactivity under ambient conditions or at the liquid-solid interface, on graphite and graphene. While it's not the ambition to give a comprehensive overview, a range of different surface reactivity concepts are highlighted.

O. Ivasenko (✉) · S. de Feyter
Department of Chemistry, KU Leuven – University of Leuven,
Celestijnenlaan 200F, 3001 Leuven, Belgium
e-mail: oleksandr.ivasenko@kuleuven.be

S. De Feyter
e-mail: steven.defeyter@kuleuven.be

D. G. de Oteyza and C. Rogero (eds.), *On-Surface Synthesis II*,
Advances in Atom and Single Molecule Machines,
https://doi.org/10.1007/978-3-319-75810-7_3

Below we outline some of the methodological aspects relevant to the topic of this chapter and considered when studying chemical reactivity at interfaces.

Substrates

Highly oriented pyrolytic graphite (HOPG) is a widely used substrate in scanning tunneling microscopy studies. Conductive, chemically inert, with large (up to hundreds of microns in size) high-quality atomically flat terraces, HOPG is perfectly suited for STM studies and represents an ideal platform for the investigation of various phenomena involving the adsorption on the basal plane (graphene sheets) of graphite. Supramolecular and covalent chemistry developed on HOPG substrates can be easily transferred onto other related materials: various types of graphene, films of reduced graphene oxide, different carbon-based adsorbents (e.g. porous graphitic carbon—PGC), rendering such studies as highly important for the development of future sensors [1, 2], electronic devices [3], catalysts [4], etc.

Molecular design

Successful STM imaging requires the localization of molecular species on the conductive substrate, preferably in the form of monolayers or ultrathin films. It is quite common to use starting materials that spontaneously form stable monolayers on surfaces. This is achieved using building blocks with extended aromatic cores (for π-π interactions with the substrates), long alkyl chains (van der Waals intermolecular interactions) or self-complementary hydrogen bonding moieties. In such cases the reaction can be detected with STM via the difference in the monolayer structures of starting materials and products (due to different molecular geometries and intermolecular interactions, see example in Fig. 1).

For the formation of highly ordered films of products strong self-assembly of starting materials could be a curse since monolayer reconstruction entails overcoming additional barriers and often results in multiple monolayer defects. A rather interesting approach is orthogonal building block design in which self-assembly function and reactivity are sufficiently independent. For example, Zimmt et al. have

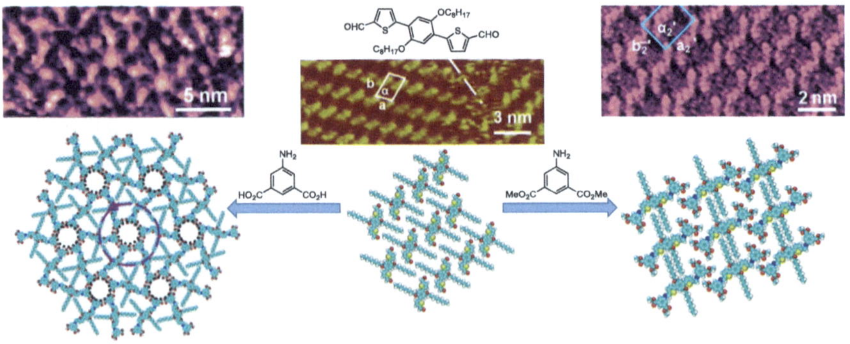

Fig. 1 Distinct monolayer structures of starting dialdehyde and two bis-imine products. Adapted with permission from [5]. Copyright © 2013 American Chemical Society

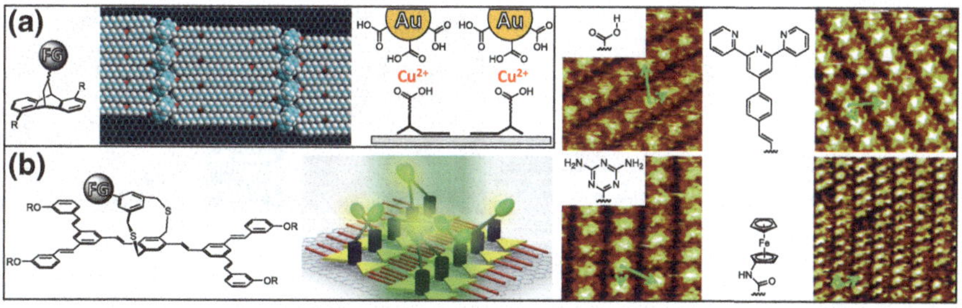

Fig. 2 Examples of building block designs in which self-assembly and reactive functionalities do not interfere with each other. **a** adapted from [6], copyright © 2012, with permission from Elsevier. **b** adapted with permission from [7, 10]. For [10] copyright © 2016 American Chemical Society, for [7] copyright © 2014 Wiley-VCH Verlag GmbH & Co. KGaA

designed carboxy-functionalized building blocks suitable for chemical binding with gold nanoparticles (Fig. 2a, [6]), while the group of Attias has developed a series of original molecular probes bearing optically, electrochemically, hydrogen-bonding and metal-coordination active units (Fig. 2b, [7–10]).

Activation methods

Similarly to solution-based syntheses, the chemical reactivity of reagents at interfaces can be activated thermally, catalytically (e.g. tuning pH in acid-catalyzed reaction), photo- or electrochemically. The exact activation method depends on the reaction and often defines the preferred synthetic setup: type of solvent, temperature, work-up, etc. For example, aqueous electrolytes are typically used in electrochemical (EC-STM) experiments, while on-surface syntheses of boroxine and imine polymers can involve solvent-free conditions and delivery of reagents (e.g. water [11], aldehyde [12]) in the form of vapors from the gas phase.

Unique to SPM studies is the use of SPM tips for localized mechanochemical and/or electrical activation of reactants. Such activation is often fully compatible with SPM measurements, and allows nanoscale manipulation and testing of surface structures (Sects. 5, 6 and 7).

Characterization

Scanning tunneling microscopy is a powerful technique for nanoscale structural characterization of conductive surfaces and monolayers adsorbed on them. In studying chemical reactions, STM can provide accurate (typical uncertainties ±0.1 nm) values for unit cells and, in the case of polymers, the length of repeat units. Together with complementary data from DFT modelling or known X-ray structure, the STM measurements can serve as a support of the intended chemical transformation. Adsorbate morphology can be obtained from atomic force microscopy (AFM) or scanning electron microscopy (SEM, e.g. 2.1 Fig. 9) investigation of dry films.

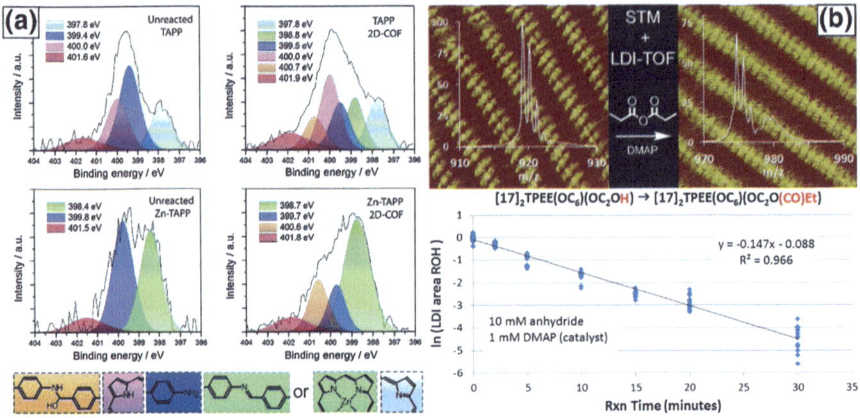

Fig. 3 Examples of STM-complementary characterizations using: **a** XPS in the synthesis of imine-based 2D COFs [18]—published by The Royal Society of Chemistry. **b** LDI-TOF in the chemical transformation of a carboxylic building block. Adapted with permission from [19]. Copyright © 2017 American Chemical Society

Unfortunately, except for special cases of STM markers [13]—functional groups that have high specific STM contrast and can be used to characterize mixtures of structurally related 'marked' and 'unmarked' molecules [14, 15], STM cannot provide detailed information about the chemical nature of the imaged adsorbates. That is why it is desirable and often mandatory to perform complementary characterization. This could be classical solution techniques, e.g. NMR, provided that there is sufficient amount of material for the analysis [16]. Otherwise, dedicated surface techniques are particularly convenient. Thus, X-ray photoelectron spectroscopy (XPS) has been used to follow formation of polyboroxines [17] and polyimines (Fig. 3a, [18]). Another interesting example is the application of laser desorption/ionization time-of-flight mass-spectrometry (LDI-TOF) and its matrix-assisted modification (MALDI-TOF) for successful characterization of chemical changes in physisorbed monolayers (Fig. 3b, [19]).

2 Chemical Reactivity

Since the formation of imines is by far the most studied organic reaction on HOPG, we discuss it in details (Sect. 2.1) covering general aspects pertinent to many other interfacial reactions, and only then provide an overview of other reactions that have been tested (Sect. 2.2).

2.1 Synthesis of Imines

Geometry of the building blocks

Basic concepts of supramolecular design that relies on the reversible formation of non-covalent interactions are also applicable to the design of new covalent architectures especially when the synthesis occurs with the reversible formation of new covalent bonds and the designed architecture represents the most stable state (thermodynamic control). Formation of imines can be reversible in a number of different regimes: (1) in aqueous solutions with carefully controlled optimal pH range, (2) at high temperatures in solid state in the presence of water vapors (often using $CuSO_4{\cdot}5H_2O$ as a source of water), (3) in certain organic solvents at room or slightly elevated temperatures. Thus, a variety of 1D polymers and 2D covalent organic frameworks (2D COFs) have been formed from suitably functionalized di-, tri- and tetra-substituted building blocks (e.g. Figure 4). If one of the building blocks is a monoaldehyde or monoamine then the product of condensation is a small molecule (Fig. 1, [5, 20]).

Single C–C bonds around imine links are free to rotate, during synthesis such rotations lead to connectivity mismatches and network deformations (defects).

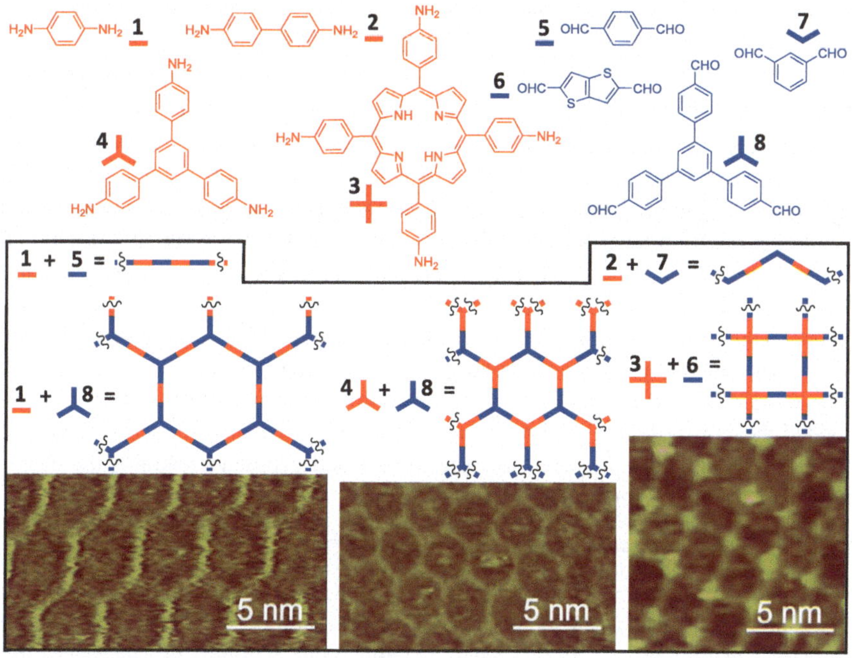

Fig. 4 Examples of the topological design of various imine-based polymeric structures. Adapted from [21], with the permission of AIP Publishing

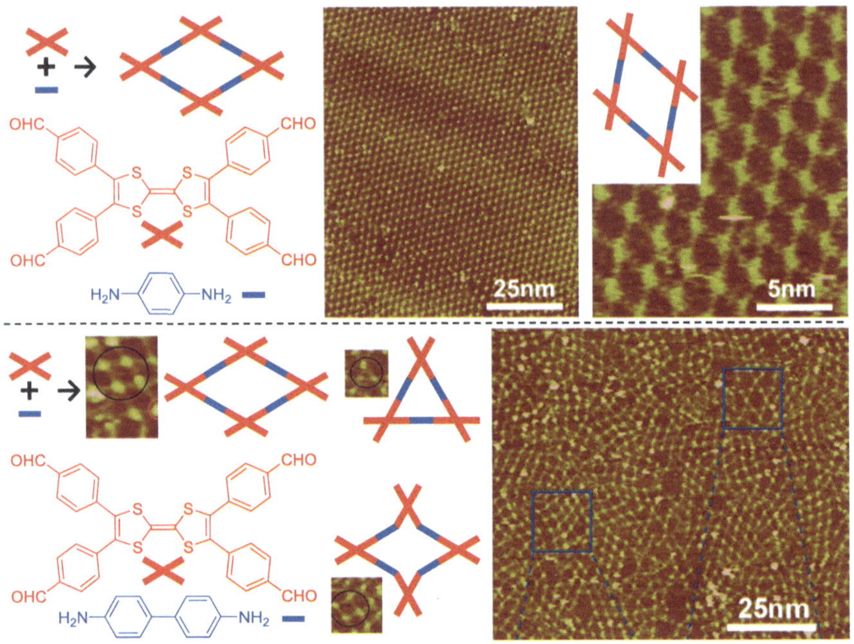

Fig. 5 Longer biphenylene linker results in a larger number of different defects in 2D COF compared to shorter p-phenylene, which forms highly crystalline COF uniform rhomboidal pores. Adapted with permission from [22]. Copyright © 2015 American Chemical Society

While every imine COF has defects, their surface density is a function of synthetic protocol (e.g. thermal annealing in the presence of water vapors often allows to overcome energy barriers and improve the quality of polymeric structures [12]) and rigidity of the building blocks (see Fig. 5 and other examples in [22–24]).

Above we have seen how small variations in molecular geometry (rotational disorder along C–C bonds) results in significant changes to the structure of reaction products. Larger changes to the building block structure may have drastic outcomes. Thus, very recently, Wang et al. demonstrated that non-planar triamine reacts with dialdehydes at gas-solid interface forming exclusively 1:3 adducts (Fig. 6a, [25]). Varying the concentration, ratio and temperature (up to 200 °C) yielded the same adducts, without formation of any other oligomeric or polymeric products. The authors proposed that the rigid tripod geometry restrains the accessibility of terminal groups for further reactions.

Stoichiometry of reagents

Stoichiometry of reagents in solution can affect various interfacial processes. An interesting example is a kinetically controlled formation of 2:1 diamine:dialdehyde non-covalent co-assembly (Fig. 6b) at room temperature at liquid-solid interface [26]. Such transient structure (life period <2 h) formed only if the diamine:dialdehyde ratio

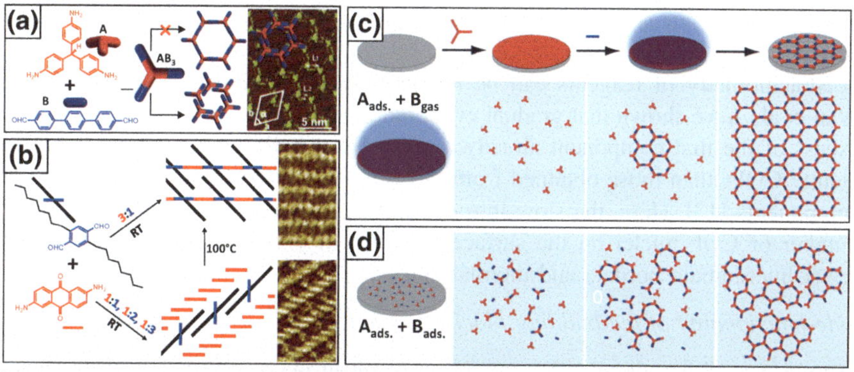

Fig. 6 Examples of: **a** a thermodynamically stable self-assembly of AB_3 adduct. Adapted with permission from [25]. Copyright © 2017 Royal Society of Chemistry. **b** metastable non-covalent co-assembly of a diamine and a dialdehyde. Adapted with permission from [26]. Copyright © 2017 American Chemical Society. **c** The controlled nucleation of 2D COF via a gradual introduction of one of the coupling components versus **d** multiple nucleation sites generated upon heating stoichiometric mixtures of both components. Adapted with permission from [12]. Copyright © 2013 American Chemical Society

in solution was equal or lower than 1:1. At 3:1 stoichiometry in solution only 1D polymeric structures were observed on the surface.

Stoichiometry of reagents is also an important parameter in any bi-component reaction. Any large deviation from the ideal stoichiometry should have a negative impact on the yield and degree of polymerization—ultimately resulting in the formation of small functional molecules (e.g. in the reaction of a triamine A and dialdehyde B large excess of B should yield new trialdehyde AB_3—analogous to the one in Fig. 6a, while large excess of A should furnish new tetraamine A_2B). Indeed such structural variations and strong dependency of the product composition on the stoichiometry of reagents were observed in surface-supported imine synthesis on Au (111) in ultra-high vacuum (UHV) [27]. In majority of the reported imine polymerizations on HOPG however, so strong influence of reagents stoichiometry was not detected. To understand this difference in behavior we should consider the following factors:

- stronger than on HOPG substrate-molecule interactions on Au(111) result in lower on-surface mobility of adsorbed precursors and small oligomers;
- UHV precludes normal pathway (i.e. hydrolysis by dissolved or vaporized water) for reversible formation/braking of imine bonds;
- in the case of reaction at liquid-solid interface there is also additional possibility for preferential adsorption of larger oligomers.

All of them render the abovementioned imine formation in UHV essentially irreversible and strongly dependent on the initial state (stoichiometry, on-surface distribution of precursors); while typical reactions on HOPG are done under reversible conditions, when polymerization is energetically favorable and once reaction began it is thermodynamically driven towards large oligomers and polymers.

Stoichiometry of reagents can be intentionally varied during synthesis. Thus, Wan et al., have shown that gradual evaporation of the second component into the excess of the first component already adsorbed on HOPG results in much higher quality COFs than those obtained from direct heating HOPG with mixture of both components [12]. Here, the slow introduction of the second component limits the number of COF nuclei on the surface (Fig. 3c, d) yielding larger domains with fewer interdomain borders and internal defects.

Side-functionality of the building blocks

Introduction of functional side-groups can result in major alternation of precursor reactivity or properties of the final product. Thus, introduction of phenolic OH group into imine polymers has been shown (depending on the structure of the building blocks and substitution pattern) to lead to unusual aggregation behavior [28], improved chemical stability (Fig. 7a, [29]) or even alter the chemical nature of the COF through additional chemical transformations (Fig. 7b, [30]).

Dynamic combinatorial chemistry

In the pioneering work by Samori et al. competitive reactivity of aliphatic diamines with a substituted benzaldehyde in the presence of HOPG was investigated [31]. The authors have shown that thermodynamically less stable self-assembly of bis-imine formed from a short diamine upon addition of longer diamine into the liquid phase undergoes quick trans-imination yielding a more stable self-assembly of bis-imine formed from the longer diamine (Fig. 8a). This simple concept of directing chemical synthesis via preferential adsorption of one of the possible products is very appealing. Unfortunately, so far other systems have shown much

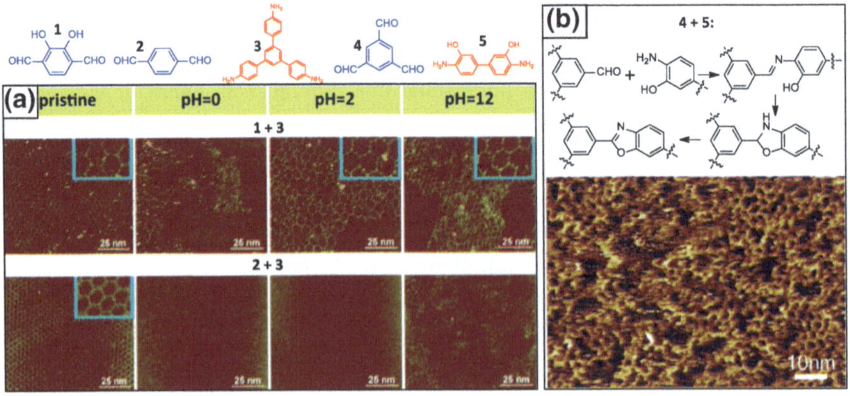

Fig. 7 Improved stability **a** and post-synthetic transformation **b** of OH-functionalized imine 2D COFs. **a** adapted with permission from [29]. Copyright © 2017 Royal Society of Chemistry. **b** adapted with permission from [30]. Copyright © 2015 Royal Society of Chemistry

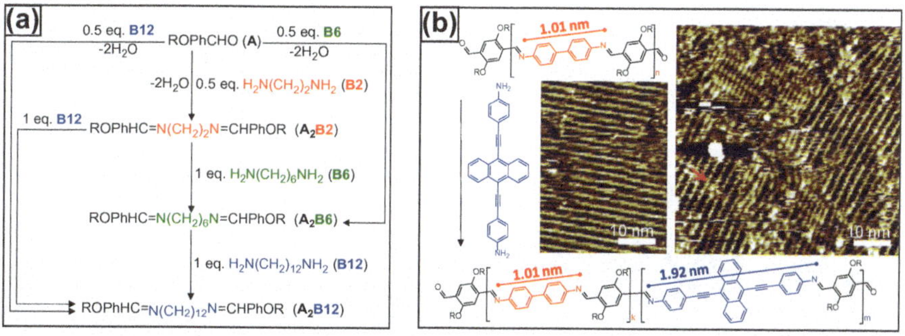

Fig. 8 Examples of dynamic covalent chemistry directed by preferential adsorption. **b** adapted from [32], copyright © 2017, with permission from Elsevier

more complex behavior (Fig. 8b, [32]) when the competition for the surface occurs not only with other reaction products but also with starting material and the selectivity in singling out one of the reaction products is unsatisfactory [16]. Interestingly, in the latter work strong heterogeneous catalysis of imine reactions even for the reaction of weakly adsorbing terephthalic aldehyde and 3,5-bis (tert-butyl)aniline was observed. Currently we investigate if adsorbents (PGC, SiO_2, etc.) can serve as general catalysts of other reactions besides formation of imines.

Formation of multilayers and thin films

Non-homogeneous deposition of precursors will result in heterogeneous polymeric films with varied thickness throughout the surface. Lei et al. developed a general and convenient methodology for the optimization of room temperature polymerizations at liquid-solid interface in which, keeping the ratios amine:aldehyde fixed, the total concentration is carefully tuned for optimal polymer quality [33]. Using high total concentrations it was also possible to grow multilayers.

We have already mentioned the formation of highly ordered COF monolayer from substituted tetrathiafulvalene (TTF) tetraaldehyde and p-phenylenediamine (Fig. 5). Surprisingly growth of the second layer appears to be templated by the first and results in eclipsed stacking. Stacking was the result of strong and specific π-π interactions between TTF cores, as convincingly demonstrated by spontaneous self-assembly of unsubstituted TTF on top of TTF sites of the COF monolayer [36]. The authors have noticed that the quality of the bilayer was higher if the polymerization was done in a stepwise fashion, otherwise monomer stacking and polymerization occur simultaneously leading to much more disordered COF films.

The morphology of the polymeric films also depends on the deposition method itself. Kunitake et al. have demonstrated that, unlike gas phase deposition leading to uniform films, deposition from liquid phase (pH-controlled synthesis in aqueous solutions) yields sponge-like structures (Fig. 9a, [34, 37]).

Finally, it should be mentioned that micropatterns of polymeric films can be deposited on any arbitrary substrate using lithographically controlled wetting or conventional ink-jet printing (Fig. 9b, [35]).

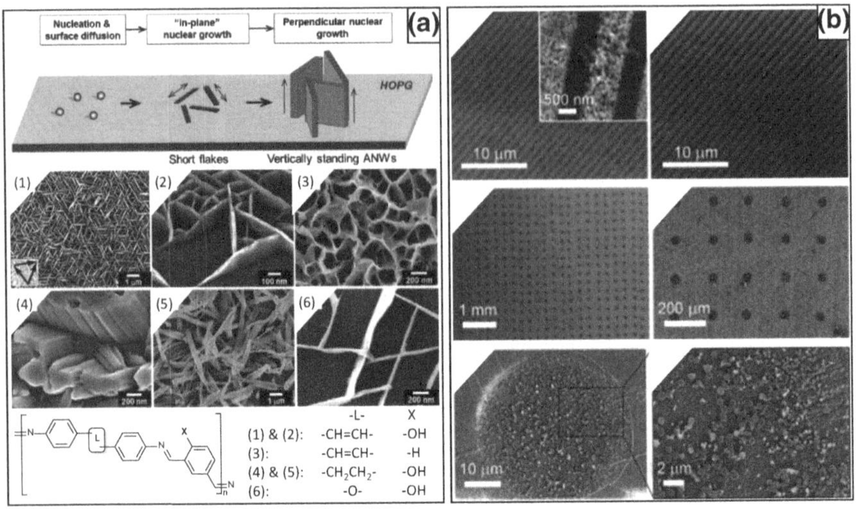

Fig. 9 Examples of **a** different growth and **b** lithographic deposition of polymeric films. **a** adapted from [34], copyright © 2013, with permission from Elsevier. **b** adapted with permission from [35]. Copyright © 2015 Wiley-VCH Verlag GmbH & Co. KGaA

2.2 *Other Reactions*

One of the simplest chemical transformation is protonation. Li et al. have shown that addition of strong acids (HNO_3, CH_3SO_3H, CF_3SO_3H) greatly accelerates the self-assembly of neutral di-pyridyl building blocks with protonated intermediates acting as active species directing on-surface self-assembly [38]. Formation of copper salt linkers was used for nanopatterned assembly of functionalized gold nanoparticles on top of strongly physisorbed monolayers on graphite (Fig. 2a, [6]).

In many aspects metal coordination is similar to protonation. In situ generation of potassium coordination complexes at HOPG-solution interface was demonstrated for valinomycin (Fig. 10a, [39]), G_4 quartets of 9-(n-octadecyl)guanine (Fig. 10b, [40]) and tetraethylene glycol derivative (mimicking the complexation of crown ethers, Fig. 10c, [41]).

Various nitrogen-containing building blocks have also been explored to create surface nanopatterns of ‘0D’ metal complexes [45], to template the nucleation of metallic nanoparticles (Fig. 10d, [42]), to synthesize 1D coordination polymers (Fig. 10e, [43]) and 2D metal-organic frameworks (MOFs, Fig. 10f, [44]).

Procedure-wise one of the simplest organic transformations are thermally induced intramolecular isomerizations. Heating is all that is necessary to consider. An interesting example of such isomerizations is tandem Claisen-Hiratani rearrangement—a [3]-sigmatropic rearrangement in which an allyl aryl ether is converted thermally to allylphenoles forming new C–C bond [46–48].

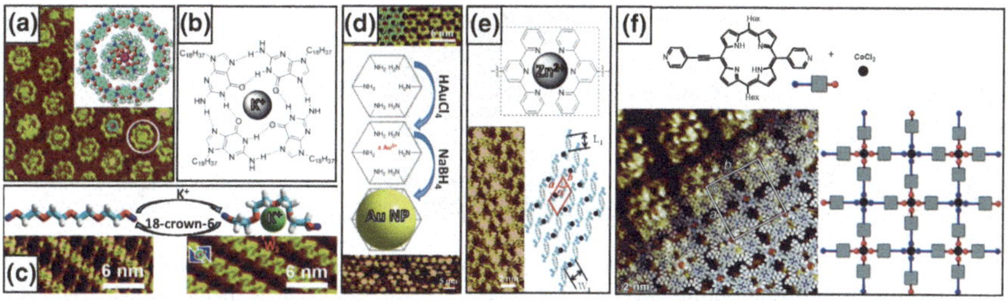

Fig. 10 Formation of coordination complexes at HOPG-solution interface. **a** Potassium-valinomycin as guest inside a macrocycle. Adapted from [39] with permission of The Royal Society of Chemistry. Copyright © 2013 Royal Society of Chemistry. **b** dynamic assembly-reassembly of substituted G4 quartets [40]. **c** Nanoscale actuation of oligoether 'springs'. Adapted from [41] with permission of The Royal Society of Chemistry. Copyright © 2014 Royal Society of Chemistry. **d** Pre-concentration of Au^{3+} with subsequent templated reduction to Au nanoparticles. Adapted from [42] with permission of The Royal Society of Chemistry. Copyright © 2015 Royal Society of Chemistry. **e** On-surface synthesis/self-assembly of 1D (adapted with permission from [43]; copyright © 2015 American Chemical Society) and 2D (adapted from [44] with permission of The Royal Society of Chemistry; copyright © 2014 Royal Society of Chemistry) metalorganic supramolecular polymers

Analogues to imine formation, C=C condensation of an aromatic dialdehyde and barbituric acid (Knoevenagel reaction) was investigated at liquid/HOPG and vapor/HOPG interfaces [49]. Here STM was successfully used for quantitative analysis of on-surface distribution of reaction products.

Also the synthesis of 2D COFs on carbon-based surfaces is not limited to imines only. A very popular reaction for the synthesis of surface-supported COFs is the formation of boroxines from different boronic acids (Fig. 11a, [17, 50, 53]). An interesting development is the use of boroxine acid esters for in situ hydrolysis and polymerization [11]. Thin films via bi-component condensations of boronic acids and hexahydroxytriphenylene (HHTP) were first prepared as early as 2011 [54], but only very recently formation of monolayer films of such 2D-COFs demonstrated to take place under mild conditions at liquid-solid interface (Fig. 11b, [51, 55]). Finally, it is possible to 'marry' different polymerization chemistries: 3- and 4-formylphenylboronic acids react with a triamine building block forming isomeric imine-boroxine hybrid 2D COFs (Fig. 11c, [52]).

3 Photochemical Reactivity

Presence of azobenzene moieties in derivatives is the calling card of a photo-switching functionality anticipated in the design. Indeed, *cis-trans* isomerization of N=N bond in azobenzenes is a well-defined process free from any side-reaction or degradation. For example isophthalic acid derivative were shown to form stable self-assembly that upon irradiation undergo reversible isomerization [56, 57]. The

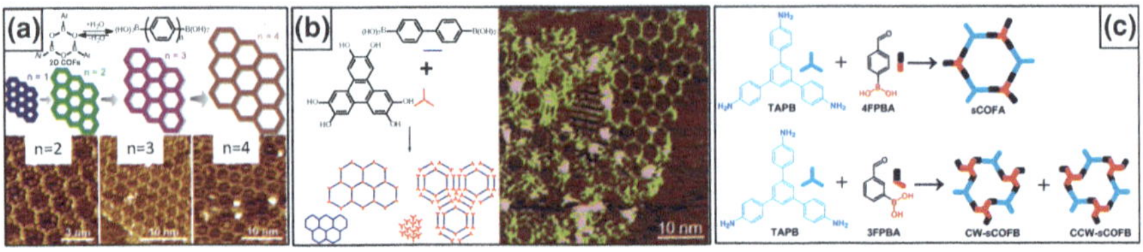

Fig. 11 Boronic acid based 2D COFs on HOPG via **a** dehydration of boronic acids into boroxines, **b** polycondensation of diboronic acids and hexahydroxytriphenylene (HHTP), **c** simultaneous formation of boroxine and imine linkages from the reaction of bifunctional building blocks. **a** adapted with permission from [50]. Copyright © 2012 American Chemical Society. **b** adapted from [51] with permission of The Royal Society of Chemistry. Copyright © 2016 Royal Society of Chemistry. **c** adapted from [52] with permission of The Royal Society of Chemistry. Copyright © 2017 Royal Society of Chemistry

assemblies of corresponding *cis*- isomers have different structure and lower stability compared to *trans*-. The latter is due to non-planarity of *cis*-azobenzene resulting in less efficient substrate molecule interaction. Alternative designs (Fig. 12a) when *cis-trans* isomerization is occurring on azobenzene units pointing away from the surface into the solution phase were also realized [58, 59].

More complex supramolecular architectures were also tested. Thus, exploring the host-guest concept, assemblies of photoactive host (Fig. 12b, [60, 63]) and guest (Fig. 12c, [61, 64]) were studied. In the former example photoisomerization was demonstrated to influence surface adsorption of a small molecular probe (coronene), while the latter example illustrated variability and responsiveness of the flexible host network to differently shaped isomeric guests. Exploring formation of supramolecular polymers, Yamauchi et al. demonstrated phototriggered supramolecular polymerization of a naphthalene-azobenzene dyad [62]. Here *cis-trans* photoisomerization allowed to produce shorter polymeric fibers with narrower polydispersity compared to those generated by thermal polymerization (Fig. 12d).

Stilbene moiety, being structurally similar to azobenzene can also undergo cis-trans photo-isomerization. In dilute solutions this is a typical outcome of photoirraditation of various building blocks with olefinic C=C bonds. However, in dense assemblies (on surface or in crystals), reactions can go further and produce covalent dimers (e.g. Fig. 13a, [65, 66]). Such transformations are often topochemical. The required close proximity and proper orientation of reacting bonds can be achieved through supramolecular design, e.g.: for efficient on-surface photocoupling supramolecular dimers of 1,2-bis(4-pyridyl)ethylene (BPE) bound by two isophthalic acid molecules were confined in a porous self-assembly on HOPG surface (Fig. 13b, [67]).

Photodimerizations can involve reconstruction of multiple bonds as in the photodimerization of anthracenes (Fig. 13c, [68]) or arylethenes (Fig. 13d, [69–71]) and in the metal-free on-surface photochemical homocoupling of 4-ethynylbenzoic acid (Fig. 13e, [72]).

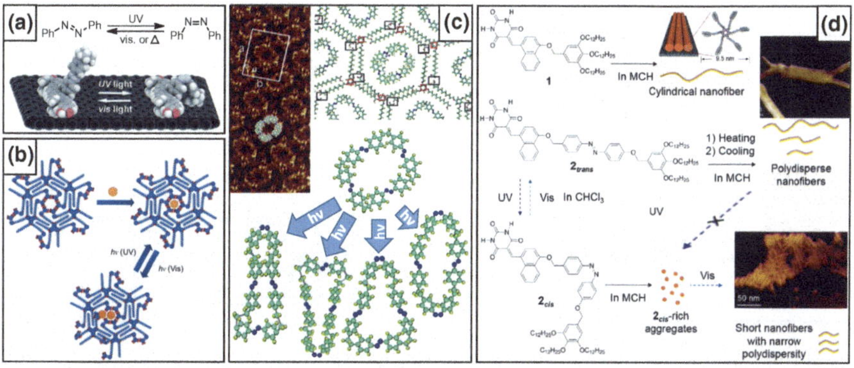

Fig. 12 The application of azobenzene units in **a** structural phototransformations of physisorbed monolayers, **b** the release/adsorption of guests; **c** tuning supramolecular network of the host in response to photoisomerization of the guest; **d** controlling the growth and morphology of nanofibers. **a** adapted with permission from [59]—published by the PCCP Owner Societies. **b** adapted with permission from [60]. copyright © 2013 Wiley-VCH Verlag GmbH & Co. KGaA. **c** adapted with permission from [61]; copyright © 2009 American Chemical Society. **d** adapted with permission from [62]; copyright © 2017 The Chemical Society of Japan

Going beyond the formation of small molecules, it is possible to design 1,3-diacetylene building blocks that undergo topochemical photopolymerization when assembled on graphite [56]. Building blocks with one diacetylene unit produce 1D polymeric chains (Fig. 14a), while introduction of two or more reactive units upon exhaustive photopolymerization can furnish non-porous 2D COFs (Fig. 14b, [73]). In the latter example formation of polydiacetylene and polyacetylene chains was assumed. Zimmt et al. demonstrated that 1,3-diacetylene units are also powerful supramolecular synthons (Fig. 14c, [74]), that, together with the possibility of photopolymerization (Fig. 14d, [75]), makes them very interesting for versatile surface nanopatterning.

Combination of azobenzene and diacetylene units within one building block resulted in reach photochemical behavior of on-surface assemblies [77]. Depending on the irradiation wavelength, diacetylene polymerization ($\nu = 254$ nm) or reversible *trans-cis* ($\nu = 365$ nm)/*cis-trans* ($\nu = 435$ nm) isomerizations of azobenzene cores can be triggered (Fig. 14e).

4 Electrochemical Reactivity

Good chemical stability and electrical conductivity of graphene sheets makes carbon-based electrodes (HOPG, graphitic rods, glassy-carbon, graphene on different supports, etc.) interesting for various electrochemical (EC) transformations.

Both inorganic [78] and organic nanostructures [79] on surfaces can be directly prepared by electrochemical synthesis. Supramolecular monolayers bearing

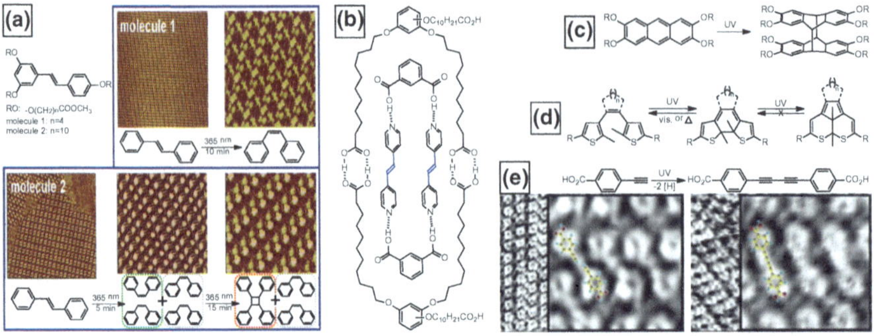

Fig. 13 Photochemical transformation of covalent bonds: **a** supramolecular network in which stilbene units undergo isomerization and partial dimerization; **b** oriented and confined inside supramolecular host pores BPE molecules readily dimerize upon photoirradiation [67]; **c**, **d** the photochemistry of anthracenes and arylethenes that has already been explored at HOPG-solution interface in [68] and [69–71], respectively; **e** photoinitiated metal-free oxidative coupling of acetylenes. **a** adapted with permission from [65]; copyright © 2014 American Chemical Society. **e** adapted with permission from [72]; copyright © 2016 American Chemical Society

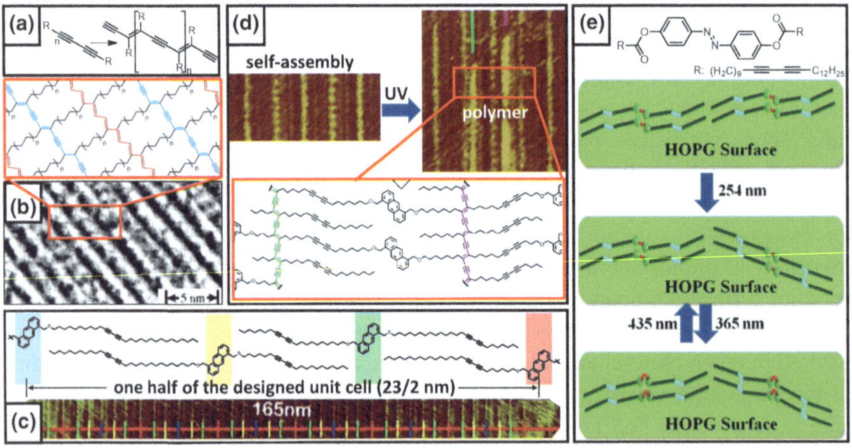

Fig. 14 Photopolymerization of diacetylene assemblies: **a** general reaction scheme; **b** model and STM of a 2D covalent frameworks; **c** design and **d** photopolymerization of four-component co-assembly; **e** building block with dual photofunctionality bearing both diacetylene and azobenzene units. **b**, **c**, **d**, **e** were adapted with permission from [73—STM image, 76—model], [74], [75] and [77], respectively. [73] copyright © Wiley-VCH Verlag GmbH & Co. KGaA. [74] copyright © 2012 American Chemical Society. [75] copyright © 2015 American Chemical Society. The model in (**b**) adapted by permission from Macmillan Publishers Ltd: Nature Chemistry [76], copyright © 2013. [77] copyright © 2012 American Chemical Society

electrochemically active groups (Fig. 15a, [80]) can be prepared by conventional self-assembly from solution and characterized using STM (structure) and cyclic voltammetry (CV, electrochemical properties).

The electrochemical STM (EC-STM, Fig. 15b) on atomically flat substrates (HOPG, graphene) is particularly suitable for structural characterization of interfacial redox processes. For example, recently we have reported in situ generation and nanoscale characterization of dicationic, monocationic and neutral assemblies of dibenzylviologen (Fig. 15c, [81]). Another recent example is the EC-STM surface characterization in a study of electrocatalytic activity of porphyrins adsorbed on graphene/Au(111) (Fig. 15d, [82]).

5 Tip-Induced Reactivity

Besides powerful imaging, scanning probe microscopies are also capable of nanoscale modification of on-surface structures. In STM, the tip can locally modify the matter because of microscopic dimensions (mechanical actuation, also possible for AFM), large 10^7–10^8 V/cm^2 electric field (electrostatic polarization), high current densities (electronic excitation), local heating or any combination of mentioned pathways. Often the exact origin of tip-induced changes is hard to determine and can only be guessed from the chemical nature of on-surface structures and type of nanoscale modifications.

Changes can be limited to non-covalent interactions in physisorbed assemblies, e.g.: (a) movement of single C_{60} molecules between the pores of a host network (Fig. 16a, [83]); (b) shattering a large domain of a supramolecular assembly into many smaller ones (Fig. 16b, [15]); (c) using electrical pulse to trigger formation of

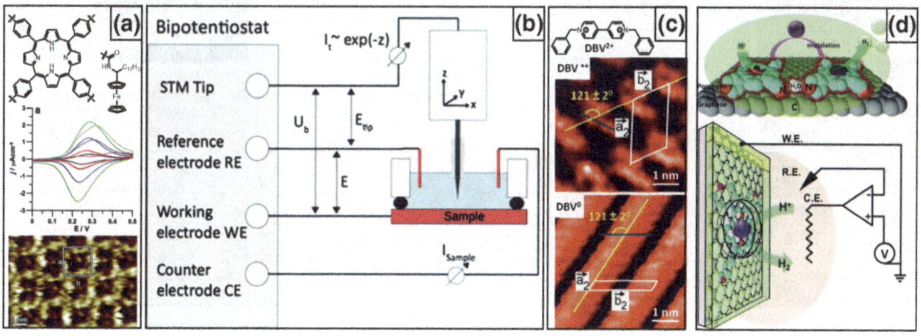

Fig. 15 **a** CV and STM of self-assembled porphyrin functionalized with electrochemically-active ferrocene moieties [80]. EC-STM operational scheme (**b**, [81]) and application for in situ generation of reduced viologens (**c**, [81]) and probing the catalytic activity of porphyrins on graphene (**d**, [82]). [80]—published by The Royal Society of Chemistry. [81]—published by The Royal Society of Chemistry. **c** adapted from [82] with permission of The Royal Society of Chemistry. Copyright © 2017 Royal Society of Chemistry

a metastable polymorph (Fig. 16c, [84]); (d) lithographic patterning of a supramolecular host-guest multilayer (Fig. 16d, [85]).

Chemical reactions on HOPG surface are also possible. STM imaging at −800 mV was sufficient for sporadic reduction of Mn(III) porphyrins to Mn(II) states and characterization of different oxygenated species thereof (Fig. 17a, [86]). An example of much more precise tip-induced reaction is [2+2]-oligomerization of C_{60} by electric pulses through its tri-and tetralayers. Nakaya et al. demonstrated unprecedented control over writing (polymerization) and erasing (depolymerization) of nanoscale structures (Fig. 17b, [87]), promoting such systems for >Tbits/in^2 aerial density non-volatile storages that can potentially be scaled to writing and erasing speed above Mbits/s or even Gbits/s.

1D polymerizations can be initiated by electrical pulsing over diacetylene self-assembly [88]. The reaction is highly localized allowing initiation of specific diacetylene rows. Monolayer defects interrupt polymerization, while presence of suitable molecules (phthalocyanine [89], C_{60} [90]) on the way of reactive chain end results in 'molecular soldering': covalent C–C coupling of small molecules and polymeric 'wire' (Fig. 18a). Finally, the layers of boroxine 2D COF and its host-guest co-assembly with C_{60} were tailor-cut using STM lithography (Fig. 18b, c, [91]). The mechanism is not known, presumably the cutting happens via the oxidation of C-B bonds.

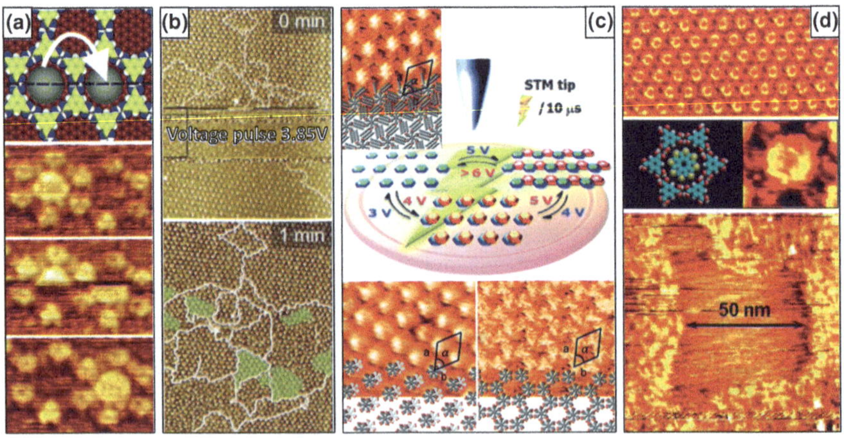

Fig. 16 STM manipulations that do not involve making/braking of covalent bonds: **a** physical displacement of single molecules [83]; pulse-induced desorption-readsorption of many molecules that leads to the formation of the same (**b**, [15]) or polymorphic (**c**, [84]) assemblies; **d** local removal of a supramolecular multilayered films [85]. **a** adapted with permission from [83]; copyright © 2017 The Chemical Society of Japan. **b** adapted with permission from [15]. Copyright © 2016 Nature Publishing Group. **c** adapted from [84] with permission of The Royal Society of Chemistry. **d** adapted from [85] with permission of The Royal Society of Chemistry

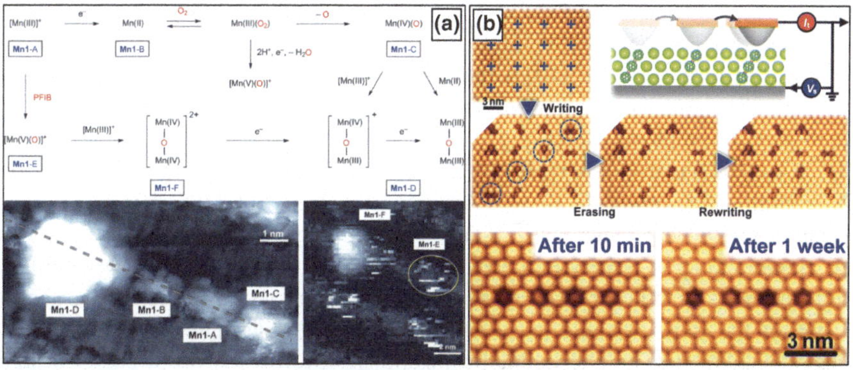

Fig. 17 STM in **a** the investigation of redox chemistry of Mn(111) porphyrin [86]; and **b** in reliable, spatially localized and precise oligomerization of C_{60} molecules in multilayered films [87]. **a** adapted with permission from [86]. Copyright © 2013 Nature Publishing Group. **b** adapted with permission from [87]. Copyright © 2016 The Japan Society of Applied Physics

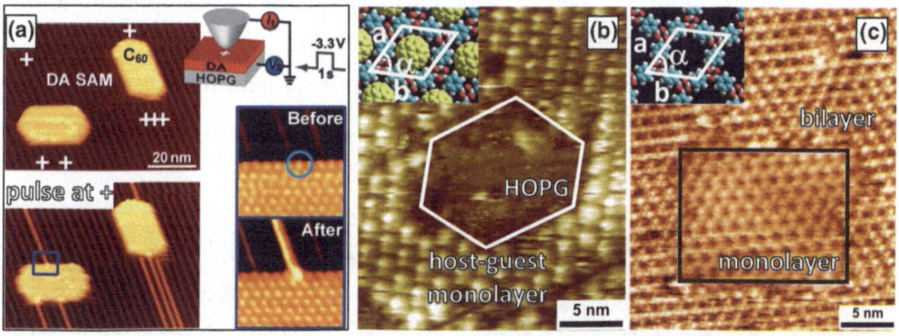

Fig. 18 STM manipulations involving reconstruction of covalent bonds: **a** pulse-induced polymerization of diacetylene assemblies can be used to covalently 'solder' polydiacetylene 'wires' to physisorbed molecules of C_{60} [90]; continuous scanning under appropriate tunneling settings was used to locally scratch boroxine 2D COF layers **c** and its host-guest assemblies with C_{60} [91]. **a** adapted with permission from [90]; copyright © 2014 American Chemical Society. **b**, **c** were adapted from [91] with permission from The Royal Society of Chemistry

6 Graphene Manipulation and Grafting

Graphene is not completely inert and under certain conditions can undergo chemical reactions. High temperature ($\geq$400 °C, e.g. [92]) or oxygen plasma oxidation [93] of HOPG in air results in random pits on the surface. Much better control over the size and localization of surface pits can be achieved via the spatially patterned ion-bombardment Fig. 19a, [94]). Pits in graphite were used as 2D corrals for studies of molecular self-assembly [95] and even for confined polymerization of diacetylenes (Fig. 19b, [96]). If the oxidation is performed on surfaces pre-covered with patterned assemblies of other materials (nanoparticles [97], wires [98], polymers [99], DNA assemblies [100], for more examples please refer to the recent review [101]) the latter act as lithographic masks producing corresponding patterns in graphene layers. Also, STM can be used to rip and fold small parts of graphene sheets (e.g. by 7 V voltage ramp with sequential scanning, Fig. 19c, [102]) or to lithographically burn-out arbitrary shapes with nm precision (by continuous scanning in air with bias $\geq$+2 V, Fig. 19d, [103, 104]).

Grafting-covalent functionalization of graphene layers via partial conversion of sp^2 carbons to sp^3—is very interesting for various applications. Here we will talk exclusively about radical arylation using aryldiazonium precursors. This is because it is one of the most versatile (easy accessibility of starting anilines) and methodologically developed graphene functionalizations. Nevertheless, the discussed design concepts, preparation strategies and SPM characterization should be also applicable to other grafting chemistries (carbenes, nitrenes, Diels-Alder adducts, etc. [105]).

Grafting chemistry is quite simple: aryldiazonium cations through one-electron reduction (usually electrochemical [106], although using reductants is also possible [107]) transform into corresponding aryl-radicals which are sufficiently reactive to attack accessible graphene surface. For characterization it is important to use combination of complementary techniques, e.g.:

- *Raman spectroscopy* to assess global uniformity of grafting (also for rough estimation of the grafting density);

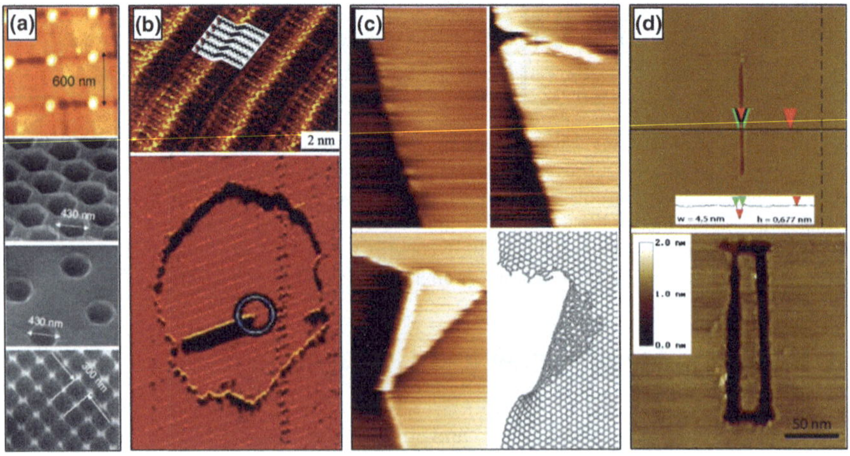

Fig. 19 **a** Advanced patterned oxidation of graphitic substrates [94]. **b** The use of corrals in HOPG for the confinement of tip-induced polymerization of diacetylene assemblies [96]. Direct exfoliation (**c**, [102]) and oxidative etching (**d**, [103]) of graphene layers with STM tip. **a** adapted with permission from [94]; copyright © 2006 IOP Publishing Ltd. **b** adapted with permission from [96]; copyright © 2005 American Chemical Society. **c** adapted from [102] with permission from the PCCP Owner Societies, copyright © 2017. **d** adapted with permission from [103]; copyright © 2017 Wiley-VCH Verlag GmbH & Co. KGaA

- *AFM* to determine film thickness (via mechanical scratching of grafted film), morphology and roughness;
- *STM* to prove nanoscale arrangement of grafted species, to distinguish them from physisorbed impurities and for spatially controlled degrafting (see below);
- *XPS* to prove chemical nature/integrity of functional groups (after grafting or any subsequent treatment of the grafts).

In STM, grafted aryls appear as bright blobs (brighter than physisorbed benzene derivatives [108, 109]) and covalent attachment to the surface underneath can be seen from characteristic local perturbation of graphene lattice (Fig. 20a). Thickness of the grafted adlayer primarily depends on the aryl structure: aryl radicals with 3-, 4-, 5-positions blocked to radical attack form well-defined non-dendritic grafts [110]. Grafting density can be easily tuned by changing grafting conditions (Fig. 20b), approaching the maximum at ~2–4 grafts/nm^2. Substituents can be varied widely including among the most unusual examples: directly graftable diazonium salts bearing free-radical moieties (Fig. 20c, [111]), metal complexes (Fig. 20d, [112]) and phosphomolybdate clusters (Fig. 20e, [113]). If necessary, functional groups on grafted aryls can be further modified by adapting surface functionalization protocols developed for organic synthesis on solid supports. Such advanced functionalization has already been successfully used for the design of various sensors [114].

Water soluble diazonium salts can be grafted from aqueous solutions (often using dilute HCl as supporting electrolyte). Water-insoluble precursors can be electrochemically grafted from organic solvents (e.g. acetonitrile, tetrahydrofuran, etc.) containing quaternary ammonium salts as supporting electrolytes. Very recently Ejigu et al. have shown that using DMSO as solvent and $CsClO_4$ as electrolyte results in simultaneous grafting and exfoliation of graphene layers [115].

Molecular adsorbates can affect grafting in two ways: by preventing the grafting of graphene lattice directly underneath the molecules, and by directing the nanostructured functionalization of self-assembly defects and interdomain borders (Fig. 21a, b, [108]). Here, n-pentacontane was chosen as the molecular adsorbate because of high chemical inertness (typical for all linear alkanes) and its strong adsorption. The latter ensured that reactive species, by- and side-products will not displace the molecular template. Using derivatives capable of ordered self-assembly (Fig. 20d) opens new possibilities for nanostructured surface functionalization during direct grafting [112] or through pre-assembly of diazonium precursor (Fig. 21c, [116]). In the last two examples it is unfortunate that STM was not used to prove the covalent nature of the observed structures and to assess the density/arrangement of molecules that are actually grafted.

Different nanostructured grafting should be accessible via soft lithography using nanoparticle assemblies as templates. While there is yet no example on graphene-based substrates, the same Ru complex (Fig. 20d) was reduced onto ITO pre-covered with a colloidal self-assembly of polystyrene beads [112]. After the removal of templates patterned thin film with circular nanowells (periodicity ~1 μm, diameter ~350 nm) was successfully prepared (Fig. 22a).

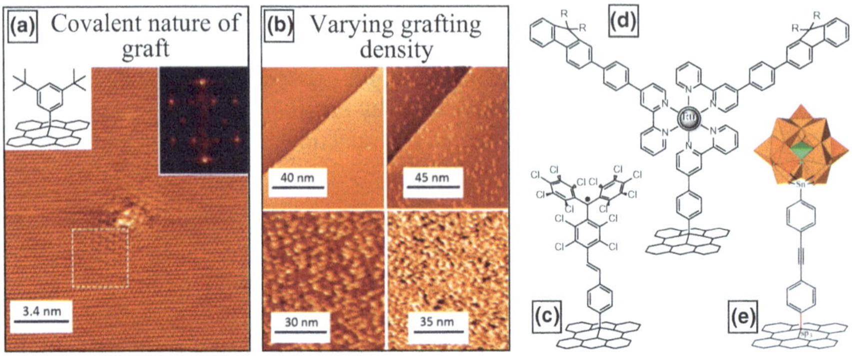

Fig. 20 Functionalization of graphene layers via covalent grafting of aryls: **a** distortion of graphene lattice in the vicinity of the graft as the evidence of the covalent nature of graft attachment (inset shows 2D FFT of the distorted lattice) [110]; **b** altering grafting conditions allows controlled tuning of the grafting density [110]. Some of the most unusual grafts feature: **c** a stable free radical [111]; **d** a metalorganic complex [112]; **e** a phosphomolybdate cluster [113]. **a**, **b** adapted with permission from [110]; copyright © 2015 American Chemical Society. **e** adapted with permission from [113]; copyright © 2016 American Chemical Society

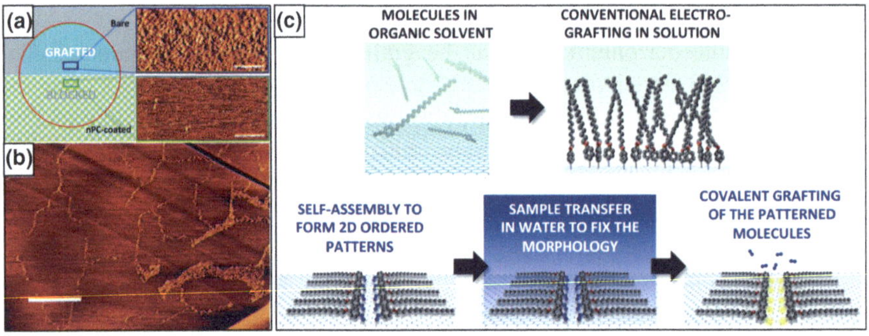

Fig. 21 Molecular self-assembly and covalent grafting: **a** the protection of graphene layers from the attack of aryl radicals by physisorbed self-assembly of n-pentacontane (nPC, n-$C_{50}H_{102}$) and **b** nanostructured grafting as the result of covalent grafting of HOPG surface at the defects and interdomain borders of nPC monolayer [108]. Preorganization of diazonium salts via on surface self-assembly leads to nanostructured grafting that follows the pattern of precursor self-assembly, while the direct grafting in organic solvent (without the self-assembly step) results in random grafting of the surface [116]. **a**, **b** were adapted from [108] with permission from The Royal Society of Chemistry. **c** adapted with permission from [116]; copyright © 2016 American Chemical Society

Nanoporous (35–200 nm) patterns (Fig. 22b) were also reported when combining the grafting and electrochemical etching of HOPG surface [117]. Performing +0.8 to −0.8 V CV cycles at 20 mV/s scan rate in 0.5, 2 or 5 M HCl solutions containing 4-carboxyphenyldiazonium salt, Cui et al. observed gradual increase of the dimension of the etched pores.

Grafted adlayers can be used to tune on-surface self-assembly. Electrodeposition of gold nanostructures on pristine HOPG substrate primarily occur at step-edges [78] while NH_2-functionalized grafted adlayer promotes uniform deposition of Au nanoparticles on top of it [118].

Examples of molecular self-assemblies on arylated HOPG are also known. For such studies the ability of STM tip to degraft covalently attached aryls selectively, under very mild conditions and with nanoscale precision opened new exciting possibilities that are only beginning to be explored. Thus, local removal of grafted species allows to direct and study ripening of 2D crystals that were kinetically trapped in the framework of a low-density grafted adlayer (Fig. 23a, b, [109]). Manually degrafted corrals can be used to localize supramolecular assemblies (Fig. 23c–e, g, [110, 120]). And finally, such corrals can be used as nanoreactors with controlled shape, size and chemical functionality. For example, we have already demonstrated the use of such corrals for self-assembly of in situ generated EC-active species under full electrochemical control [81], and for the investigation of tip-induced and photochemical polymerization of diacetylenes under nanoconfinement (Fig. 23f, h, [119]).

7 Summary and Conclusions

In this contribution, we have summarized a variety of molecule-based surface chemistry approaches on graphene and graphite. Several of the chemistries rely on the formation of well-ordered monolayers that undergo a postfunctionalization, be it orthogonal to the surface or in the monolayer plane. Nowadays, the formation of two-dimensional polymers is one of the most popular applications of on-surface chemistry, driven by the prospect of new and interesting properties that may arise for these new materials. Several types of protocols can induce on-surface reactions

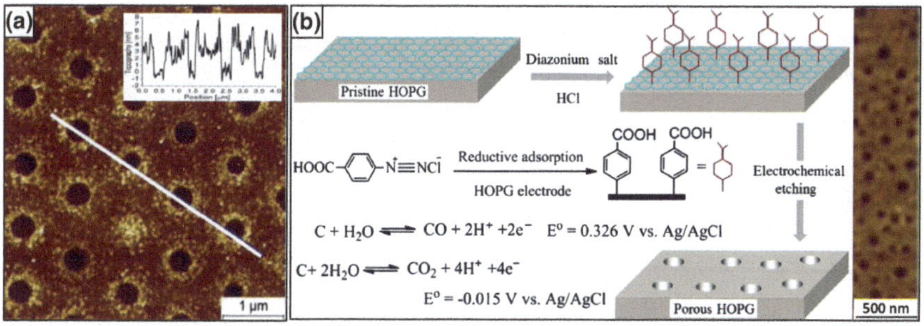

Fig. 22 Nanopatterns formed via: **a** electrochemical reduction of a diazonium salt on ITO electrode covered with polystyrene beads [112]; **b** simultaneous aryl grafting and graphite oxidation [117]. **a** adapted with permission from [112]; copyright © 2016 American Chemical Society. **b** adapted from [117], copyright © 2016, with permission from Elsevier

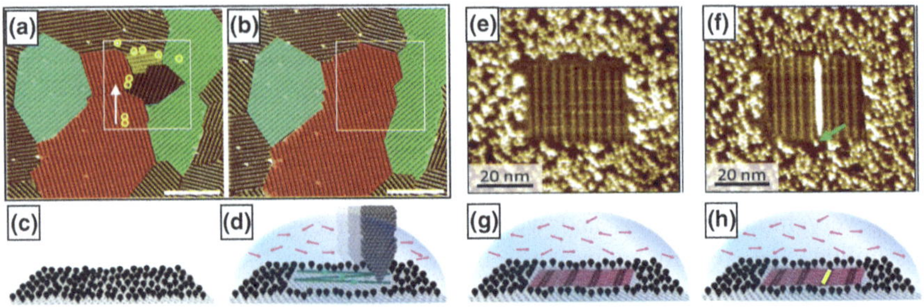

Fig. 23 The power of controlled degrafting: **a** application of manual removal of grafts (marked with yellow circles) to promote **b** local (marked with the white rectangle) Ostwald ripening of physisorbed supramolecular assembly [109]. In situ STM lithography **d** of a dense adlayer of covalent grafts **c** for directed self-assembly (**e**, **g**) and tip induced polymerization (**f**, **h**) of 10,12-pentacosadiynoic acid [119]. **a**, **b** were adapted from [109] with permission from The Royal Society of Chemistry. **c**, **d**, **g**, **h** adapted with permission from [120]; copyright © 2016 American Chemical Society. **e**, **f** were adapted from [119] with permission, copyright © 2017, from The Royal Society of Chemistry

such as thermal, light, or electrochemical activation. One aspect that has not received as much attention yet is the use of surfaces to affect the product distribution of a chemical reaction that occurs in solution. Initial reports suggest that such combination of on-surface self-assembly and combinatorial chemistry might be complementary to existing synthetic strategies. Also, the complexity and, hence, the potential of covalent functionalization of carbon surfaces is not yet fully realized. Although such chemistries are known for more than two decades, their investigation using scanning probe microscopy techniques and the nanoscale characterization will lead to new patterning strategies. Combining different on-surface reactivity protocols may lead to a new generation of smart surfaces and materials.

Acknowledgements This work is supported by the Fund of Scientific Research–Flanders (FWO), KU Leuven—Internal Funds, Belgian Federal Science Policy Office (IAP-7/05), European Research Council under the European Union's Seventh Framework Program (FP7/2007–2013)/ ERC Grant Agreement No. 340324, and from the European Union Framework Program for Research and Innovation Horizon 2020 as a Future and Emerging Technologies Action (FET Open) under Grant Agreement No. 664878 (2D-INK).

References

1. Hughes, G., Westmacott, K., Honeychurch, K., Crew, A., Pemberton, R., Hart, J.: Biosensors **6**, 50 (2016)
2. Shivananju, B.N., Yu, W., Liu, Y., Zhang, Y., Lin, B., Li, S., Bao, Q.: Adv. Funct. Mater. n/a (2016)
3. Ferrari, A.C., Bonaccorso, F., Fal'ko, V., Novoselov, K.S., Roche, S., Boggild, P., Borini, S., Koppens, F.H.L., Palermo, V., Pugno, N., Garrido, J.A., Sordan, R., Bianco, A., Ballerini, L., Prato, M., Lidorikis, E., Kivioja, J., Marinelli, C., Ryhanen, T., Morpurgo, A.,

Coleman, J.N., Nicolosi, V., Colombo, L., Fert, A., Garcia-Hernandez, M., Bachtold, A., Schneider, G.F., Guinea, F., Dekker, C., Barbone, M., Sun, Z., Galiotis, C., Grigorenko, A. N., Konstantatos, G., Kis, A., Katsnelson, M., Vandersypen, L., Loiseau, A., Morandi, V., Neumaier, D., Treossi, E., Pellegrini, V., Polini, M., Tredicucci, A., Williams, G.M., Hee Hong, B., Ahn, J.-H., Min Kim, J., Zirath, H., van Wees, B.J., van der Zant, H., Occhipinti, L., Di Matteo, A., Kinloch, I.A., Seyller, T., Quesnel, E., Feng, X., Teo, K., Rupesinghe, N., Hakonen, P., Neil, S.R.T., Tannock, Q., Lofwander, T., Kinaret, J.: Nanoscale **7**, 4598 (2015)

4. Huang, H., Wang, X.: Journal of Materials Chemistry A **2**, 6266 (2014)
5. Hu, F.-Y., Zhang, X.-M., Wang, X.-C., Wang, S., Wang, H.-Q., Duan, W.-B., Zeng, Q.-D., Wang, C.: ACS Appl. Mater. Interfaces. **5**, 1583 (2013)
6. Wei, X., Tong, W., Fidler, V., Zimmt, M.B.: J. Colloid Interface Sci. **387**, 221 (2012)
7. Du, P., Jaouen, M., Bocheux, A., Bourgogne, C., Han, Z., Bouchiat, V., Kreher, D., Mathevet, F., Fiorini-Debuisschert, C., Charra, F., Attias, A.-J.: Angew. Chem. Int. Ed. **53**, 10060 (2014)
8. Bléger, D., Mathevet, F., Kreher, D., Attias, A.-J., Bocheux, A., Latil, S., Douillard, L., Fiorini-Debuisschert, C., Charra, F.: Angew. Chem. Int. Ed. **50**, 6562 (2011)
9. Du, P., Bléger, D., Charra, F., Bouchiat, V., Kreher, D., Mathevet, F., Attias, A.-J.: Beilstein J. Nanotechnol. **6**, 632 (2015)
10. Le Liepvre, S., Du, P., Kreher, D., Mathevet, F., Attias, A.-J., Fiorini-Debuisschert, C., Douillard, L., Charra, F.: ACS Photonics **3**, 2291 (2016)
11. Spitzer, S., Rastgoo-Lahrood, A., Macknapp, K., Ritter, V., Sotier, S., Heckl, W.M., Lackinger, M.: Chem. Commun. (2017)
12. Liu, X.-H., Guan, C.-Z., Ding, S.-Y., Wang, W., Yan, H.-J., Wang, D., Wan, L.-J.: J. Am. Chem. Soc. **135**, 10470 (2013)
13. Giancarlo, L.C., Flynn, G.W.: Acc. Chem. Res. **33**, 491 (2000)
14. Schull, G., Ness, H., Douillard, L., Fiorini-Debuisschert, C., Charra, F., Mathevet, F., Kreher, D., Attias, A.-J.: J. Phys. Chem. C **112**, 14058 (2008)
15. Fang, Y., Ghijsens, E., Ivasenko, O., Cao, H., Noguchi, A., Mali, K.S., Tahara, K., Tobe, Y., De Feyter, S.: Nat. Chem. **8**, 711 (2016)
16. Plas, J., Waghray, D., Adisoejoso, J., Ivasenko, O., Dehaen, W., De Feyter, S.: Chem. Commun. **51**, 16338 (2015)
17. Dienstmaier, J.F., Gigler, A.M., Goetz, A.J., Knochel, P., Bein, T., Lyapin, A., Reichlmaier, S., Heckl, W.M., Lackinger, M.: ACS Nano **5**, 9737 (2011)
18. Hu, Y., Goodeal, N., Chen, Y., Ganose, A.M., Palgrave, R.G., Bronstein, H., Blunt, M.O.: Chem. Commun. **52**, 9941 (2016)
19. He, J., Fang, C., Shelp, R.A., Zimmt, M.B.: Langmuir **33**, 459 (2017)
20. Li, Y., Wan, J., Deng, K., Han, X., Lei, S., Yang, Y., Zheng, Q., Zeng, Q., Wang, C.: J. Phys. Chem. C **115**, 6540 (2011)
21. Liu, X.-H., Guan, C.-Z., Zheng, Q.-N., Wang, D., Wan, L.-J.: J. Chem. Phys. **142**, 101905 (2015)
22. Dong, W.-L., Wang, L., Ding, H.-M., Zhao, L., Wang, D., Wang, C., Wan, L.-J.: Langmuir **31**, 11755 (2015)
23. Sun, X., Fan, L., Zhou, X., Tian, W.Q., Guo, Z., Li, Z., Li, X., Lei, S.: Chem. Commun. **51**, 5864 (2015)
24. Xu, L., Yu, Y., Lin, J., Zhou, X., Tian, W.Q., Nieckarz, D., Szabelski, P., Lei, S.: Nanoscale **8**, 8568 (2016)
25. Yue, J.-Y., Markoulides, M., Regan, A.C., Li, S.-Y., Chronakis, N., Gourdon, A., Chen, T., Yan, H.-J., Wang, D.: Chem. Commun. **53**, 428 (2017)
26. Yu, Y., Zheng, Y., Lei, S.: J. Phys. Chem. C **121**, 593 (2017)

27. Gong, Z., Yang, B., Lin, H., Tang, Y., Tang, Z., Zhang, J., Zhang, H., Li, Y., Xie, Y., Li, Q., Chi, L.: ACS Nano **10**, 4228 (2016)
28. Yu, Y., Yang, L., Liu, C., Tian, W.Q., Wang, Y., Lei, S.: Chem. Commun. **52**, 8317 (2016)
29. Mo, Y.-P., Liu, X.-H., Sun, B., Yan, H.-J., Wang, D., Wan, L.-J.: Phys. Chem. Chem. Phys. **19**, 539 (2017)
30. Xu, L., Cao, L., Guo, Z., Zha, Z., Lei, S.: Chem. Commun. **51**, 8664 (2015)
31. Ciesielski, A., El Garah, M., Haar, S., Kovaříček, P., Lehn, J.-M., Samorì, P.: Nat Chem **6**, 1017 (2014)
32. Yu, Y., Lin, J., Lei, S.: RSC Adv. **7**, 11496 (2017)
33. Yu, Y., Lin, J., Wang, Y., Zeng, Q., Lei, S.: Chem. Commun. **52**, 6609 (2016)
34. Higuchi, R., Tanoue, R., Sakaguchi, K., Yanai, K., Uemura, S., Kunitake, M.: Polymer **54**, 3452 (2013)
35. de la Peña Ruigómez, A., Rodríguez-San-Miguel, D., Stylianou, K.C., Cavallini, M., Gentili, D., Liscio, F., Milita, S., Roscioni, O.M., Ruiz-González, M.L., Carbonell, C., Maspoch, D., Mas-Ballesté, R., Segura, J.L., Zamora, F.: Chem.—A Eur. J. **21**, 10666 (2015)
36. Dong, W.-L., Li, S.-Y., Yue, J.-Y., Wang, C., Wang, D., Wan, L.-J.: Phys. Chem. Chem. Phys. **18**, 17356 (2016)
37. Higuchi, R., Tanoue, R., Enoki, N., Miyasato, Y., Sakaguchi, K., Uemura, S., Kimizuka, N., Kunitake, M.: Chem. Commun. **48**, 3103 (2012)
38. Li, H., Xu, X., Shang, J., Li, J., Hu, X., Teo, B.K., Wu, K.: J. Phys. Chem. C **116**, 21753 (2012)
39. Li, Y., Liu, C., Xie, Y., Li, X., Fan, X., Yuan, L., Zeng, Q.: Chem. Commun. **49**, 9021 (2013)
40. Ciesielski, A., Lena, S., Masiero, S., Spada, G.P., Samorì, P.: Angew. Chem. Int. Ed. **49**, 1963 (2010)
41. Luo, D., Zhang, X., Shen, Y., Xu, J., Shu, L., Zeng, Q., Wang, C.: Chem. Commun. **50**, 9369 (2014)
42. Geng, Y., Liu, M., Xue, J., Xu, P., Wang, Y., Shu, L., Zeng, Q., Wang, C.: Chem. Commun. **51**, 6820 (2015)
43. Gong, Y., Zhang, S., Geng, Y., Niu, C., Yin, S., Zeng, Q., Li, M.: Langmuir **31**, 11525 (2015)
44. Garah, M.E., Ciesielski, A., Marets, N., Bulach, V., Hosseini, M.W., Samori, P.: Chem. Commun. **50**, 12250 (2014)
45. Kikkawa, Y., Koyama, E., Aoyagi, M., Schneider, N., Takahashi, M., Fujiwara, K., Kanesato, M.: Supramol. Chem. **23**, 9 (2011)
46. Omori, K., Kikkawa, Y., Tokuhisa, H., Kanesato, M., Hiratani, K.: Colloids Surf., A **356**, 58 (2010)
47. Omori, K., Kikkawa, Y., Kanesato, M., Hiratani, K.: Chem. Commun. **46**, 8008 (2010)
48. Yoshihiro, K., Kazuhiro, O., Masatoshi, K., Kazuhisa, H.: Chem. Lett. **41**, 1196 (2012)
49. Geng, Y., Dai, H., Chang, S., Hu, F., Zeng, Q., Wang, C.: ACS Appl. Mater. Interfaces. **7**, 4659 (2015)
50. Dienstmaier, J.F., Medina, D.D., Dogru, M., Knochel, P., Bein, T., Heckl, W.M., Lackinger, M.: ACS Nano **6**, 7234 (2012)
51. Yu, L., Li, Z.-B., Wang, D.: Chem. Commun. **52**, 13771 (2016)
52. Yue, J.-Y., Mo, Y.-P., Li, S.-Y., Dong, W.-L., Chen, T., Wang, D.: Chem. Sci. **8**, 2169 (2017)
53. Ono, K., Johmoto, K., Yasuda, N., Uekusa, H., Fujii, S., Kiguchi, M., Iwasawa, N.: J. Am. Chem. Soc. **137**, 7015 (2015)
54. Colson, J.W., Woll, A.R., Mukherjee, A., Levendorf, M.P., Spitler, E.L., Shields, V.B., Spencer, M.G., Park, J., Dichtel, W.R.: Science **332**, 228 (2011)

55. Liu, C., Yu, Y., Zhang, W., Zeng, Q., Lei, S.: Chem.—A Eur. J. **22**, 18412 (2016)
56. Miura, A., De Feyter, S., Abdel-Mottaleb, M.M.S., Gesquière, A., Grim, P.C.M., Moessner, G., Sieffert, M., Klapper, M., Müllen, K., De Schryver, F.C.: Langmuir **19**, 6474 (2003)
57. Grim, P.C.M., Vanoppen, P., Rücker, M., Feyter, S.D., Valiyaveettil, S., Moessner, G., Müllen, K., Schryver, F.C.D.: J. Vac. Sci. Technol. B: Microelectron. Nanometer Struct. Process. Meas. Phenom. **15**, 1419 (1997)
58. Bléger, D., Ciesielski, A., Samorì, P., Hecht, S.: Chem.—A Eur. J. **16**, 14256 (2010)
59. Jacob, H., Ulrich, S., Jung, U., Lemke, S., Rusch, T., Schutt, C., Petersen, F., Strunskus, T., Magnussen, O., Herges, R., Tuczek, F.: Phys. Chem. Chem. Phys. **16**, 22643 (2014)
60. Tahara, K., Inukai, K., Adisoejoso, J., Yamaga, H., Balandina, T., Blunt, M.O., De Feyter, S., Tobe, Y.: Angew. Chem. Int. Ed. **52**, 8373 (2013)
61. Shen, Y.-T., Guan, L., Zhu, X.-Y., Zeng, Q.-D., Wang, C.: J. Am. Chem. Soc. **131**, 6174 (2009)
62. Yamauchi, M., Kanao, N., Adhikari, B., Karatsu, T., Yagai, S.: Chem. Lett. **46**, 111 (2017)
63. Tahara, K., Nakatani, K., Iritani, K., De Feyter, S., Tobe, Y.: ACS Nano **10**, 2113 (2016)
64. Shen, Y.-T., Deng, K., Zhang, X.-M., Feng, W., Zeng, Q.-D., Wang, C., Gong, J.R.: Nano Lett. **11**, 3245 (2011)
65. Liao, L.-Y., Li, Y.-B., Zhang, X.-M., Geng, Y.-F., Zhang, J.-Y., Xie, J.-L., Zeng, Q.-D., Wang, C.: J. Phys. Chem. C **118**, 15963 (2014)
66. Abdel-Mottaleb, M.M.S., De Feyter, S., Gesquière, A., Sieffert, M., Klapper, M., Müllen, K., De Schryver, F.C.: Nano Lett. **1**, 353 (2001)
67. Xue, J., Xu, J., Hu, F., Liao, L., Li, M., Duan, W., Zeng, Q., Wang, C.: Phys. Chem. Chem. Phys. **16**, 25765 (2014)
68. Kikkawa, Y., Kihara, H., Takahashi, M., Kanesato, M., Balaban, T.S., Lehn, J.-M.: J. Phys. Chem. B **114**, 16718 (2010)
69. Yokoyama, S., Hirose, T., Matsuda, K.: Langmuir **31**, 6404 (2015)
70. Maeda, N., Hirose, T., Yokoyama, S., Matsuda, K.: J. Phys. Chem. C **120**, 9317 (2016)
71. Bonacchi, S., El Garah, M., Ciesielski, A., Herder, M., Conti, S., Cecchini, M., Hecht, S., Samorì, P.: Angew. Chem. **127**, 4947 (2015)
72. Colazzo, L., Sedona, F., Moretto, A., Casarin, M., Sambi, M.: J. Am. Chem. Soc. **138**, 10151 (2016)
73. Takami, T., Ozaki, H., Kasuga, M., Tsuchiya, T., Ogawa, A., Mazaki, Y., Fukushi, D., Uda, M., Aono, M.: Angew. Chem., Int. Ed. Engl. **36**, 2755 (1997)
74. Xue, Y., Zimmt, M.B.: J. Am. Chem. Soc. **134**, 4513 (2012)
75. Yang, Y., Zimmt, M.B.: Langmuir **31**, 12408 (2015)
76. Colson, J.W., Dichtel, W.R.: Nat. Chem. **5**, 453 (2013)
77. Zhang, X.-M., Xu, S.-D., Li, M., Shen, Y.-T., Wei, Z.-Q., Wang, S., Zeng, Q.-D., Wang, C.: J. Phys. Chemi. C **116**, 8950 (2012)
78. Gómez, J.J.A., Zubieta, C., Ferullo, R.M., García, S.G.: Appl. Surf. Sci. **363**, 356 (2016)
79. Tang, Z., Liu, S., Wang, Z., Dong, S., Wang, E.: Electrochem. Commun. **2**, 32 (2000)
80. El Garah, M., Santana Bonilla, A., Ciesielski, A., Gualandi, A., Mengozzi, L., Fiorani, A., Iurlo, M., Marcaccio, M., Gutierrez, R., Rapino, S., Calvaresi, M., Zerbetto, F., Cuniberti, G., Cozzi, P.G., Paolucci, F., Samori, P.: Nanoscale **8**, 13678 (2016)
81. Huynh, T.M.T., Phan, T.H., Ivasenko, O., Mertens, S.F.L., De Feyter, S.: Nanoscale **9**, 362 (2017)
82. Seo, S., Lee, K., Min, M., Cho, Y., Kim, M., Lee, H.: Nanoscale **9**, 3969 (2017)
83. Griessl, S.J.H., Lackinger, M., Jamitzky, F., Markert, T., Hietschold, M., Heckl, W.M.: J. Phys. Chem. B **108**, 11556 (2004)
84. Lee, S.-L., Hsu, Y.-J., Wu, H.-J., Lin, H.-A., Hsu, H.-F., Chen, C.-H.: Chem. Commun. **48**, 11748 (2012)

85. Ivasenko, O., MacLeod, J.M., Chernichenko, K.Y., Balenkova, E.S., Shpanchenko, R.V., Nenajdenko, V.G., Rosei, F., Perepichka, D.F.: Chem. Commun. **10**, 1192 (2009)
86. den Boer, D., Li, M., Habets, T., Iavicoli, P., Rowan, A.E., Nolte, R.J.M., Speller, S., Amabilino, D.B., De Feyter, S., Elemans, J.A.A.W.: Nat Chem **5**, 621 (2013)
87. Masato, N., Masakazu, A., Tomonobu, N.: Jpn. J. Appl. Phys. **55**(11), 4 (2016)
88. Okawa, Y., Akai-Kasaya, M., Kuwahara, Y., Mandal, S.K., Aono, M.: Nanoscale **4**, 3013 (2012)
89. Okawa, Y., Mandal, S.K., Hu, C., Tateyama, Y., Goedecker, S., Tsukamoto, S., Hasegawa, T., Gimzewski, J.K., Aono, M.: J. Am. Chem. Soc. **133**, 8227 (2011)
90. Nakaya, M., Okawa, Y., Joachim, C., Aono, M., Nakayama, T.: ACS Nano **8**, 12259 (2014)
91. Plas, J., Ivasenko, O., Martsinovich, N., Lackinger, M., De Feyter, S.: Chem. Commun. **52**, 68 (2016)
92. Stevens, F., Kolodny, L.A., Beebe, T.P.: J. Phys. Chem. B **102**, 10799 (1998)
93. Fredriksson, H., Chakarov, D., Kasemo, B.: Carbon **47**, 1335 (2009)
94. Artur, B., Moritz, H., Ninette, S., Stefan, S.J., Sharali, M., Fabián, P.-W., Patrice, B., Dagmar, G., Manfred, M.K.: Nanotechnology **17**, 5889 (2006)
95. Stevens, F., Buehner, D., Beebe, T.P.: J. Phys. Chem. B **101**, 6491 (1997)
96. Sullivan, S.P., Schnieders, A., Mbugua, S.K., Beebe, T.P.: Langmuir **21**, 1322 (2005)
97. Cong, C.X., Yu, T., Ni, Z.H., Liu, L., Shen, Z.X., Huang, W.: J. Phys. Chem. C **113**, 6529 (2009)
98. Xu, W., Seo, H.-K., Min, S.-Y., Cho, H., Lim, T.-S., Oh, C.-Y., Lee, Y., Lee, T.-W.: Adv. Mater. **26**, 3459 (2014)
99. Son, J.G., Son, M., Moon, K.-J., Lee, B.H., Myoung, J.-M., Strano, M.S., Ham, M.-H., Ross, C.A.: Adv. Mater. **25**, 4723 (2013)
100. Jin, Z., Sun, W., Ke, Y., Shih, C.-J., Paulus, G.L.C., Hua Wang, Q., Mu, B., Yin, P., Strano, M.S.: Nat. Commun. **4**, 1663 (2013)
101. Zheng, Y., Wang, H., Hou, S., Xia, D.: Adv. Mater. Technol. **2**, 1600237 (2017)
102. Rubio-Verdu, C., Saenz-Arce, G., Martinez-Asencio, J., Milan, D.C., Moaied, M., Palacios, J.J., Caturla, M.J., Untiedt, C.: Phys. Chem. Chem. Phys. **19**, 8061 (2017)
103. Dobrik, G., Tapasztó, L., Nemes-Incze, P., Lambin, P., Biró, L.P.: Phys. Status Solidi B **247**, 896 (2010)
104. Tapaszto, L., Dobrik, G., Lambin, P., Biro, L.P.: Nat Nano **3**, 397 (2008)
105. Park, J., Yan, M.: Acc. Chem. Res. **46**, 181 (2013)
106. Downard, A.J.: Electroanalysis **12**, 1085 (2000)
107. Barrière, F., Downard, A.J.: J. Solid State Electrochem. **12**, 1231 (2008)
108. Li, Z., Van Gorp, H., Walke, P., Phan, T.H., Fujita, Y., Greenwood, J., Ivasenko, O., Tahara, K., Tobe, Y., Uji-i, H., Mertens, S.F.L., De Feyter, S.: Nanoscale **9**, 5188 (2017)
109. Braganca, A.M., Greenwood, J., Ivasenko, O., Phan, T.H., Mullen, K., De Feyter, S.: Chem. Sci. **7**, 7028 (2016)
110. Greenwood, J., Phan, T.H., Fujita, Y., Li, Z., Ivasenko, O., Vanderlinden, W., Van Gorp, H., Frederickx, W., Lu, G., Tahara, K., Tobe, Y., Uji-i, H., Mertens, S.F.L., De Feyter, S.: ACS Nano **9**, 5520 (2015)
111. Seber, G., Rudnev, A.V., Droghetti, A., Rungger, I., Veciana, J., Mas-Torrent, M., Rovira, C., Crivillers, N.: Chem.—A Eur. J. **23**, 1415 (2017)
112. Nguyen, V.Q., Sun, X., Lafolet, F., Audibert, J.-F., Miomandre, F., Lemercier, G., Loiseau, F., Lacroix, J.-C.: J. Am. Chem. Soc. **138**, 9381 (2016)
113. Huder, L., Rinfray, C., Rouchon, D., Benayad, A., Baraket, M., Izzet, G., Lipp-Bregolin, F., Lapertot, G., Dubois, L., Proust, A., Jansen, L., Duclairoir, F.: Langmuir **32**, 4774 (2016)
114. Cao, C., Zhang, Y., Jiang, C., Qi, M., Liu, G.: ACS Appl. Mater. Interfaces. **9**, 5031 (2017)
115. Ejigu, A., Kinloch, I.A., Dryfe, R.A.W.: ACS Appl. Mater. Interfaces. **9**, 710 (2017)

116. Xia, Z., Leonardi, F., Gobbi, M., Liu, Y., Bellani, V., Liscio, A., Kovtun, A., Li, R., Feng, X., Orgiu, E., Samorì, P., Treossi, E., Palermo, V.: ACS Nano **10**, 7125 (2016)
117. Cui, L., Xu, Y., Liu, B., Yang, W., Song, Z., Liu, J.: Carbon **102**, 419 (2016)
118. Gonzalez, M.C.R., Orive, A.G., Salvarezza, R.C., Creus, A.H.: Phys. Chem. Chem. Phys. **18**, 1953 (2016)
119. Verstraete, L., Hirsch, B.E., Greenwood, J., De Feyter, S.: Chem. Commun. **53**, 4207 (2017)
120. Verstraete, L., Greenwood, J., Hirsch, B.E., De Feyter, S.: ACS Nano **10**, 10706 (2016)

Dehydrogenative and Dehalogenative Homocoupling Reactions of C–X Groups on Metal Surfaces

Liangliang Cai, Qiang Sun and Wei Xu

Abstract Surface-assisted synthesis involving hydrocarbons has aroused great attention due to its remarkable potential in constructing novel carbon nanostructures. Particularly, C–C coupling between reactants by cleaving the pre-defined C–X groups (X stands for hydrogen and halogens) followed by forming new carbon-carbon bonds, represents one of the best choices for controllable fabrication of advanced carbon nanostructures. In this chapter, we reviewed the recent achievements of the on-surface reactions of C–X groups activations and C–C couplings, where different carbon species including alkynyl (sp^1), alkenyl (sp^2), aryl (sp^2) and alkyl (sp^3) groups are studied.

1 Introduction

Recently, the fabrication of nanostructures/nanomaterials through covalent interlinking of molecular building blocks on metal surfaces had been an enormously active research field [1–8]. Various chemical reactions have been employed on surfaces, including, for example, the Ullmann reaction [9, 10], Bergman cyclization [11, 12], Glaser coupling [13, 14], imine formation [15, 16], decarboxylative polymerization [17], azide–alkyne click reaction [18, 19], and dimerization of N-heterocyclic carbenes [20]. Among others, on-surface chemistry involving hydrocarbons has aroused great attention due to its fascinating potential in constructing novel carbon-rich nanostructures/nanomaterials. In particular, C–C coupling between reactants by cleaving the pre-defined C–X groups (X stands for hydrogen and halogens) followed by forming new carbon-carbon bonds, represents one of the best choices for controllable fabrication of advanced carbon nanostructures/nanomaterials.

L. Cai · Q. Sun · W. Xu (✉)
Interdisciplinary Materials Research Center, Tongji-Aarhus Joint Research Center for Nanostructures and Functional Nanomaterials, College of Materials Science and Engineering, Tongji University, Shanghai 201804, People's Republic of China
e-mail: xuwei@tongji.edu.cn

D. G. de Oteyza and C. Rogero (eds.), *On-Surface Synthesis II*,
Advances in Atom and Single Molecule Machines,
https://doi.org/10.1007/978-3-319-75810-7_4

In particular, the dehalogenative homocouplings of aryl halides (sp^2-carbon) have been explored most intensely [21–27]. This type of reaction proved to be a promising strategy for the construction of graphene-derived carbon nanostructures including graphene nanoribbons with different edge structures and widths [21, 28–31], porous graphene and hyperbenzene [27, 32], nanographene as well as two-dimensional polymers consisting of porphyrin units [9, 10, 33]. Most recently, the interests in constructing novel or advanced carbon nanostructures such as carbyne, graphyne and graphdiyne have drawn emerging attention of C–X groups involving other types of hybridized carbon [13, 14, 34–36]. In this research field, means of real-space direct visualization by scanning probe microscopy under ultra-high vacuum provides unambiguous identification of the formed nanostructures, and invaluable insights into the reaction scheme have also been obtained from complementary density functional theory (DFT) calculations.

In this chapter, we begin with a short review of the on-surface Ullmann type reactions, then we review the most recent achievements of the on-surface reactions of C–X groups, which involve: (1) different carbon species including alkynyl (sp^1), alkenyl (sp^2) and alkyl (sp^3) groups. (2) dehydrogenation and dehalogenation. Scheme 1 shows the overview illustration of the functional groups that are used for on-surface investigations. In the last part we describe some examples of the sequential C–X groups activation on surfaces. These findings have given us a general understanding of the different chemical properties between different C–X bonds on metal surfaces, and provided fruitful strategies for synthesis of novel complexes and construction of advanced carbon nanostructures on surfaces.

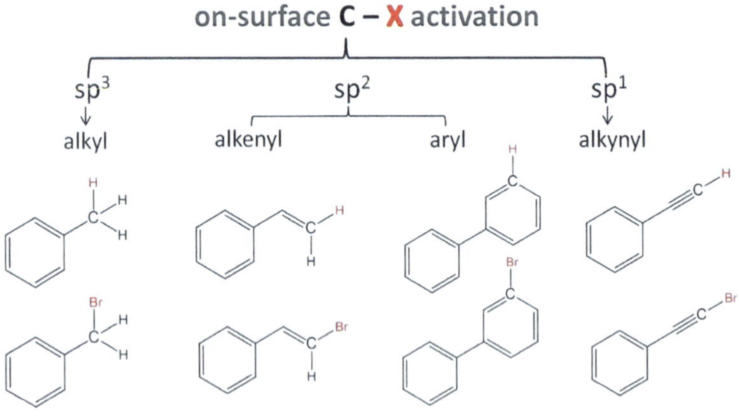

Scheme 1 On-surface C–X activation of sp^3, sp^2 and sp^1 hybridized carbons

2 Results and Discussions

2.1 *On-Surface Dehalogenative Homocoupling*

Dates back to 1901, Fritz Ullmann et al. discovered that the aromatic carbon-halogen bond can be cleaved to form diaryls with the help of the Cu powder catalyst [37]. Ever since, the Ullmann reaction has become a textbook case and it has been widely used in organic synthesis [38, 39]. About a century later, Saw-Wai Hla et al. introduced this well-known reaction onto the metal surface [40]. In their early work, they demonstrated that the Ullmann reaction could be induced by the scanning tunneling microscopy (STM) tip manipulating single molecules. After deposition of iodobenzene on the Cu(111) substrate, the STM tip was utilized to inject electrons into the molecule inducing the C–I bond cleavage to give a phenyl radical and an iodine atom. The phenyl radical is believed to be stabilized by π interaction with the terrace and σ bonding to a Cu step edge atom. After the initial iodine abstraction process, the authors further manipulated one phenyl bringing it closer to another phenyl by STM. Finally, by excitation of two phenyl radicals with inelastic tunneling, covalent coupling of the phenyl radicals was triggered and the biphenyl could be formed and identified by following STM imaging.

This breakthrough opens up a promising vista for on-surface synthesis. However, a clear drawback of the approach mentioned above is that electron-injection by the STM tip is a time-consuming and very inefficient way to induce a chemical reaction. This reaction did not draw remarkable attention until several years later Leonhard Grill et al. discovered a much simpler and more straightforward way to activate the C–Br bond in on-surface synthesis, which later boomed the emerging field of on-surface chemistry. Meanwhile, on-surface Ullmann-type coupling has been investigated by researchers across the world and still amazingly ranks among the most studied on-surface reactions so far. They demonstrated that the aryl C–Br bond cleavage and the subsequent covalent C–C bond coupling can be achieved upon thermal treatment [9]. Three kinds of brominated tetraphenylporphyrins (Br_n–TPP, n = 1, 2, 4) were selected as building blocks (Fig. 1). Firstly, the brominated TPP molecules were deposited onto a clean Au(111) substrate, which was kept at room temperature during deposition. If the evaporator temperature was 550 K or lower during the deposition, large highly ordered self-assembly islands of intact brominated TPP molecules were observed. Subsequently, annealing the molecule-covered surface to 610 K induced a C–Br bond cleavage to form the corresponding aryl radicals, followed by coupling to TPP-connections. Alternatively, raising the evaporation temperature of the molecular source to 610 K would cause the C–Br bond cleavage already in the evaporator, which led to deposition of TPP-type radicals onto the gold surface, and then these activated radicals react with each other to form intermolecular chemical bonds upon thermal diffusion (kept at room temperature). By varying the TPP-based molecules with one, two or four Br substituents, the authors succeeded in forming different nano-architectures, that is, for BrTPP only dimerization was observed,

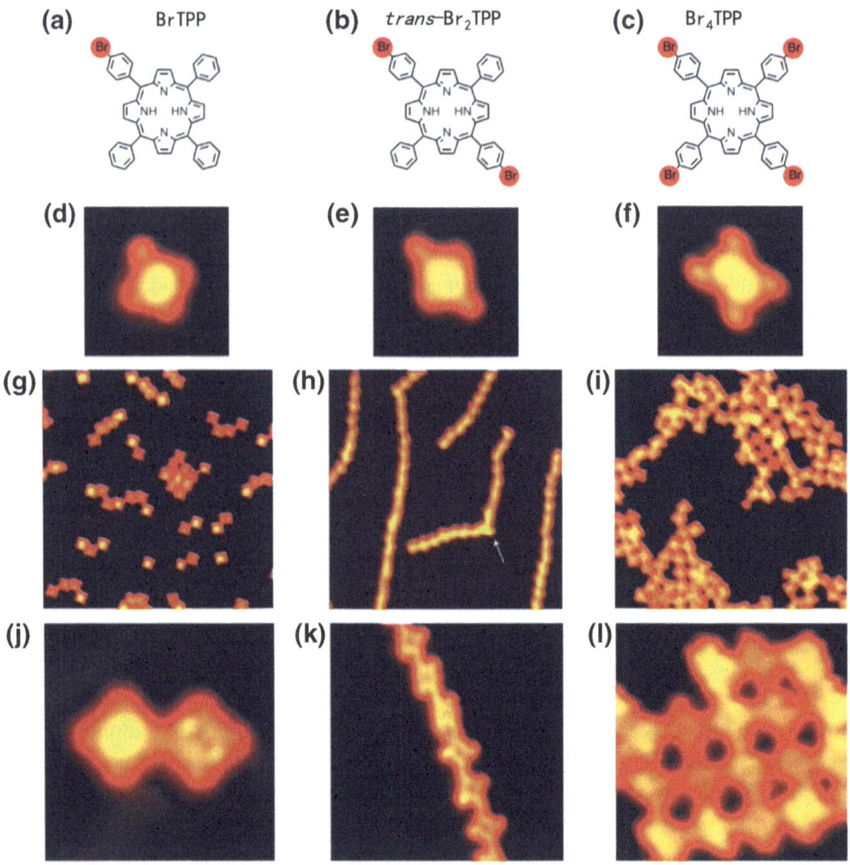

Fig. 1 Thermally induced on-surface Ullmann-type coupling [9]. Building nanoarchitectures using different monomer building blocks with one (left column), two (middle column) and four (right column) Br substituents. **a–c** Chemical structures. **d–f** STM images of single intact molecules (all 3.5 × 3.5 nm^2), where the brominated legs appear larger than the ones without bromines. **g–i** Overview STM images (all 30 × 30 nm^2) of the nanostructures after activation and connection. **j–l** Zoom-in STM images of the formed nanostructures (**j** 5 × 5 nm^2, **k** 10 × 10 nm^2 and **l** 8.5 × 8.5 nm^2) (Adapted with permission from Ref. [9]. Copyright 2007 by the Nature Publishing Group)

whereas *trans*-Br_2TPP led to the formation of long linear chains. Br_4TPP enabled the fabrication of two-dimensional networks (Fig. 1).

Later on, massive works based on the Ullmann-type reactions have been carried out. For instance, more knowledge comes from the works reported by the group of Lackinger and Beton [23, 24]. Both teams independently studied the synthesis of 2-D, polymeric hexagon structures by utilizing 1,3,5-tris(4-bromophenyl)benzene (TBPB) as precursors *via* on-surface Ullmann reaction. After deposition of TBPB molecules onto Cu(111) and Au(111) respectively, comparable network structures

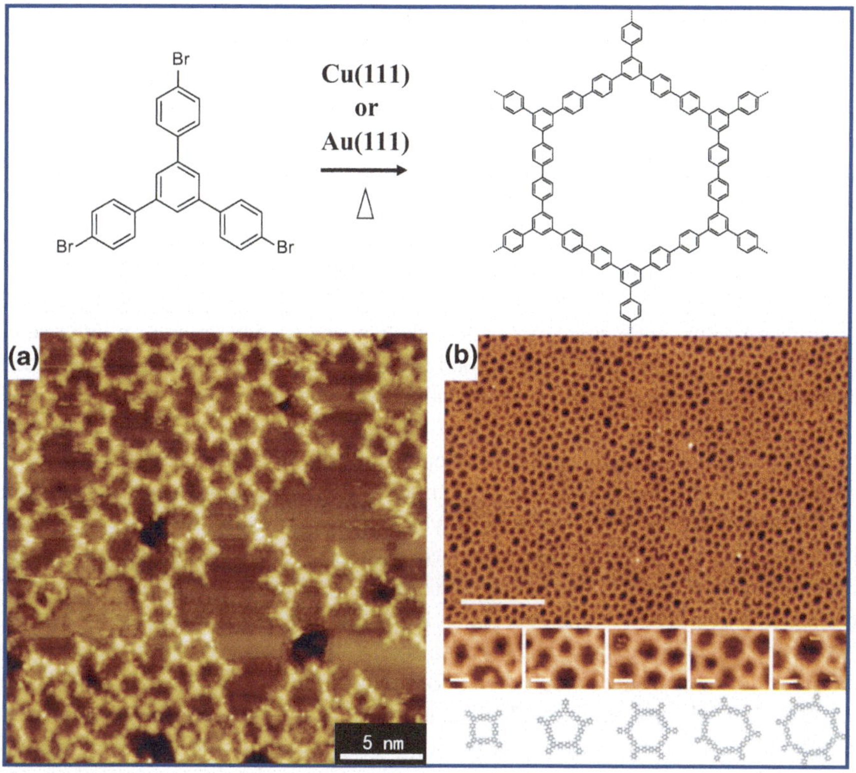

Fig. 2 Network formation via Ullmann-type coupling by **a** Lackinger et al. [23] and **b** Beton et al. [24] (Adapted with permission from Refs. [23, 24]. Copyright 2009 and copyright 2010 by The Royal Society of Chemistry)

were constructed. The precursor molecules exhibited high conformational flexibility during the growth of network structures, which results in the construction of inhomogeneous network structures (Fig. 2a, b). Later, researches using building blocks with high conformational rigidity showed that more uniform network structures can be accomplished [25, 26].

J. Michael Gottfrid and co-workers revealed the reactivity of a surface-assisted system could be affected as a function of the substrate temperature and reported a Ullmann coupling reaction where the regulation between a cyclic product and an open-chain structure could be achieved by varying the surface temperature during deposition of precursor molecules [27]. Deposition of 4,4″-dibromo-*m*-terphenyl (DMTP) onto the Cu(111) substrate at 300 K led to the fabrication of a highly ordered zigzag-chain structure consisting of C–Cu–C bridged organometallics. While depositing DMTP onto Cu(111) at 550 K resulted in the construction of hyperbenzene, a hexagonal macrocycle consisting of 18 connected phenyl rings.

Owing to the success in thermal induced on-surface Ullmann-type reaction, the synthesis of graphene nanoribbons with atomic precise width and edge structure has been achieved. Roman Fasel et al., were the first to report a simple bottom-up method for the growth of atomically precise graphene nanoribbons of different topologies and width [28]. The authors applied on-surface Ullmann-type coupling and subsequent cyclodehydrogenation reactions of 10,10′-dibromo-9,9′-bianthryl (DBBA) precursor to synthesize the prototypical armchair ribbon of width N = 7. Thermal sublimation of DBBA onto the Au(111) substrate removes their halogen substituents, thus forming the biradical species. During the first thermal activation step, the biradical species diffuse along the surface and undergo radical addition reactions to form single covalent C–C bonds between each monomer, giving rise to linear polymer chains. Then, in a second thermal activation step, a surface-assisted cyclodehydrogenation builds up an extended fully aromatic system (Fig. 3a). In the bottom-up approach, a careful design of the precursors can give access to a wide range of different structures. For example, the authors fabricated chevron-type graphene nanoribbons (GNRs) with alternating widths of N = 6 and N = 9 using 6,11-dibromo-1,2,3,4-tetraphenyltriphenylene precursor and a threefold GNR junction with a tri-halogen-functionalized monomer. Michael F. Crommie et al. constructed N = 13 armchair GNRs utilizing the precursor shown in Fig. 3b [29]. Patrick Han et al. and Roman Fasel et al. found that the coupling selectivity usually introduced by halogen substitution was overwhelmed by the structural and catalytic properties of the substrate when the DBBA molecules were introduced to Cu(111) substrate [30, 31] and obtained identical chiral GNRs as shown in Fig. 3c. Shigeki Kawai and co-workers presented boron-doped graphene nanoribbons with widths of N = 7, 14 and 21 by on-surface reactions with an organoboron precursor (Fig. 3d) [41]. Besides, Ingmar Swart and coworkers found that for the aryl chloride, cyclodehydrogenation occurs before dehalogenation and polymerization [42]. The planar bisanthene radicals displayed a different coupling behavior compared to the staggered bianthryl radicals, which resulted in the fabrication of polybisanthenes.

The C–C homocoupling reaction of precursors with predefined C–X (X stands for halogens) groups has proven to be one of the most efficient ways for controllable fabrication of different carbon scaffoldings. Nevertheless, dehalogenative homocouplings have been only limited to aryl halides (sp^2-carbon). More recently, alkyl (sp^3-carbon), alkenyl (sp^2-carbon) and alkynyl (sp^1-carbon) halides have also been demonstrated on surfaces by the group of Wei Xu (Fig. 4) [34, 43, 44]. In these systems, C–Br activations are surprisingly found to occur at room temperature, which are much lower than their aryl bromide counterparts. For alkenyl and alkynyl halides organometallic intermediates could be identified (Fig. 4b, c). The identification of these organometallic intermediates in the processes demonstrates the role of the metal surface as a catalyst in the homocoupling reaction. It is also worth mentioning that by introducing the dehalogenation homocoupling of alkynyl halides to an inert Au(111) surface, the authors demonstrate an effective approach for the construction of carbon nanostructures involving acetylenic scaffoldings. Three organic molecules with one, two and three alkynyl bromide groups have been delicately designed and synthesized. As expected, the formation of dimer structures,

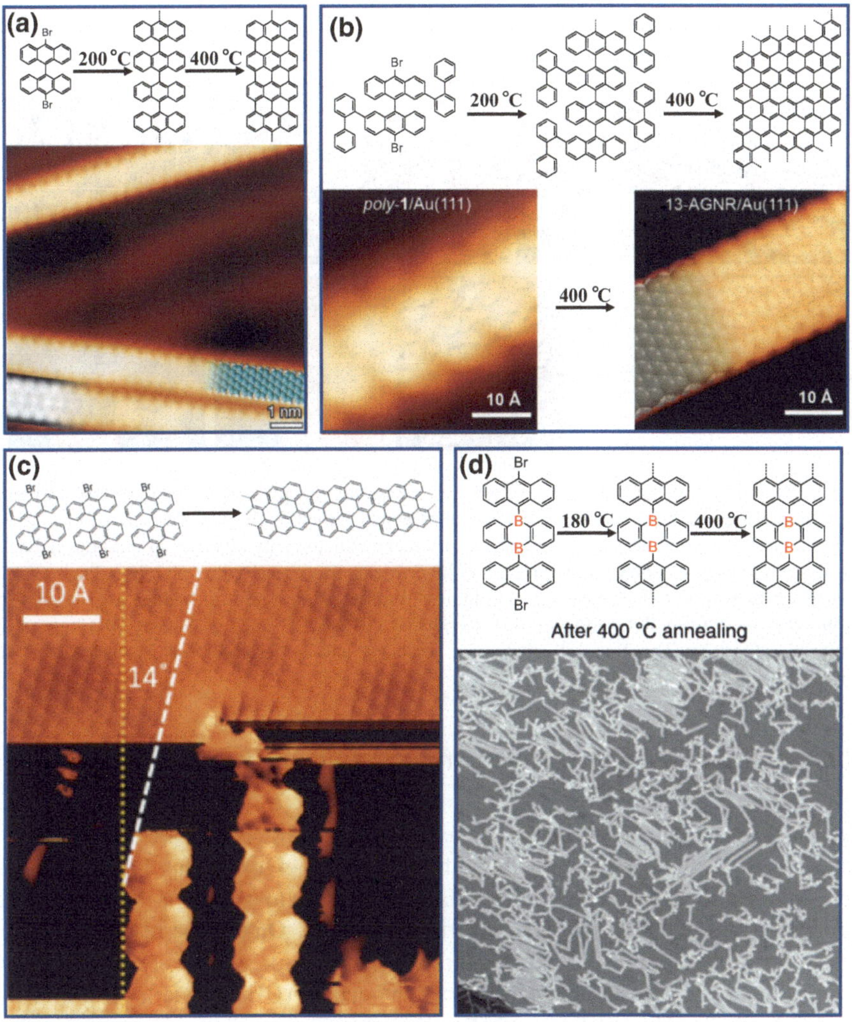

Fig. 3 Bottom-up synthesis of graphene nanoribbons (GNRs). Dehalogenation followed by cyclodehydrogenation to give fully aromatic GNRs by **a** Fasel et al. [28] and **b** Crommie et al. [29] (Adapted with permission from Refs. [28, 29]. Copyright 2010 by the Nature Publishing Group and copyright 2013 by the American Chemical Society). **c** Dehalogenation followed by hydrogen passivation and finally gives identical chiral GNRs [30] (Adapted with permission from Ref. [30]. Copyright 2014 by the American Chemical Society) **d** Atomically controlled substitutional boron-doping of graphene nanoribbons [41] (Adapted with permission from Ref. [41]. Copyright 2015 by the Nature Publishing Group)

one-dimensional (1-D) molecular wires and two-dimensional (2-D) molecular networks with acetylenic scaffoldings are successfully achieved on Au(111) substrate (Fig. 4c).

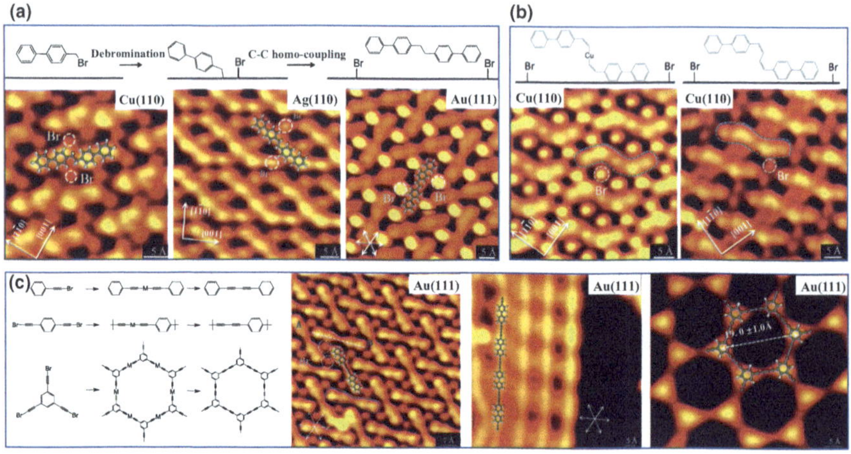

Fig. 4 On-surface dehalogenative reactions involving alkyl (sp^3-carbon), alkenyl (sp^2-carbon) and alkynyl (sp^1-carbon) halides. **a** Formation of dimer structures by (sp^3-carbon) halides [43]. **b** Formation of organometallic intermediates and diene structures by alkenyl (sp^2-carbon) halides [44]. **c** Formation of dimer structures, one-dimensional molecular wires and two-dimensional molecular networks with acetylenic scaffoldings by alkynyl (sp^1-carbon) halides [34] (Adapted with permission from Refs. [34, 43, 44]. Copyright 2016 by The Royal Society of Chemistry, copyright 2016 by The Royal Society of Chemistry, and copyright 2016 by the American Chemical Society)

2.2 *On-Surface Dehydrogenative Homocoupling*

Dehalogenative homocoupling reactions after the cleavage of C–X (X stands for halogens, typically for Br and I) bond have a drawback of leaving halogen atoms remaining as contaminations on the surface. A solution to overcome this problem is offered by the so-called "clean" on-surface reactions. On-surface reactions where by-products are completely avoided or only gaseous or volatile by-products are formed that leave the surface under ultrahigh vacuum (UHV) conditions would all be good options. Thus, dehydrogenative homocoupling reactions represent elegant ways to realize this target. During the reaction process, only inert hydrogen gas is formed along with the targeted products. The challenge lies in the selective C–H activation of only specific C–H bonds in the chosen organic building blocks with the presence of many other C–H bonds.

One of the very first on-surface C–H bond activation was reported by Raval et al. [7]. They found that deposition of tetra(mesityl)porphyrin (TMP) molecules onto a Cu(110) surface with a low coverage and then annealing the molecule-covered sample at between 150 and 200 °C led to the fabrication of covalently connected C–C bonds (Fig. 5a). The close-up STM image unraveled that the formed linear, angular and grid-like structures consisted of edge-connected TMP units. The authors drew a conclusion that C–H activation of the *para* methyl groups occurred,

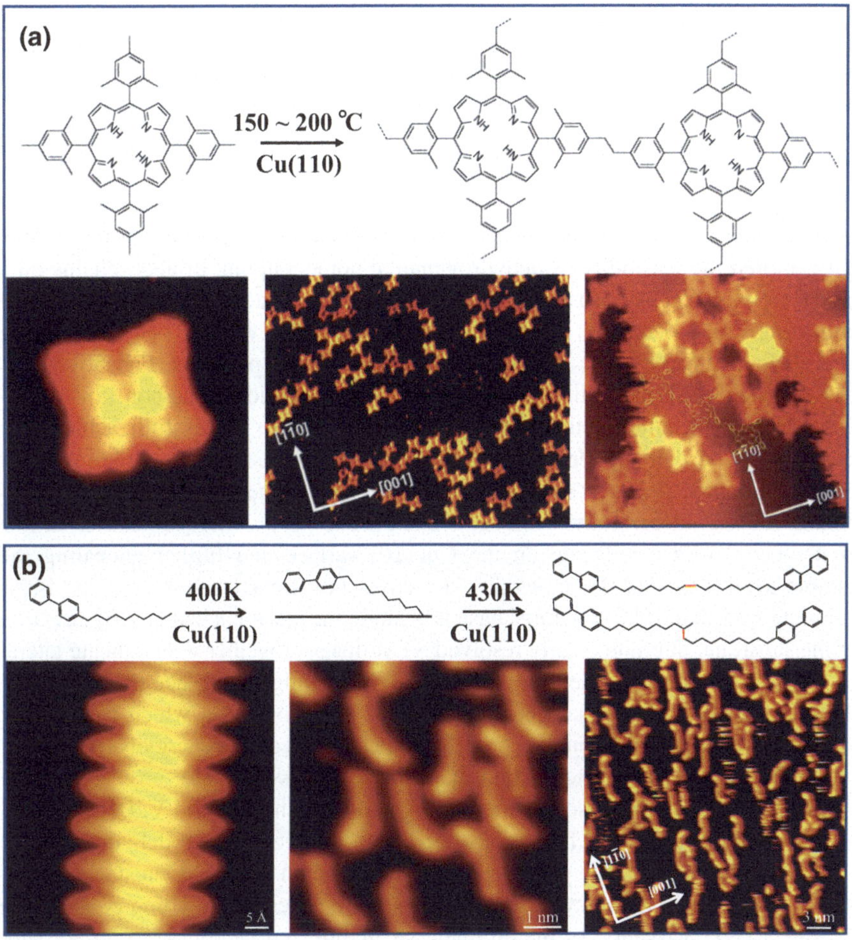

Fig. 5 **a** Dehydrogenative homocoupling of TMP molecules via C–H activation of benzylic sp^3 hybridized carbon atoms [7] (Adapted with permission from Ref. [7]. Copyright 2008 by The Royal Society of Chemistry). **b** Dehydrogenative homocoupling of alkyl chains on Cu(110) [46] (Adapted with permission from Ref. [46]. Copyright 2016 by Wiley-VCH)

followed by C–C coupling of benzyl radicals to eventually fabricate the observed nanostructures.

Not only benzylic but also alkyl sp^3 hybridized C–H groups can be activated on surfaces. This example was demonstrated by Lifeng Chi and co-workers in 2011 [45]. They reported a highly selective C–H activation and subsequent dehydrogenative C–C coupling reaction of long-chain linear alkanes on an anisotropic Au (110) surface. Mass spectrometric analysis identified the formation of H_2 as the only by-product, which further verifies the C–H activated reaction. The analogous measurement with the deuterated alkane was also performed, where the formation of D_2 gas was experimentally detected as well. For further confirmation of the dehydrogenative C–C coupling, the authors explored 1,4-di(eicosyl)benzene and eicosylbenzene under the same conditions on the Au(110) substrate. They found that the reaction took place exclusively at specific sites (terminal CH_3 or penultimate CH_2 groups) in the chains at the temperature between 420 and 470 K, and selected for aliphatic C–H activation rather than aromatic C–H activation. This work has demonstrated the huge potential offered by on-surface chemistry as well as the catalytic activity of the metal surface. To be more general, the group of Wei Xu explored the feasibility of dehydrogenative homocoupling of alkyl chains on a much simpler Cu(110) surfaces, and demonstrated a stepwise dehydrogenative homocoupling proceeding from the intact alkyl chain via the dehydrogenative intermediates finally to the formation of divers coupling products (Fig. 5b) [46].

On-surface C–H activation is not limited to sp^3 hybridized carbon. Sp^2 hybridized carbon (including both alkenes and arenes) has been successfully achieved in surface-assisted dehydrogenative homocoupling reactions. For example, Wei Xu et al. demonstrated that a selective aryl-aryl coupling *via* direct C–H bond activation can be successfully achieved on Cu(110) [47]. After deposition of the quaterphenyl (4Ph) molecule on the Cu(110) surface and further annealing the sample to 500 K, the 4Ph molecules are covalently linked together *via* aryl-aryl coupling into chain-like structures basically along the close-packed [1–10] direction of the substrate. Submolecularly resolved STM images together with delicate lateral STM manipulations revealed that the aryl-aryl coupling took place exclusively at the *para* site of the terminal phenyl groups of 4Ph molecules, which was further verified by density functional theory (DFT) calculations on the energy barriers of all possible C–H activation sites within the 4Ph molecules (Fig. 6a).

Porphyrin-based precursors were investigated by both Johannes V. Barth as well as Wei Xu groups, which demonstrated the surface-assisted dehydrogenative homocoupling to fabricate 2-D network structures. In these circumstances, porphyrin and cobalt-phthalocyanine molecules were successfully homo-coupled *via* the direct C–H activation of the aromatic sp^2 hybridized carbon atom on Ag substrates, respectively (Fig. 6b, c) [48, 49].

Along with the surface-assisted dehydrogenative homocoupling of sp^2 hybridized carbon atom, the group of Wei Xu investigated on-surface C–H activation of terminal alkenes [50]. 4-vinyl-1,1′-biphenyl (VBP) was deposited onto copper substrates. After annealing the molecule-covered Cu(110) to 425 K, VBP molecules dimerized *via* covalent C–C coupling at the vinylic position with a quite high yield (>80%). Contrary to the dimerization products of alkenes in solutions, that is, butane moieties, VBP molecules unexpectedly couple to give diene compounds (Fig. 7). Notably, at this low temperature, activation of the aromatic C–H bonds could be fully suppressed.

Glaser coupling, the homocoupling of terminal alkynes *via* C_{sp}–H activation, is very well established in solutions. Over the past few years, dehydrogenative homocoupling of terminal alkynes has been introduced to metal surfaces, evidencing a more complicated scenario. Diverse products including diynes [13, 14, 51–54], organometallic motifs [35, 36], phenyl groups [55, 56], and others could be

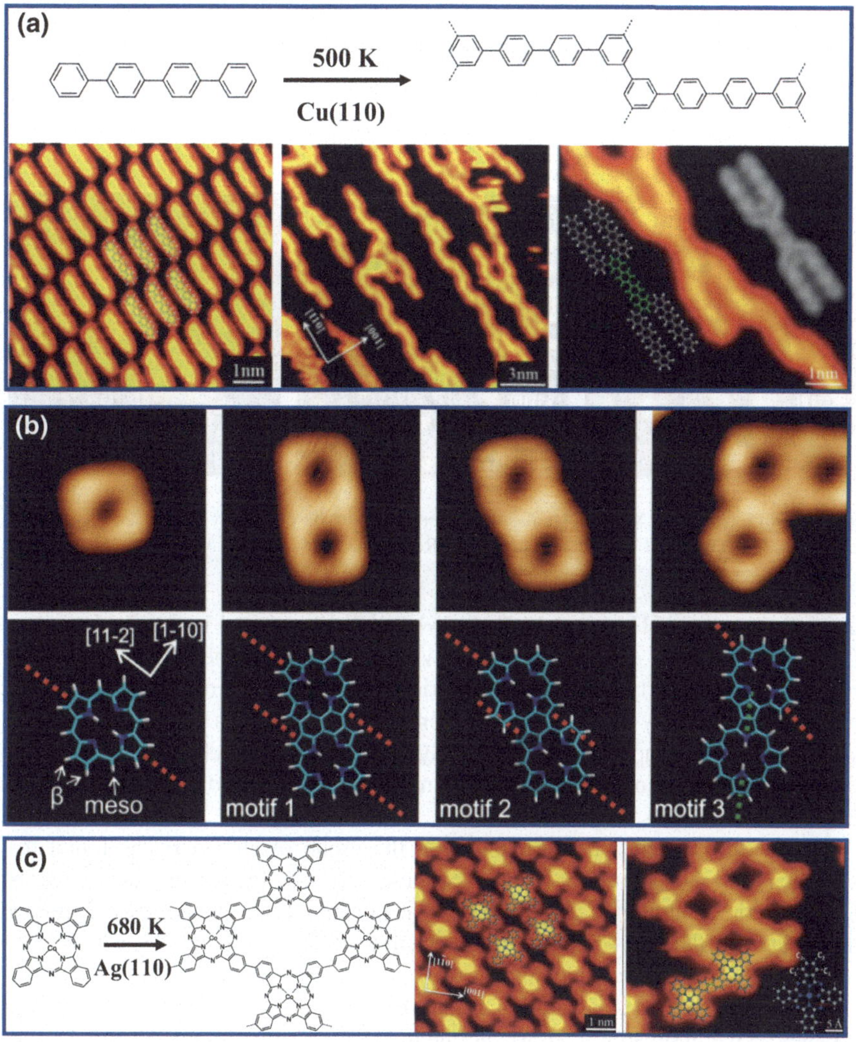

Fig. 6 C–H activation of aromatic sp^2 hybridized carbon atoms [47–49] (Adapted with permission from Refs. [47–49]. Copyright 2014 by The Royal Society of Chemistry, copyright 2014 by the American Chemical Society, and copyright 2015 by The Royal Society of Chemistry)

formed on different surfaces [14, 57], and usually byproducts are inevitable when triggering the reaction at elevated temperatures.

Mario Ruben and Johannes V. Barth et al. made first attempts to construct 2-D frameworks based on the surface-assisted Glaser coupling [13]. 1,3,5-tris-(4-ethynylphenyl)benzene (Ext-TEB) was deposited onto a flat Ag(111) surface. After annealing to 400 K, 2-D network structures with butadiyne connection moieties were fabricated (Fig. 8a). They also attempted to synthesize extended

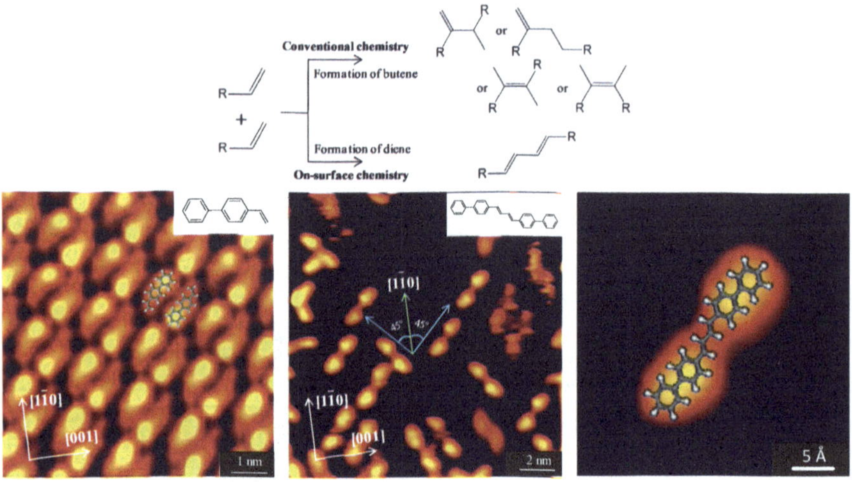

Fig. 7 Dimerization of VBP molecules at the terminal vinylic position via surface-assisted C–H activation [14] (Adapted with permission from Ref. [50]. Copyright 2015 by Wiley-VCH)

graphdiyne wires [51] by a 4,4″-diethynyl-1.1′:4′,1”terphenyl molecule. However, On the flat Ag(111) surface thermal activation triggers a variety of side-reactions resulting in irregularly-branched polymeric networks. To increase the chemo- and regioselectivity, they confined the terminal alkyne reactants to step edges of a vicinal Ag(877) template, and fortunately the formation of chemoselective butadiyne was preferred over branching side-reactions encountered on the flat Ag(111) (Fig. 8b).

Besides the vicinal surface templating, the group of Harald Fuchs found that selectivity of the on-surface Glaser coupling can be dramatically increased by simple steric shielding of the active alkyne moieties [14]. Hence, they decorated alkyl chains at the *ortho*-position of the reacting alkyne functionalities, which led to the suppression of side reactions (Fig. 8c).

The on-surface Glaser-coupling generally takes place by thermal activation of alkynes on metal substrates. However, the interaction between molecular precursors and surface metal atoms often induces unwanted reaction pathways once excessive thermal energy is provided. Harald Fuchs and Mauro Sambi together with their co-workers showed light-induced homocoupling of terminal alkynes. Fuchs et al. showed the formation of aryl-alkyne dimers on Ag(111) induced by UV irradiation at a wavelength of 375 nm at room temperature [53]. The light-induced Glaser coupling with a bisalkynylarene as a building block mainly provides dimeric structures instead of long polymer chains in contrast to the thermal-induced one discussed above. Further reaction towards long polymer chains is also suppressed because of the low surface mobility of the dimers at room temperature. Recently, Sambi et al. also reported a light-induced metal-free homocoupling of terminal alkynes on highly oriented pyrolytic graphite (HOPG) [54].

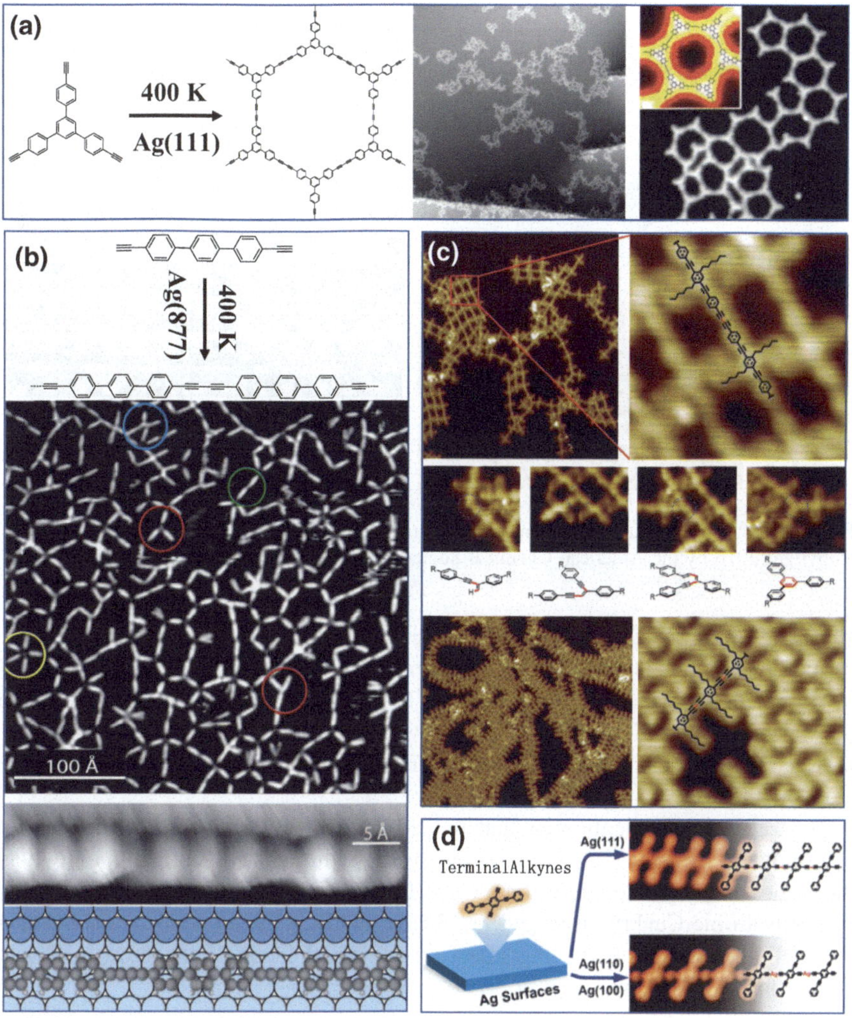

Fig. 8 On-surface Glaser-type coupling reactions of terminal arylalkynes: **a**, **b** by Ruben/Barth et al. [13, 51], **c** by Studer/Fuchs et al. [14], and by **d** Wu et al. [35] (Adapted with permission from Refs. [13, 14, 35, 51]. Copyright 2012 by the Nature Publishing Group, copyright 2014 by the American Chemical Society, copyright 2013 by Wiley-VCH, and copyright 2015 by the American Chemical Society)

Two major strategies have been adopted to control the on-surface reaction of terminal alkynes including changing the precursor backbone or employing different substrates. Kai Wu and co-workers studied the role of different lattice planes of the

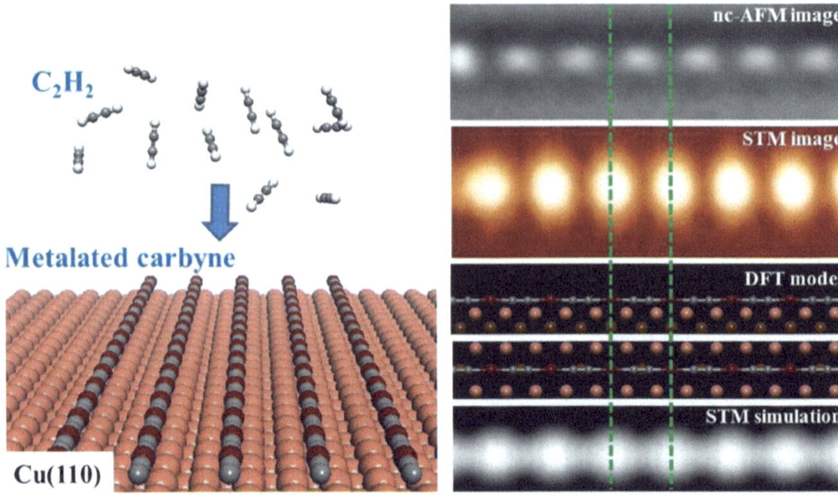

Fig. 9 Synthesis of metalated carbyne by dehydrogenative coupling of ethyne on Cu(110) [36] (copyright 2016 by the American Chemical Society)

silver substrate and found that 1-D molecular wires were formed by Glaser-type coupling on Ag(111) whereas 1-D organometallic nanostructures were fabricated involving the surface metal atoms on Ag(110) and Ag(100) (Fig. 8d). Such a lattice dependent reaction of the terminal alkyne was originated from the matching degree between the periodicities of the constructed molecular wires and the surface lattice structures [35].

It is also worth mentioning that the group of Wei Xu reported a successful synthesis of metalated carbyne chains by dehydrogenative coupling of ethyne molecules and copper adatoms on a Cu(110) substrate under UHV conditions [36]. The anisotropic Cu(110) was selected for its relatively high chemical activity and the well-known 1-D templating effect which could facilitate the growth of 1-D carbon-based nanomaterials (Fig. 9).

2.3 Sequential C–X Groups Activation

The sequential on-surface reactions combining dehalogenative with dehydrogenative homocoupling are reviewed in this section. The groups of Hecht/Grill and Lackinger reported the analogous sequential Ullmann-type coupling reaction [10, 58]. At lower temperature, selective activation of the weaker C–I bonds was

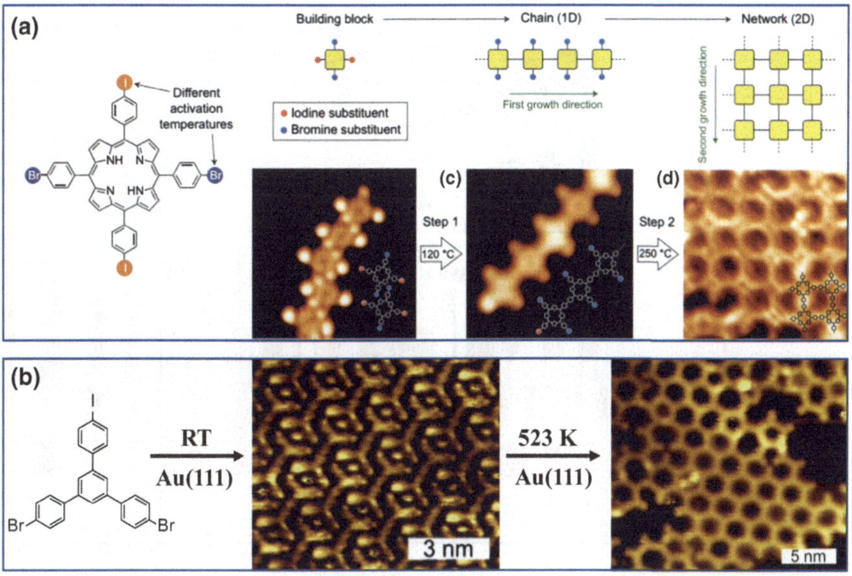

Fig. 10 Sequential combination of on-surface Ullmann-type coupling reactions **a** by the group of Hecht/Grill [10] and **b** by Lackinger and co-workers [58] (Adapted with permission from Refs. [10, 58]. Copyright 2012 by the Nature Publishing Group and copyright 2014 by the American Chemical Society)

accomplished, resulting in covalent bonded substructures. At higher temperatures, C–Br bond activation could be also achieved, finally realizing the construction of 2-D network structures (Fig. 10).

The group of Wei Xu illustrated on-surface formation of polyphenyl chains through a sequential reaction pathway involving Ullmann-type coupling and subsequent dehydrogenative coupling [59]. The *para*-bromobiphenyl molecules reacted *via* Ullmann-type coupling to quaterphenyl on Cu(110) at 465 K. Further increasing the temperature to 500 K induced intermolecular dehydrogenative coupling at the *meta*-positions of the terminal phenyl moieties in the formed quaterphenyl to give polyphenyl chains (Fig. 11a). Mauro Sambi and co-workers applied the Ullmann coupling to generate polyphenyl wires on Au(111) at 400 K by using 4,4″-dibromo-*p*-terphenyl (DBTP) as building block. Further annealing to 650 K led to intermolecular dehydrogenation to give graphene nanoribbons (Fig. 11b) [60].

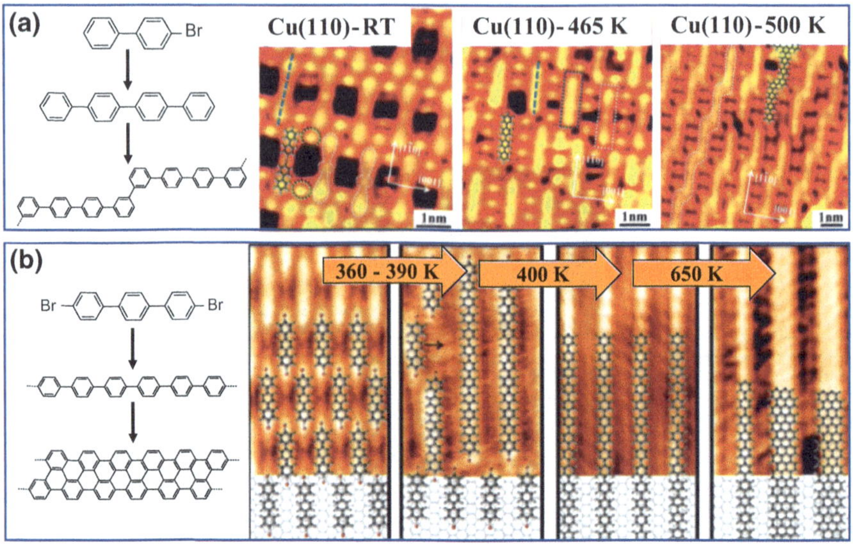

Fig. 11 On-surface Ullmann-type coupling and subsequent intermolecular dehydrogenation **a** by Xu and co-workers [59] and **b** by Basagni, Sedona and co-workers [60] (Adapted with permission from Refs. [59, 60]. Copyright 2014 by The Royal Society of Chemistry and copyright 2015 by the American Chemical Society)

3 Summary and Outlook

On-surface C–X activation and the following C–C coupling undoubtedly represent one of the most promising and efficient methods for syntheses of novel molecules and advanced carbon-based nanostructures, which are hard to obtain or even unachievable by traditional solution chemistry. During the last decade which can be called the infant era of on-surface synthesis, we have witnessed the successful formation of a vast number of graphene derived nanostructures, among which graphene nanoribbon belongs to one of the most representative ones. To that end, Ullmann-type on-surface reaction, that is, dehalogenative homocoupling of aryl halide represents the most intensely studied topic across the worldwide surface science community.

A few forward-looking on-surface studies have been showing the growing interests on the synthesis of prospective low-dimensional carbon nanostructures including but not being limited to carbyne and graphyne, which were proposed long ago but yet to be synthesized. The development of on-surface synthesis arouses our attention and preliminary movements on the synthesis of these promising materials again. In particular, on-surface C–X activations followed by C–C coupling might be the best reaction of choice. In light of the requirements of carbon centers other than aryl sp^2 hybridized one, we may foresee flourishing studies of sp^1 and sp^3 hybridized carbon center, or ever heteroatoms other than carbon involved in the reactions in the near future. Along with a long pathway to these long-standing goals, it is of high relevance and urgency to study the fundamental reaction products, pathways as well as their mechanisms. In addition to the Ullmann-type coupling on surfaces, lots of works are yet to be carried out in the future.

References

1. Björk, J., Hanke, F.: Chem. Eur. J. **20**, 928 (2014)
2. Lindner, R., Kihnle, A.: ChemPhysChem **16**, 1582–1592 (2015)
3. Fan, Q.T., Gottfried, J.M., Zhu, J.F.: Acc. Chem. Res. **48**, 2484–2494 (2015)
4. Dong, L., Liu, P.N., Lin, N.: Acc. Chem. Res. **48**, 2765–2774 (2015)
5. Klappenberger, F., Zhang, Y.-Q., Bjçrk, J., Klyatskaya, S., Ruben, M., Barth, J.V.: Acc. Chem. Res. **48**, 2140–2150 (2015)
6. Haq, S., Hanke, F., Sharp, J., Persson, M., Amabilino, D.B., Raval, R.: ACS Nano **8**, 8856–8870 (2014)
7. In't Veld, M., Iavicoli, P., Haq, S., Amabilino, D.B., Raval, R.: Chem. Commun. **13**, 1536–1538 (2008)
8. Lackinger, M., Heckl, W.M.: J. Phys. D **44**, 464011 (2011)
9. Grill, L., Dyer, M., Lafferentz, L., Persson, M., Peters, M.V., Hecht, S.: Nat. Nanotechnol. **2**, 687 (2007)
10. Lafferentz, L., Eberhardt, V., Dri, C., Africh, C., Comelli, G., Esch, F., Hecht, S., Grill, L.: Nat. Chem. **4**, 215–220 (2012)
11. de Oteyza, D.G., Gorman, P., Chen, Y.C., Wickenburg, S., Riss, A., Mowbray, D.J., Etkin, G., Pedramrazi, Z., Tsai, H.Z., Rubio, A., Crommie, M.F., Fischer, F.R.: Science **340**, 1434 (2013)
12. Sun, Q., Zhang, C., Li, Z., Kong, H., Tan, Q., Hu, A., Xu, W.: J. Am. Chem. Soc. **135**, 8448 (2013)
13. Zhang, Y.Q., Kepcija, N., Kleinschrodt, M., Diller, K., Fischer, S., Papageorgiou, A.C., Allegretti, F., Bjork, J., Klyatskaya, S., Klappenberger, F., Ruben, M., Barth, J.V.: Nat. Commun. **3**, 1286 (2012)
14. Gao, H.Y., Wagner, H., Zhong, D., Franke, J.H., Studer, A., Fuchs, H.: Angew. Chem. Int. Ed. **52**, 4024 (2013) Angew. Chem. **125**, 4116 (2013)
15. Weigelt, S., Busse, C., Bombis, C., Knudsen, M.M., Gothelf, K.V., Strunskus, T., Wçll, C., Dahlbom, M., Hammer, B., Lægsgaard, E., Besenbacher, F., Linderoth, T.R.: Angew. Chem. Int. Ed. **46**, 9227 (2007) Angew. Chem. **119**, 9387 (2007)
16. Liu, X.H., Guan, C.Z., Ding, S.Y., Wang, W., Yan, H.J., Wang, D., Wan, L.J.: J. Am. Chem. Soc. **135**, 10470 (2013)
17. Gao, H.Y., Held, P.A., Knor, M., Mìck-Lichtenfeld, C., Neugebauer, J., Studer, A., Fuchs, H.: J. Am. Chem. Soc. **136**, 9658 (2014)
18. Bebensee, F., Bombis, C., Vadapoo, S.R., Cramer, J.R., Besenbacher, F., Gothelf, K.V., Linderoth, T.R.: J. Am. Chem. Soc. **135**, 2136 (2013)
19. Díaz Arado, O., Mçnig, H., Wagner, H., Franke, J.H., Langewisch, G., Held, P.A., Studer, A., Fuchs, H.: ACS Nano **7**, 8509 (2013)
20. Matena, M., Stçhr, M., Riehm, T., BjÅrk, J., Martens, S., Dyer, M.S., Persson, M., Lobo-Checa, J., Mìller, K., Enache, M., Wadepohl, H., Zegenhagen, J., Jung, T.A., Gade, L. H.: Chem. Eur. J. **16**, 2079 (2010)
21. Talirz, L., Ruffieux, P., Fasel, R.: On-surface synthesis of atomically precise graphene nanoribbons. Adv. Mater. **28**, 6222–6231 (2016)
22. Wang, W., Shi, X., Wang, S., Van Hove, M.A., Lin, N.: J. Am. Chem. Soc. **133**, 13264 (2011)

23. Gutzler, R., Walch, H., Eder, G., Kloft, S., Heckl, W.M., Lackinger, M.: Chem. Commun. **29**, 4456–4458 (2009)
24. Blunt, M.O., Russell, J.C., Champness, N.R., Beton, P.H.: Chem. Commun. **46**, 7157–7159 (2010)
25. Bieri, M., Nguyen, M.T., Gröning, O., Cai, J., Treier, M., Aït-Mansour, K., Ruffieux, P., Pignedoli, C.A., Passerone, D., Kastler, M., Mllen, K., Fasel, R.: J. Am. Chem. Soc. **132**, 16669–16676 (2010)
26. Bieri, M., Blankenburg, S., Kivala, M., Pignedoli, C.A., Ruffieux, P., Mllen, K., Fasel, R.: Chem. Commun. **47**, 10239–10241 (2011)
27. Fan, Q., Wang, C., Han, Y., Zhu, J., Hieringer, W., Kuttner, J., Hilt, G., Gottfried, J.M.: Angew. Chem. **125**, 4766–4770 (2013)
28. Cai, J., Ruffieux, P., Jaafar, R., Bieri, M., Braun, T., Blankenburg, S., Muoth, M., Seitsonen, A.P., Saleh, M., Feng, X., Mullen, K., Fasel, R.: Nature **466**, 470–473 (2010)
29. Chen, Y.C., de Oteyza, D.G., Pedramrazi, Z., Chen, C., Fischer, F.R., Crommie, M.F.: ACS Nano **7**, 6123–6128 (2013)
30. Han, P., Akagi, K., Canova, F.F., Mutoh, H., Shiraki, S., Iwaya, K., Weiss, P.S., Asao, N., Hitosugi, T.: ACS Nano **8**, 9181–9187 (2014)
31. Sánchez-Sánchez, C., Dienel, T., Deniz, O., Ruffieux, P., Berger, R., Feng, X., Müllen, K., Fasel, R.: ACS Nano **10**, 8006–8011 (2016)
32. Bieri, M., Treier, M., Cai, J., Aït-Mansour, K., Ruffieux, P., Gröning, O., Gröning, P., Kastler, M., Rieger, R., Feng, X., Müllen, K., Fasel, R.: Chem. Commun. **45**, 6919–6921 (2009)
33. Simonov, K.A., Vinogradov, N.A., Vinogradov, A.S., Generalov, A.V., Zagrebina, E.M., Svirskiy, G.I., Cafolla, A.A., Carpy, T., Cunniffe, J.P., Taketsugu, T., Lyalin, A., Mårtensson, N., Preobrajenski, A.B.: ACS Nano **9**, 8997–9011 (2015)
34. Sun, Q., Cai, L., Ma, H., Yuan, C., Xu, W.: ACS Nano **10**, 7023–7030 (2016)
35. Liu, J., Chen, Q., Xiao, L., Shang, J., Zhou, X., Zhang, Y., Wang, Y., Shao, X., Li, J., Chen, W., Xu, G.Q., Tang, H., Zhao, D., Wu, K.: ACS Nano **9**, 6305–6314 (2015)
36. Sun, Q., Cai, L., Wang, S., Widmer, R., Ju, H., Zhu, J., Li, L., He, Y., Ruffieux, P., Fasel, R., Xu, W.: J. Am. Chem. Soc. **138**, 1106–1109 (2016)
37. Ullmann, F., Bielecki, J.: Ber. Dtsch. Chem. Ges. **34**, 2174–2185 (1901)
38. Ullmann, F., Meyer, G.M., Loewenthal, O., Gilli, O.: Justus Liebig's Annalen der Chemie **331**, 38–81 (1904)
39. Fanta, P.E.: The Ullmann synthesis of biaryls. Synthesis **01**, 9–21 (1974)
40. Hla, S.W., Bartels, L., Meyer, G., Rieder, K.H.: Phys. Rev. Lett. **85**, 2777–3780 (2000)
41. Kawai, S., Saito, S., Osumi, S., Yamaguchi, S., Foster, A.S., Spijker, P., Meyer, E.: Atomically controlled substitutional boron-doping of graphene nanoribbons. Nat. Commun. **6**, 8098 (2015)
42. Jacobse, P.H., van den Hoogenband, A., Moret, M.E., Gebbink, R.J.M.K., Swart, I.: Angew. Chem. Int. Ed. **55**, 13052–13055 (2016)
43. Sun, Q., Cai, L., Ding, Y., Ma, H., Yuan, C., Xu, W.: Phys. Chem. Chem. Phys. **18**, 2730–2735 (2016)
44. Sun, Q., Cai, L., Ma, H., Yuan, C., Xu, W.: Chem. Commun. **52**, 6009–6012 (2016)
45. Zhong, D., Franke, J., Podiyanachari, S.K., Blömker, T., Zhang, H., Kehr, G., Erker, G., Fuchs, H., Chi, L.: Science **334**, 213–216 (2011)
46. Cai, L., Sun, Q., Zhang, C., Ding, Y., Xu, W.: Chem. Eur. J. **22**, 1918–1921 (2016)
47. Sun, Q., Zhang, C., Kong, H., Tan, Q., Xu, W.: Chem. Commun. **50**, 11825–11828 (2014)
48. Wiengarten, A., Seufert, K., Auwärter, W., Ecija, D., Diller, K., Allegretti, F., Bischoff, F., Fischer, S., Duncan, D.A., Papageorgiou, A.C., Klappenberger, F., Acres, R.G., Ngo, T.H., Barth, J.V.: J. Am. Chem. Soc. **136**, 9346–9354 (2014)
49. Sun, Q., Zhang, C., Cai, L., Xie, L., Tan, Q., Xu, W.: Chem. Commun. **51**, 2836–2839 (2015)
50. Sun, Q., Cai, L., Ding, Y., Xie, L., Zhang, C., Tan, Q., Xu, W.: Angew. Chem. Int. Ed. **54**, 4549–4552 (2015)
51. Cirera, B., Zhang, Y.Q., Bjçrk, J., Klyatskaya, S., Chen, Z., Ruben, M., Barth, J.V., Klappenberger, F.: Nano Lett. **14**, 1891–1897 (2014)

52. Gao, H.Y., Franke, J.H., Wagner, H., Zhong, D., Held, P.A., Studer, A., Fuchs, H.: J. Phys. Chem. C **117**, 18595–18602 (2013)
53. Gao, H.Y., Zhong, D., Mçnig, H., Wagner, H., Held, P.A., Timmer, A., Studer, A., Fuchs, H.: J. Phys. Chem. C **118**, 6272–6277 (2014)
54. Colazzo, L., Sedona, F., Moretto, A., Casarin, M., Sambi, M.: J. Am. Chem. Soc. **138**, 10151–10156 (2016)
55. Zhou, H., Liu, J., Du, S., Zhang, L., Li, G., Zhang, Y., Tang, B.Z., Gao, H.J.: J. Am. Chem. Soc. **136**, 5567–5570 (2014)
56. Liu, J., Ruffieux, P., Feng, X., Müllen, K., Fasel, R.: Chem. Commun. **50**, 11200–11203 (2014)
57. Eichhorn, J., Heckl, W.M., Lackinger, M.: Chem. Commun. **49**, 2900–2902 (2013)
58. Eichhorn, J., Nieckarz, D., Ochs, O., Samanta, D., Schmittel, M., Szabelski, P.J., Lackinger, M.: ACS Nano **8**, 7880–7889 (2014)
59. Zhang, C., Sun, Q., Chen, H., Tan, Q., Xu, W.: Chem. Commun. **51**, 495–498 (2015)
60. Basagni, A., Sedona, F., Pignedoli, C.A., Cattelan, M., Nicolas, L., Casarin, M., Sambi, M.: J. Am. Chem. Soc. **137**, 1802–1808 (2015)

On-Surface Ullmann Reaction for the Synthesis of Polymers and Macrocycles

Qitang Fan, Junfa Zhu and J. Michael Gottfried

Abstract Compared to organic synthesis in solution, the benefits of on-surface synthesis are apparent especially when it involves insoluble reactants or when in situ characterizations of products with large molecular weight are required. This article covers the on-surface synthesis of hydrocarbon polymers and macrocycles via an Ullmann type reaction of haloarenes on metal single-crystal surfaces. The related formation of stable organometallic reaction intermediates with carbon-metal-carbon bonds is also discussed. Regarding the on-surface synthesis of polymers, deposition onto hot metal surfaces (in contrast to post-annealing) as well as high diffusion rates of the monomers lead to the increase of their chain lengths. To obtain high yields of macrocycles, it is important to enable the ring closure at the stage of the organometallic species with reversible carbon-metal-carbon bonds and to apply high-dilution conditions. In contrast, the formation of macrocycles by cyclisation of polymer chains with only covalent bonds is of low probability due to the particular mechanism of the ring-closure process in two-dimensional confinement.

1 Introduction

Since Friedrich Wöhler reported the first organic synthesis in 1828 with the conversion of ammonium cyanate to urea, [1] the field of synthetic organic chemistry has developed mainly by performing reactions in the homogeneous liquid phase and much less in heterogeneous systems, e.g., at solid-liquid interfaces. A wide

Q. Fan · J. M. Gottfried (✉)
Fachbereich Chemie, Philipps-Universität Marburg, Hans-Meerwein-Str, 35032 Marburg, Germany
e-mail: michael.gottfried@chemie.uni-marburg.de

Q. Fan · J. Zhu
National Synchrotron Radiation Laboratory and Collaborative Innovation Center of Suzhou Nano Science and Technology, University of Science and Technology of China, Hefei 230029, People's Republic of China

D. G. de Oteyza and C. Rogero (eds.), *On-Surface Synthesis II*,
Advances in Atom and Single Molecule Machines,
https://doi.org/10.1007/978-3-319-75810-7_5

range of reaction types enables the synthesis of diverse chemicals such as pharmaceuticals or synthetic polymer materials including silicon, plastic, artificial fibers, etc. [2]. The fact that most of the available reactions require solution conditions represents a considerable restriction, because sufficient solubility of the reactants in a suitable solvent is required. In the synthesis of shape-persistent nano-sized macrocycles, which have been considered as important intermediates for two-dimensional (2D) polymers, [3] sufficient solubility of reactants and products can only be achieved with long alkyl side chains at the (often conjugated) molecular backbones. These side chains are potentially disadvantageous in the further reactions of the macrocycles. This example illustrates how solution-based synthesis requires compromises regarding the structure of the molecules and that not all desired structures can be synthesized in the homogeneous liquid environment.

With the emergence of the on-surface synthetic approach, in which the reactions are performed at the solid/vacuum interface in the absence of solvents, this limitation can be overcome. Concerning the above-mentioned example of shape-persistent macrocycles, cyclic oligophenylenes (honeycombenes) without alkyl side chains have been synthesized on metal surfaces under ultrahigh-vacuum (UHV) conditions [4–6]. One of the most robust and versatile reactions currently used in on-surface synthesis is the Ullmann C–C coupling reaction [7, 8] which is a modified version of a long-known heterogeneous (solid/liquid) reaction originally developed for the synthesis of biphenyls from aryl halides [9]. It has been employed to synthesize various, mostly conjugated, polymers including polyphenylene [10–14] polythiophene [15], graphene nanoribbons [16], and others [17]. One of the intrinsic benefits of on-surface synthesis is the possibility of in situ characterization of the products with powerful surface science techniques, especially scanning probe microscopy/spectroscopy with sub-molecular spatial resolution. This approach allows for the sophisticated control over the electronic properties of the synthesized polymers via modification of the custom designed precursor monomers. Another advantage of on-surface synthesis comes into play when monolayers of large organic molecules, metal complexes or organometallic compounds are considered. The typically used preparation method of such layers, physical vapor deposition under vacuum conditions, is often not applicable, because the molecules thermally decompose before they develop a practically usable vapor pressure. In such cases, an alternative is the deposition of smaller precursor molecules, which react on the surface to form the desired larger species. An example is the above-mentioned synthesis of oligophenylene macrocycles [4–6]. In general, the on-surface synthesis of cyclic polymers or macrocycles is challenging, because the 2D confinement reduces the degrees of freedom of the precursor chain and hinders the encounter of its ends as the final step of the cyclization. Nevertheless, there are several successful examples including [18]- and [30]-honeycombene, [4, 5] cyclothiophene [15], and sexiphenylene [10].

In this chapter, we review the present state of the on-surface Ullmann reaction for the synthesis of both open-chain polymers and macrocycles. In the first part, the formation, assembly, and stabilization of organometallic chains with carbon-metal-carbon bonds from haloarenes containing two C–X (X = halogen) groups on metal surfaces are discussed. These organometallic chains are important intermediates for the formation of covalently bonded polymers. Two procedures for the on-surface synthesis of polymers are compared: post-annealing of precursor layers and deposition of precursor molecules on metal surfaces kept at elevated temperatures ("hot" deposition). The "hot" deposition along with a high diffusion rate of the monomers have been found to favor the formation of long polymer chains. In addition, control of the electronic properties of the synthesized polymers such as poly-*para*-phenylene and graphene nanoribbons by modification of precursor monomer is illustrated with examples from the recent literature. In the second part, important prerequisites for the on-surface synthesis of macrocycles are discussed, including the design of monomers with bent carbon skeletons such as 4,4″-dibromo-*meta*-terphenyl (DMTP), the use of surface templates, and the application of the high-dilution principle. Mechanisms of the on-surface formation of macrocycles are described using the example of [18]-honeycombene (hyperbenzene). It is shown that high yields of macrocycles are only obtained if the cyclization occurs at the stage of intermediates with reversible bonds, e.g., an organometallic phase with carbon-metal-carbon bonds, whereas all-covalent intermediate chains have a lower probability for cyclization. Finally, properties and applications of in situ synthesized conjugated polymers and macrocycles on surfaces are summarized.

2 Synthesis of Polymers via Surface Ullmann Reaction

2.1 The Emergence of the On-Surface Ullmann Reaction

The Ullmann reaction, first described in 1901, [18] is one of the oldest heterogeneous reactions in organic synthesis. It was originally proposed for the synthesis of biphenyls from aromatic halides in solution with the assistance of copper powder. The copper does not function as a catalyst, but is needed in stoichiometric amounts. A first link between Ullmann reaction and surface chemistry was established in 1992 when Xi and Bent studied the reaction of iodobenzene on Cu(111) with surface science techniques including temperature programmed reaction spectroscopy [19, 20]. It was reported that the reaction is initiated by the insertion of a copper atom into the C–I bond. This process leads to the formation of an organometallic intermediate, which then reacts with an additional monomer into biphenyl and copper iodide. It is should be noted that the organometallic intermediate could be phenyl bonded to a surface copper adatom or a lattice copper atom (Scheme 1). The reaction pathways then were found to be coverage dependent because of the different adsorption behaviors of the monomer. The first application of the Ullmann reaction to the on-surface fabrication of covalent nanostructures was reported in 2007 by Grill et al., who successfully implemented the coupling of tetraphenylporphyrins (TPP) with different numbers and positions of peripheral Br substituents into covalently linked dimers, chains, and 2D arrays on a Au(111)

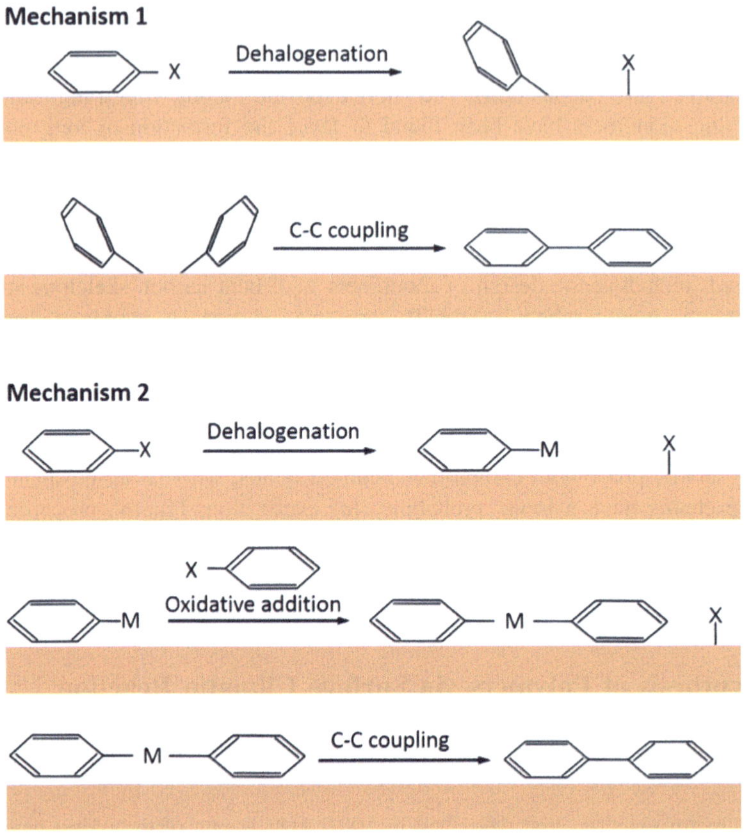

Scheme 1 Reaction mechanisms for the Ullmann reaction of a phenyl halide on a surface

surface [21]. Since then, various aromatic halides were employed for the successful synthesis of zero-dimensional (0D), 1D, and 2D nanostructures on solid surfaces in ultra-high vacuum (UHV) [7]. Looking back on the historic development of this research field, the rather late arrival of this breakthrough is mainly due to the complex mechanism of the surface Ullmann reaction, which requires an optimal balance of the energy barriers for monomer diffusion, desorption, aggregation as well as C–X, C–H and C–C bonds scission. This poses various restraints on the design of the precursor monomers, for example their molecular weights and chemical constitution. As a simple example, the C–X bond activation barrier must be lower than the desorption barrier of the precursor, otherwise the heating procedure will lead to desorption of the precursor monomers before their coupling into larger products. This case commonly occurs to precursors with small molecular weights, which have weak interactions with the metal substrates. Otherwise, alternative approaches are required to trigger the surface Ullmann reaction without simultaneous desorption of the precursor. For instance, Rieder et al. induced the surface Ullmann reaction of iodobenzene on Cu(111) by electron injection through a metal tip [22]. In awareness of this issue, monomers with moderate molecular weights were employed, such as the 4,4″-dibromo-*para*-terphenyl (DBPTP) and 10,10′-dibromo-9,9′-bianthryl (DBBA), which result in the successful on-surface synthesis of polyphenylene [13, 14] and graphene nanoribbons, [23] respectively. Eventually, the optimization of the parameters for on-surface Ullmann reaction to obtain 1D/2D nanostructures with higher degree of order was studied in detail, [24–26] which prompted the prosperity of the research field of on-surface synthesis using reactions such as Ullmann coupling, [7, 8, 27, 28] Glaser coupling, [29] dehydrogenative coupling, [30–33] etc.

2.2 *Identification of Organometallic Intermediates on Surfaces*

In the Ullmann reaction, the structure of the product polymers and (cyclic) oligomers are closely related to the structure of the precursor monomer. Precursor monomers with two halogen substituents typically lead to the formation of one-dimensional (1D) polymer chains after their C–C coupling, as was demonstrated by the formation of tetraphenylporphyrin chains from dibromo-tetraphenylporphyrin in the work by Grill et al. [21]. Noteworthy, as mentioned in the section above, one possible mechanism for the Ullmann reaction is to initially form the organometallic intermediate, which then undergoes demetalation forming the final C–C bonded covalent species. Since these intermediates have significant impact on the structure of the final C–C bonded products, numerous studies were conducted with the organometallic species to explore mechanistic details of the on-surface Ullmann reaction [8]. Commonly, on copper and silver surfaces, room-temperature stable organometallic polymeric chains with C–Cu(Ag)–C bonds are formed before their transformation into the final covalent nanostructures by further annealing treatment. The rare observations of organometallic species with C–Au–C bonds is due to their intrinsically lower stability or shorter lifetime compared to those of C–Cu(Ag) bonds [34]. The first detailed study of an organometallic nanostructure in the context of the Ullmann reaction was reported by Wang et al., who identified the copper atoms in organometallic polymeric chains via scanning tunneling spectroscopy (STS) in the Ullmann reaction of dibromo-*para*-terphenyl (DBPTP) on Cu(111) [13].

2.3 *Stabilization of Organometallic Chains*

In contrast to 2D organometallic networks, which are stabilized in two dimensions by organometallic bonds, the 1D organometallic chains are stabilized by

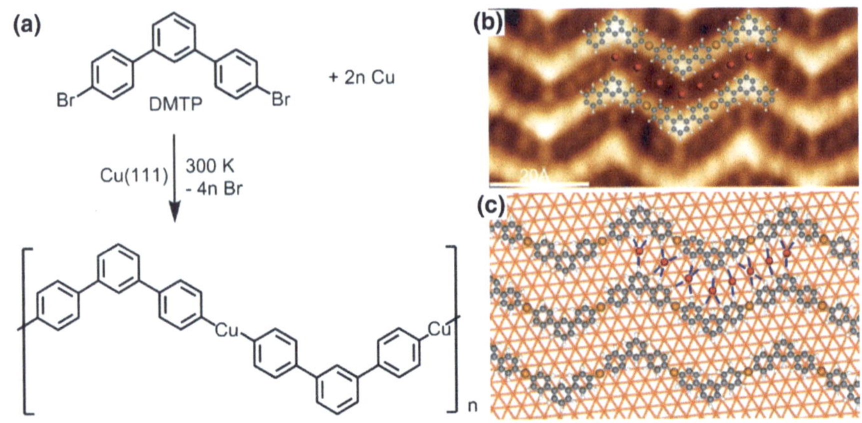

Fig. 1 **a** Reaction scheme for the formation of zigzag shaped organometallic chains from 4,4″-dibromo-*meta*-terphenyl (DMTP). **b** Molecularly resolved STM image of a small region of the one-dimensional (1D) zigzag organometallic chain island. The chemisorbed Br adatoms between the chains are resolved and labeled with red spheres. **c** Adsorption model for the structure in panel (**b**). Grey spheres represent carbon atoms; white, hydrogen; brown, copper; red, bromine. Adapted with permission from Ref. [35], © (2014) American Chemical Society

organometallic bonds along the chain and by weak intermolecular interactions in the direction perpendicular to the chains [11, 13, 34–36]. Figure 1a shows the formation of a zigzag shaped organometallic chain from 4,4″-dibromo-*meta*-terphenyl (DMTP). The combination of two different types of interactions between the precursor monomers along two different directions leads to their aggregation into ordered domains as shown by the STM image in Fig. 1b. On the basis of the apparent heights of the species and the lateral distances, the type of the bond such as C–C and C–Ag–C bonds or other, weaker interactions can be determined. As a complementary approach, low energy electron diffraction (LEED) is typically employed to confirm the periodicity of the formed ordered organometallic chain islands [11, 37]. In addition, laterally integrating surface-analytical techniques such as X-ray photoelectron spectroscopy (XPS) give valuable insight into the chemical composition of the adsorbates.

In most studies of the formation of 1D organometallic chains, the Br adatoms were typically identified by indirect evidence, because an effective local, chemically sensitive probe is generally not available. Although various intermediates of the surface Ullmann reaction have been successfully studied using vibrational spectroscopy via inelastic tunneling spectroscopy, [38] this approach is normally too time-consuming for general surface analysis. As a more practical alternative, XPS has been used especially to study the chemical state of the Br atoms and to monitor the dissociation of the C–Br bond as the initial step of the surface Ullmann reaction [11]. The XPS chemical shifts suggest that the Br atoms released from the precursor molecules bind directly to the underlying metal surface. In combination with the fact that the surface metal adatoms were typically incorporated into the organometallic chains forming carbon-metal bonds, the excess protrusions with different apparent heights were thus reasonably attributed to chemisorbed Br adatoms. These Br adatoms tend to adsorb between the organometallic chains, where they are stabilized by Br···H bonds formed with the peripheral hydrogen atoms of the monomers as illustrated in Fig. 1c [35]. The spot-like protrusions assigned to chemisorbed Br atoms between the chains can be seen in Fig. 1b. Similar results have been reported in other related studies [11, 13]. (Contrasting observations were reported for organometallic chains formed from 1,4-dibromobenzene (DBB) by Giovannantonio et al., who found that some of the Br atoms could sit on top of the bridging copper atoms in the organometallic chains [11]. The reason for this deviating behavior most likely lies in the limited space available between the phenylene units of the DBB derived chain, as compared to the terphenylene units in the DBPTP derived chains (see above). The excess Br atoms may then adsorb on the bridging copper atoms in the organometallic chains rather than on the surface lattice copper atoms.) According to the derived adsorption model, the distances between Br and the periphery hydrogen atoms fit well to the range of typical Br···H bond lengths [39]. The formation of Br···H bonds has also proved to exert considerable influence on the final adsorption configuration of the organometallic chains. For instance, 9,10-dibromoanthracene monomers form, after dissociation of the C–Br bonds, organometallic chains on Ag(111) consisting of non-coplanar anthryl biradical moieties. Parallel arrangement is not possible because of the alternative formation of Br···H bonds at the two sides of the chain [40]. In other words, the adsorbate structure represents the energetically most favorable configuration for the co-adsorption of organometallic chains and Br adatoms. If the precursor contains a C–I reactive group instead of C–Br, the resulting domains of organometallic chains are expected to be different, because iodine atoms are less electronegative than Br atoms and therefore have weaker interactions (I···H hydrogen bonds) with the organometallic chains. This was demonstrated by the different unit cells of the domains of organometallic chains formed by 1,4-dibromobenzene and 1,4-iodobenzene on Cu(110) [11].

2.4 Dynamics of the Formation of Organometallic Chains

The growth dynamics of organometallic chains is influenced by their stabilization in 2D islands. These islands grow by attachment of individual dehalogenated monomers and metal adatoms or complete oligomeric organometallic chains that have preformed on a terrace. By reversal of these processes, the chain islands can also shrink. This has been observed for DMTP on Cu(111) at 300 K [35]. As shown in Fig. 2a, the organometallic chain islands initially loses a whole chain and then regains a whole chain over a period of several minutes. The large aspect ratio of the islands indicates that the attachment of additional chains is slower than the chain growth process. An alternative explanation is based on the fact that the organometallic bond is stronger than the Br···H hydrogen bonds. Therefore, the detachment of *meta*-terphenyl (MTP) units by C–Cu–C bond scission is more difficult than the detachment of chains by dissociation of Br···H bonds. This also finally leads to the formation of island that are elongated in the direction of the organometallic chains.

Another important aspect for the formation of islands of organometallic chains is the source of metal adatoms. It has been proposed that the metal adatoms incorporated into the organometallic chains originate from the 2D gas of metal atoms present on a metal surface [41, 42]. However, if this source is not sufficient, metal atoms can also be removed directly from the lattice. For example, deposition of DMTP on Cu(111) at 300 K leads to the formation of vacancy islands of monoatomic depth, especially if the DMTP coverage is in the high submonolayer range (Fig. 2b). The vacancy islands do not occur if the substrate temperature is higher (440 K) during the deposition of DMTP (Fig. 2c). This strongly implies that the concentration of copper adatoms on Cu(111) at 300 K is not sufficient (or the

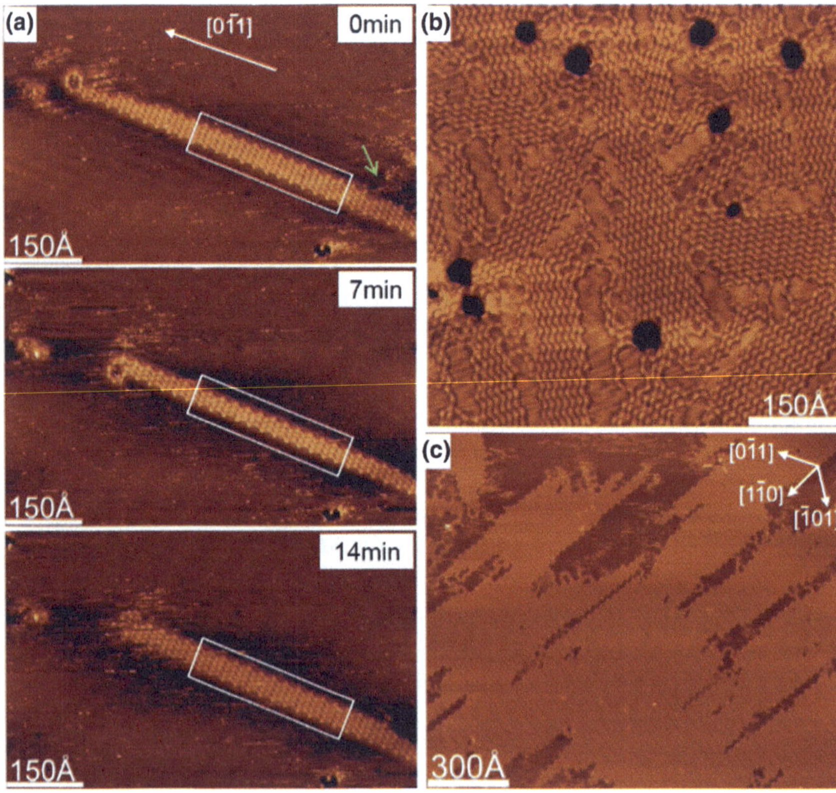

Fig. 2 **a** Sequence of STM images of the same region of the sample prepared by deposition of 0.2 ML DMTP onto Cu(111) at 300 K, taken with an interval of 7 min. **b**, **c** Overview STM images obtained after deposition of 0.77 ML DMTP onto Cu(111) held at 300 K and 440 K, respectively. Adapted with permission from Ref. [35], © (2014) American Chemical Society

process of their formation is too slow) to compensate the consumption during the formation of organometallic chains. Therefore, the extraction of copper atom from the terraces comes into play as an additional supply. These observations indicate that the concentration of surface adatoms can be one of the rate-determining factors for the formation of longer organometallic chains, as is exemplified by the comparison between Fig. 2c (440 K) and Fig. 2b (300 K).

2.5 Formation of Covalent Polymers

The covalently bonded polymers form in two ways: either by post-annealing of the preformed organometallic polymeric chains or by the deposition of precursor monomers onto a surface kept at a sufficiently high temperatures to induce direct C–C coupling [7]. The major differences between these two procedures concern the dynamics of assembly during the polymerization via C–C coupling. In the post-annealing procedure, the monomers have already arranged into compact ordered islands of organometallic chains, in which the repeat units are in the right configuration for C–C coupling. However, during this process, large domains of organometallic chains typically split into smaller domains consisting of covalent polymer chains [12], which are again stabilized by the hydrogen bonds formed with chemisorbed halogen adatoms. One reason for the reduced long-range order is that the elimination of metal atoms reduces the lattice constant in the chain direction, which makes the whole island shrink. This often leads to cracks in the islands, which cannot heal because of the large sliding diffusion barrier of the islands. The low mobility also prevents the formation of larger domains by encounter and C–C coupling between smaller domains of polymer chains. In contrast, the procedure of deposition onto the "hot" surface leads to direct reaction of the monomers and formation of polymer chains, which subsequently aggregate to form islands stabilized by hydrogen bonds. This approach typically results in the formation of longer polymer chains compared to those obtained by the post-annealing procedure. The main reason for better performance of the "hot deposition" approach is that the increased diffusion rate leads to a higher probability for the attachment of single monomer units to a preformed chain, compared to the post-annealing approach. The enclosure of monomers in the organometallic chain islands prevents the monomers from diffusing freely on the surface.

The lengths of the formed polymers depends on the rates of deposition, diffusion and reaction of the monomers during C–C coupling. In the case of high deposition rates, the high stationary monomer coverage makes the initial formation of dimers most likely, while longer chains form subsequently by the coupling of dimers and other short oligomeric chains. However, for low deposition rates, dimer formation has a lower probability and adsorbing precursor monomers are more likely to couple with a preformed oligomeric chain. This favors the further growth of preformed long polymer chains rather than the formation of a new short chain. These considerations are also supported by the fact that longer chains are formed by "hot" deposition compared to post-annealing, because the post-annealing procedure is in most aspects equivalent to "hot" deposition of the precursor monomer with infinitely high deposition rate.

High diffusion rates of monomers or preformed oligomeric chains leads to a high probability for their encounters, resulting in the formation of longer chains. The diffusion rates are not only influenced by temperature, but also by the choice of substrate. Experimental and theoretical studies indicate that the diffusion rates of aromatic molecules on coinage metal substrates decrease in the order Au(111) > Ag (111) > Cu(111) [26, 43]. As can be expected by the above considerations, the lengths of the covalent chains formed from the same precursor on these three surfaces follow the same tendency. Figure 3b–d show STM images taken after deposition of DMTP onto Cu(111), Ag(111) and Au(111) at the indicated temperatures. As can be seen, the most probable lengths of the zigzag shaped oligophenylene chains formed on the three surfaces decrease in the order Au(111) > Ag (111) > Cu(111). Additional evidence for this reasoning is provided by the fact that longer poly-*para*-phenylene chains are formed on Au(111) [14] than on Cu(111) [13] from the DBPTP monomer. Besides, longer chains are also formed at higher

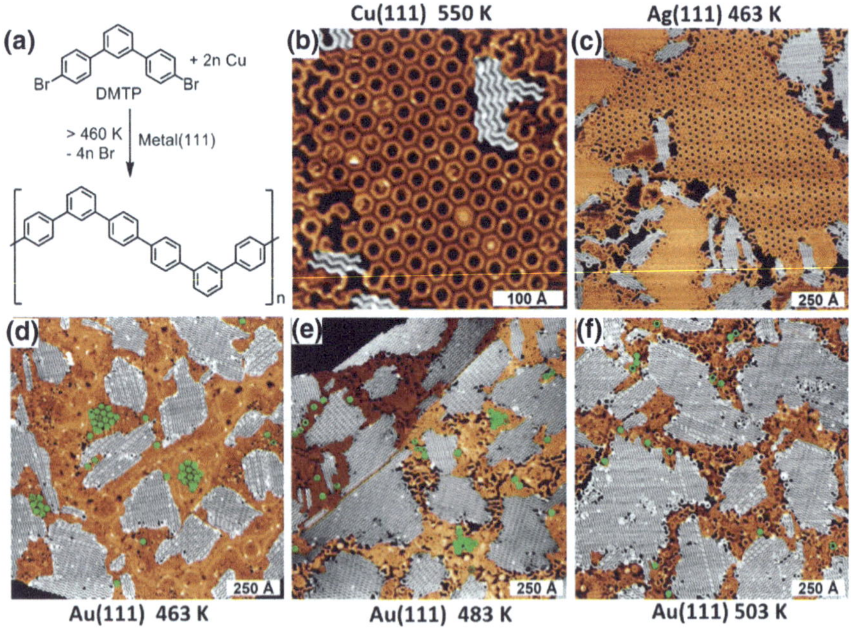

Fig. 3 **a** Reaction scheme for the formation of zigzag oligophenylene chain from DMTP on coinage metal (111) surfaces. **b–f** show the STM images taken after deposition of DMTP onto Cu (111), Ag(111), and Au(111) held at the indicated temperatures. The employed vapor deposition rates of DMTP for panel (**b**) and (**c–f**) are 50 monolayer per hour (ML/h) and 5 ML/h, respectively. Grey and green coloring is used to mark the zigzag polyphenylene chains and hyperbenzene molecules, respectively. Adapted with permission from Refs. [4, 6], © (2013) John Wiley & Sons

substrate temperatures during precursor deposition. As shown by Fig. 3d–f, deposition of DMTP onto Au(111) at increasing temperatures leads to the obvious enhancement of the most probable length of the zigzag polyphenylene chains.

By considering these aspects, diverse polymer chains were synthesized from different two-fold halogenated arenes. The topology of the polymer chain is determined by both the positions of the halogen atoms in the molecule and the shape of its carbon skeleton. For instance, linear oligophenylene chains can be synthesized from DBPTP [13], 1,4-dibromobenzene, [11, 12] or 1,4-diiodobenzene [10]. For these precursors, the C–X bonds are oriented along one of their two axes of symmetry. When this is not the case, e.g., for 1,3-diiodobenzene, zigzag shaped rather than straight oligophenylene chains are formed [10]. Similarly, using DMTP monomers with bent carbon skeleton also leads to the formation of zigzag shaped oligophenylene chains [4]. Furthermore, the properties of polymers can be modified by functionalization of the precursor monomers. Basagni et al. reported that the band structure of the parallelly aligned poly-*para*-phenylene chains formed from DBPTP can be tuned by doping with different numbers of N atoms in the precursor [44]. Another example is the formation of flexible molecular wires with high conductance from bis(5-bromo-2-thienyl)-benzobis(1,2,5-thiadiazole), in which alternative donor and acceptor units were introduced [45]. For comparison, the polyfluorene chain formed from dibromoterfluorene (DBTF) monomers has a conductance decreasing with its length [46]. It was concluded that the introduction of donor and acceptor units in the precursor monomer enhances the conductance of the polymer chains [45].

Graphene nanoribbons represent another interesting class of polymer chains. The synthesis of these ribbons is commonly achieved via Ullmann coupling of haloarenes with wide carbon skeletons followed by cyclo-dehydrogenative C–C coupling [16]. One exception without the cyclo-dehydrogenative C–C coupling is the one-step Ullmann coupling of 1,4,5,8-tetrabromonaphthalene (TBN) into the arm-chaired graphene nanoribbon with a width of five carbon atoms [34]. The design of different precursors leads to the formation of graphene nanoribbons with different widths and edge topologies [16]. Besides the arm-chaired edge [23], graphene nanoribbons with coved [47] or zigzag edges [48] have been reported, some of which are based on monomers with rather sophisticated design. The widths of graphene nanoribbons is usually determined by the size of precursor monomers. Up to now, arm-chaired graphene nanoribbons with widths of 3, [10, 13, 14] 5, [34, 49] 7, [23] 9, [50, 51] and 13 [52] carbon atoms were achieved. With the precise tailoring of the topologies and widths of graphene nanoribbons, their corresponding band structure (especially the band gap) can be controlled. Additional modification of the band structure of graphene nanoribbons can be achieved by doping with heteroatoms in the precursor monomers, e.g., boron [53, 54] and nitrogen [44, 55]. These examples illustrate the major benefits of the synthesis of polymers or graphene nanoribbons through on-surface Ullmann coupling: The highly controllable substituents and topologies pave the way to the synthesis of structurally well-defined polymers with desired functionalities for applications in molecular electronics and other fields of current technology.

3 Synthesis of Macrocycles via Surface Ullmann Reaction

3.1 Cyclisation via Ring-Closure on a Surface

In solution-based chemistry, one of the typical strategies for the synthesis of macrocycles or cyclic polymers is the ring-closure method, which involves the encounter and coupling of the reactive end groups of a polymer chain. Similar to the development of on-surface polymerization for the synthesis of diverse polymers, the application of the ring-closure method to reactions on solid surfaces would enable the on-surface synthesis of macrocycles with desired properties. A major benefit of the on-surface approach is apparent: Because of the solvent-free conditions, it can serve as an alternative method for the cyclisation of insoluble reactants, which cannot be used in solution [5]. The simplest strategy for macrocycle synthesis is the ring-closure of homodifunctional linear oligomers. Due to the uniform reaction product of surface Ullmann reaction, one would also expect that the C–X groups (X = halogen) are highly feasible for the ring-closure of linear oligomers on a surface. However, one crucial aspect of the on-surface case, compared to synthesis in solution, is the deposition of the reactant oligomer with high molecular weight onto solid surfaces in UHV. Physical vapor deposition through thermal sublimation is of limited use, because the large organic molecules usually decompose before they reach a practically usable vapor pressure. Even if more sophisticated techniques such as laser-induced desorption [56] or electrospray deposition [57] could help to overcome this limitation, the entanglement of the long oligomer chains and their arrangement in a conformation favorable for cyclization may not be possible because of energy barriers induced by the adsorption potential.

To solve this deposition-related problem, the hierarchical approach can be used: in the first step, small precursor monomers that can easily be deposited oligomerize into linear oligomer chains, which then undergo unimolecular ring-closure forming macrocycles in the second step. However, due to the 2D confinement of the reactants in the surface reaction, there are differences to ring-closure reactions in solution. Most importantly, the two terminal groups of the linear oligomer have more degrees of freedom in solution than in the adsorbed state on a surface. As a result, the diffusion behavior of the terminal groups, which need to meet for ring closure, is different. For instance, the cyclisation of the oligo-*para*-phenylene chain shown in Scheme 2a could be achieved in solution by bending of the C–C σ bonds towards the direction perpendicular to the plane of phenylene units. In other words, this ring-closure process is achieved by a conformational change of the chain. On a surface, however, the 2D confinement makes it much more difficult to reach the cyclization geometry, because the in-plane bending of the C–C bonds leads to much higher strain and steric hindrance due to the proximity of hydrogen atoms of the neighboring phenylene units (red oval in Scheme 2a). Only if the chain is long enough and therefore the bending angle per phenylene unit is small, it may be possible for the oligo-*para*-phenylene to form rings on the surface due to the reduced steric repulsion between the involved hydrogen atoms. We will see later

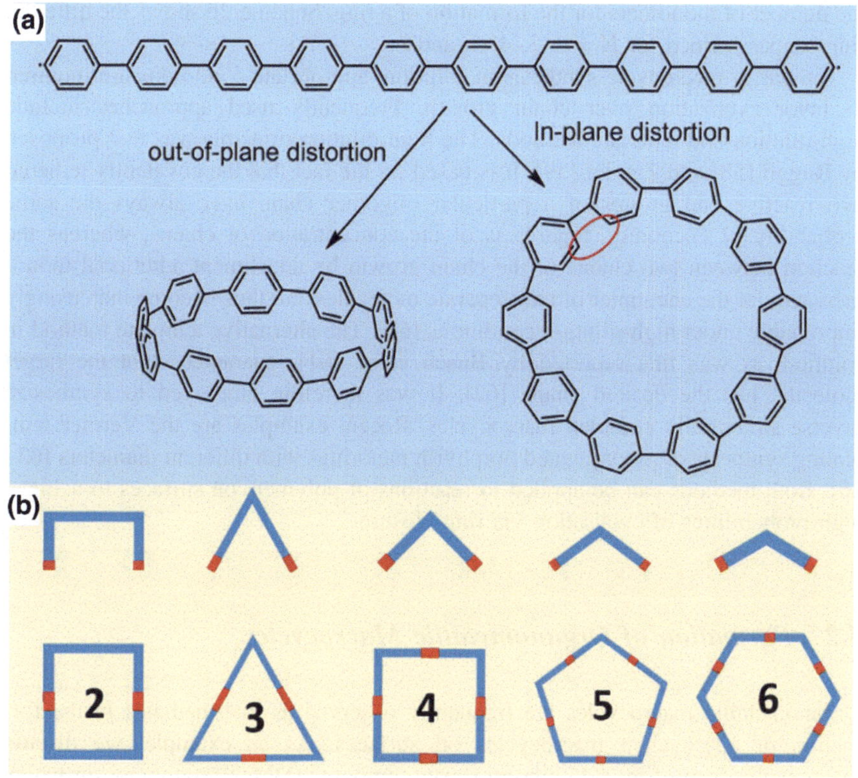

Scheme 2 **a** Reaction scheme for the ring-closure of an oligo-*para*-phenylene chain with 10 phenylene units. The out-of-plane and in-plane cyclization leads to the formation of macrocycles with different conformations, with much higher ring strain for the in-plane conformation. **b** The design of building blocks consisting of C–X bonds with different angles for the synthesis of macrocycles with different shapes

that low-barrier *conformational* changes of a chain in solution may not be possible in the 2D confinement on a surface, because they would require the flipping and thus the partial desorption of the chain at high energy costs. In such cases, a lower-energy pathway may be available that involves the reorganization of chemical bonds, i.e., a *constitutional* modification of the chain. In other words, the diffusion and successful approach of the two terminal groups of an oligomer on a surface could rely more on an isomerization of the chain rather than its conformational change. Ring-closure on a surface would then require an isomerization of an open oligomer chain into a transient isomer that resembles the final macrocycle. These considerations also provide directions regarding the design of the corresponding precursor monomers. In order to allow for the isomerization of the initially formed oligomer chain into the shape resembling a strain-free ring, the monomers should contain two C–X bonds with an angle of $180 - \frac{360}{N}$, where N is the number of monomers for the formation of a ring. Scheme 2b shows the different ring shapes formed for N = 2, 3, 4, 5, and 6.

Similar to macrocycle synthesis in solution, appropriate conditions are required to favor cyclization over chain growth. Frequently used approaches include high-dilution and template methods. The high-dilution principle was first proposed by Ruggli [58] and Ziegler [59]. It is based on the fact that the covalently tethered two reactive end groups of a particular oligomer chain have always the same probability of encounter, regardless of the concentration of chains, whereas the reaction between two chains or the chain growth by attachment additional monomers require the encounter of two separate molecules and thus become increasingly improbable under high-dilution conditions [60]. The alternative template method in solution, as was first reported by Busch et al. [61], guarantees that the target molecule has the desired shape [62]. It was therefore employed to synthesize diverse structurally complex macrocycles. Recent examples are the Vernier templating synthesis of π-conjugated porphyrin nanorings with different diameters [63–66]. Both methods can be applied to reactions of polymers on surfaces to achieve high probabilities of cyclisation via ring-closure.

3.2 Formation of Organometallic Macrocycles

Organometallic macrocycles are frequently observed as intermediates in the formation of all-covalent macrocycles on surfaces. As an example, we discuss macrocycles formed by 4,4″-dibromo-*meta*-terphenyl (DMTP) monomer, in which the two C–Br bonds enclose an angle of 120°. DMTP fulfils the conditions for the formation of non-linear open chains, which could undergo ring-closure into macrocycles. The smallest ideal (i.e., non-strained) ring formed from DMTP molecules is a hexagon, according to Scheme 2b. On the stage of the organometallic phase, the major product after deposition (with a deposition rate $f = 5$ ML/h) of DMTP onto metal surfaces, e.g., Cu(111), consist of organometallic zigzag chains, as shown by the large-scale STM image in Fig. 4a. Macrocycles are only formed as minority species and include the hexagonal $(MTP\text{-}Ag)_6$ as well as the larger rings $(MTP\text{-}Ag)_{10}$, $(MTP\text{-}Ag)_{14}$, $(MTP\text{-}Ag)_{16}$, and $(MTP\text{-}Ag)_{18}$ [35]. The predominant formation of chains is not surprising, because the islands of zigzag organometallic chains are more compact than those of organometallic macrocycles with cavities and thus allow for a higher local coverage. The more compact packing of the chain islands also leads to stronger intermolecular interactions per unit surface area and thus makes the chain islands thermodynamically more stable. When the chains are the thermodynamically favored product, kinetic reaction control is necessary to favor the formation of macrocycles. Alternatively, the thermodynamic stability of the macrocycles (or of the transition states leading to these macrocycles) can be increased by suitable templates.

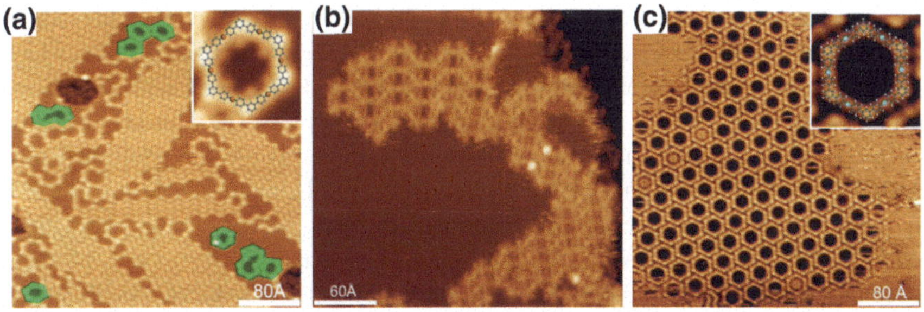

Fig. 4 STM images of organometallic species obtained by deposition of submonolayers of DMTP onto Cu(111), Cu(110) and Ag(111) surfaces held at 300, 300, and 443 K, respectively. The deposition rates *f* of DMTP are 50 ML/h for the panels (**a**) and (**b**) and 0.5 ML/h for panel (**c**). The insets in panel (**a**) and (**c**) show the magnified view of the $(MTP\text{-}Cu)_6$ and $(MTP\text{-}Ag)_6$ organometallic macrocycles. The organometallic macrocycles in panel (**a**) are green colored. Black spheres in the insets represent carbon atoms; white, hydrogen; orange, copper, cyan, silver. Adapted with permission from Refs. [6, 77], © (2016) John Wiley & Sons

As mentioned above, the template and high-dilution methods are established for macrocycle synthesis in the homogeneous liquid phase. The template method typically involves template molecules with the "negative" shape of the desired rings [67]. In addition, reversible interactions are employed between the formed macrocycle and the template monomer, because it is necessary to finally remove the template. The template method is also applicable in 2D confinement, as has been shown for numerous examples in on-surface supramolecular chemistry [68]. Assembly of small tecton molecules into large supramolecular structures has been achieved through weak interactions such as π-π stacking [69, 70] as well as hydrogen [71, 72] and coordination bonds [73–75]. One example for the application of molecular templates for steering the on-surface Ullmann reaction was reported by Lin et al., who employed the 5,15-bis(4′-bromophenyl)-10,20-bis(4′-pyridyl) porphyrin as precursor in a surface Ullmann reaction on Au(111) [76]. The co-deposition of this precursor with copper leads to the formation of covalently bonded porphyrin dimer or trimers rather than porphyrin chains with lengths obeying a Γ type distribution. Therefore, the template effects of coordination chains with pyridyl–Cu–pyridyl bonds on the surface Ullmann reaction are revealed unambiguously.

Beside the possible usage of molecular templates, the substrate itself can serve as a template in on-surface synthesis and enhance the probability of ring-closure. This has been demonstrated by the higher yield of cyclic species observed on Cu(111) compared to Cu(110) after deposition of DMTP at 300 K, as shown in Fig. 4a and b (macrocycles marked in green color). The template effect posed by the substrate stems mainly from the symmetry matching between the adsorbates and the surface lattice. Therefore, the $(MTP\text{-}Ag)_6$ macrocycle with six-fold symmetry has a higher probability of formation on the six-fold symmetric Cu(111) surface than on two-fold symmetric Cu(110) surface [77]. In fact, the synthesis of organometallic macrocycles on the Cu(110) surface from DMTP could only be achieved by introduction of an additional template, namely, the more corrugated "piano keyboard" pattern on the partially oxidized Cu(110)–(2 × 1)O surface with alternating stripes of bare Cu surface and bundled oxygen rows [37]. The Ullmann reaction proceeds predominantly on the stripes of bare Cu(110), while the adjacent Cu–O chain bundles impose strong lateral confinement. The Ullmann reaction of DMTP on the Cu stripes is influenced both by steric effects and attractive interactions between the molecular backbones and the Cu–O chains. The extent of these influences increases when the width of the bare Cu stripes is reduced. As illustrated in Fig. 5a–e, organometallic macrocycles are the major species on Cu stripes narrower than 5.6 nm. Conversely, organometallic chains with uniform length are formed when the width of the Cu stripes exceeds 5.6 nm (Fig. 5f, g). It is proposed that the formation of macrocycles on the narrow Cu stripes is due to the alignment of the straight parts of DMTP along the orientation of Cu–O chain, which is mainly achieved by the formation of hydrogen bonds between the oxygen atoms in the Cu–O chains and the peripheral hydrogen atoms of the MTP units [37].

The high-dilution principle is another important method to increase the yield of organometallic macrocycles in the surface Ullmann reaction. For reactions on surfaces, the high-dilution condition is equivalent to a low stationary coverage of the precursor monomers. This can be achieved by low deposition rates f for the

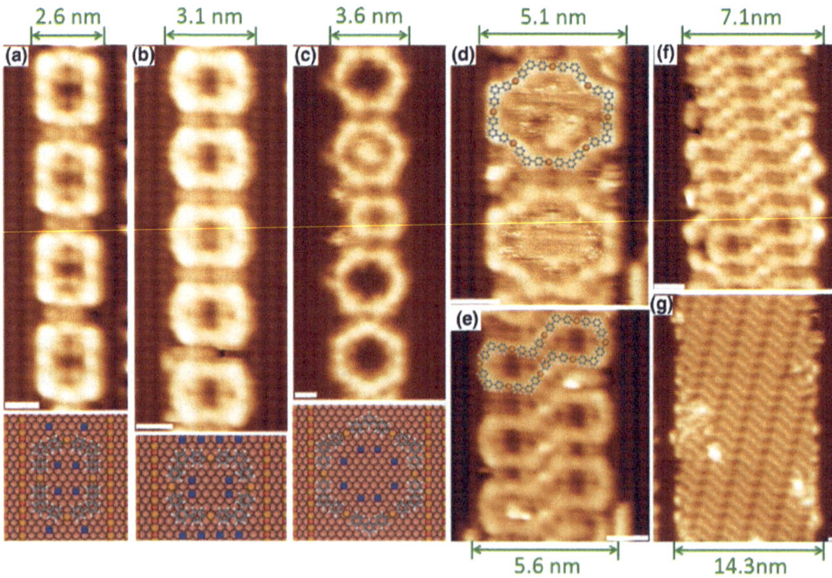

Fig. 5 High-resolution STM images of organometallic species, including **a–e** macrocycles and **f**, **g** oligomeric chains, formed by deposition of submonolayers of DMTP onto Cu(110)–(2 × 1)O held at 383 K. The corresponding adsorption models are shown underneath the panels (**a–c**). The widths of the bare Cu stripes in the STM images are indicated by the green numbers. Black spheres represent carbon atoms; white, hydrogen; blue, bromine; red, oxygen; orange, copper. Adapted with permission from Ref. [37], © (2016) American Chemical Society

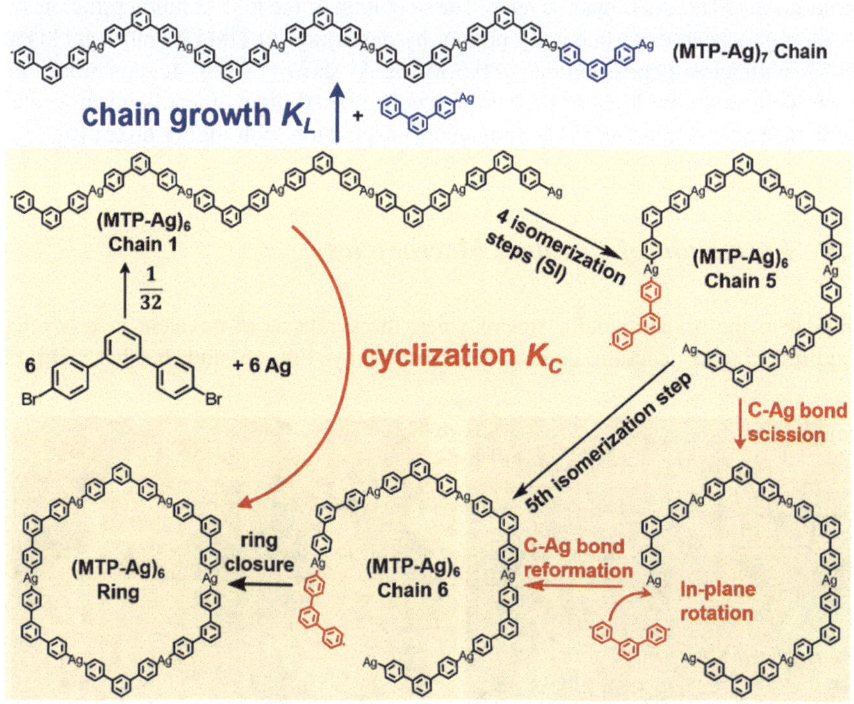

Scheme 3 Chain growth (light blue shaded) and chain cyclisation (light brown shaded) of the all-*trans* configured organometallic $(MTP\text{-}Ag)_6$ Chain 1. The isomerization of Chain 1 into Chain 6 is achieved by three steps: detachment, in-plane rotation, re-attachment of MTP units. K_C and K_L denote the rate constants for the cyclization and chain growth processes, respectively

precursor. As exemplified by the formation of $(MTP\text{-}Ag)_6$ organometallic macrocycles from DMTP on Ag(111) under high-dilution condition, the mechanism can be described as shown in Scheme 3. The high-dilution condition ensures that sufficient time is available for the isomerization of the preformed $(MTP\text{-}Ag)_6$ open chain into the transient intermediate state resembling a ring, which then undergoes the final ring-closure process into a closed macrocycle. Due to the reversibility of C–Cu organometallic bonds at the reaction temperature, the isomerization of organometallic chains proceeds mainly by the C–Cu bond scission and reformation. As illustrated in Scheme 3, the detached chain segment can undergo in-plane rotation and subsequently re-attach to the chain again, resulting in its isomerization. (Note that the isomerization is necessary because of the 2D confinement, where in 3D a conformational change would lead to the same result.) Additionally, once the $(MTP\text{-}Ag)_6$ macrocycle has formed, its decomposition into two short chains is energetically more unfavorable than that of an open chain. This is because the splitting of a ring into two moieties requires the dissociation of at least two C–Ag bonds, compared to the scission of only one C–Ag bond dissociation necessary for the detachment of a chain segment. This selection process eventually favors the build-up of $(MTP\text{-}Ag)_6$ macrocycles. The operation of the high-dilution principle on surfaces has been experimentally proven by deposition of DMTP onto Ag(111) at 443 K with a low deposition rate of 0.5 ML/h. As shown by Fig. 4c, this procedure gives a considerably higher yield of the six-membered organometallic macrocycles on both Ag(111) and Cu(111), compared to deposition with higher fluxes [6].

3.3 *Formation of Covalent Macrocycles*

Similar to the organometallic macrocycles, the synthesis of covalent macrocycles requires kinetic reaction control. As shown by Fig. 6a and b, the yield of

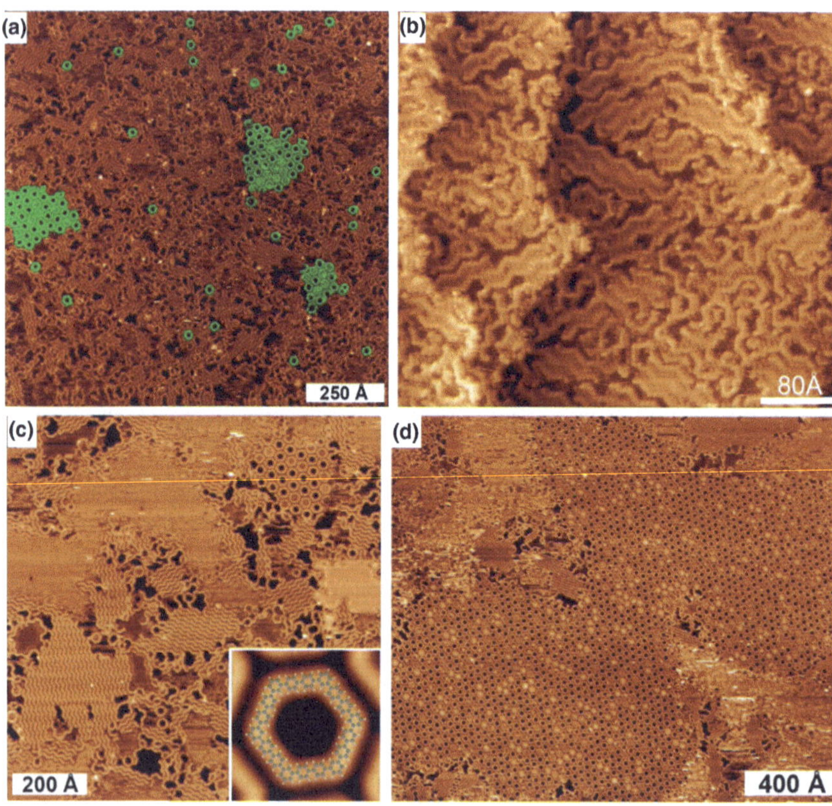

Fig. 6 **a–c** Overview STM images taken after deposition of DMTP onto Cu(111) at 300 K, onto Cu(110) at 383 K, and onto Ag(111) at 353 K, followed by annealing to 550, 458, and 463 K, respectively. **d** Large-scale STM image taken after deposition of DMTP onto Ag(111) held at 463 K with an extremely low deposition rate 0.05 ML/h. The hyperbenzene macrocycles in panel (**a**) are marked by green color. The molecular model of hyperbenzene is shown in inset of panel (**c**). Grey spheres represent carbon atoms; white, hydrogen. Adapted with permission from Refs. [6, 77], © (2016) John Wiley & Sons

hyperbenzene obtained by direct C–C coupling is higher on Cu(111) than on Cu (110) [4, 77]. In line with the discussion above, this is attributed to the template effect of the six-fold symmetric Cu(111) surface, which favors the formation of the hexagonal hyperbenzene [4], while the two-fold symmetric Cu(110) surface favors the formation of polyphenylene zigzag chains [77]. The operation of this template effect is based on the interaction between the Cu(111) surface and the organic adsorbates. Since this interaction consists mainly of physical adsorption, it is not strong enough alone to guarantee a high probability of the formation of hyperbenzene. An additional increase of the yield of hyperbenzene can be obtained by application of the high-dilution principle. This is achieved by deposition of DMTP with an extremely low deposition rate *f* of 0.05 ML/h on a Ag(111) surface held at 463 K. By this procedure, the two requirements for the pseudo-high dilution are satisfied: First, the consumption of DMTP monomers is faster than their addition. This is proven by the absence of intact molecules on the surface. Second, the final cyclic product, hyperbenzene, is stable under the applied reaction conditions, because no side reactions by C–C or C–H bond activation are observed. The rigorous application of this pseudo-high dilution condition results in the formation of hyperbenzene with a high yield of 83%, as shown in Fig. 6d. This yield is much higher than that obtained by the conventional post-annealing procedure (Fig. 6c), in which a submonolayer reservoir of densely packed MTP units has been deposited before the C–C coupling is induced by the temperature increase. In the latter case, the high initial precursor concentration undermines the high-dilution condition by favoring chain growth processes with second-order kinetics.

Regarding the mechanism for the formation of hyperbenzene on Ag(111), it has been proposed that that cyclization happens on the stage of the organometallic phase, resulting in the formation of the organometallic $(MTP\text{-}Ag)_6$ macrocycle, which then undergoes demetalation around 440–450 K to form hyperbenzene $(MTP)_6$. Accordingly, deposition of DMTP onto Ag(111) held at 443 K leads to the coexistence of both organometallic $(MTP)_6Ag_x$ macrocycles and hyperbenzene, as is illustrated in Fig. 7a–d. The simultaneous appearance of macrocycles with six, two, one and zero Ag atoms strongly indicates that hyperbenzene forms from the $(MTP\text{-}Ag)_6$ macrocycle through stepwise elimination of Ag atoms without destruction of the ring shape as shown Fig. 7e. Therefore, the yield of hyperbenzene should be proportional to the yield of $(MTP\text{-}Ag)_6$ macrocycles, which can be derived according to Scheme 3: Six DMTP monomers initially react with six Ag atoms into the six-membered $(MTP\text{-}Ag)_6$ organometallic chain by forming five C–Ag–C bridges. This chain then undergoes cyclization into the $(MTP\text{-}Ag)_6$ macrocycle or further grow into a seven-membered $(MTP\text{-}Ag)_7$ chain by attachment of another monomer unit. The competition between cyclization and chain growth determines the yield ratio of $(MTP\text{-}Ag)_6$ macrocycles and $(MTP\text{-}Ag)_7$ chains. Based on the conventional Jacobson-Stockmayer (J-S) model [78, 79], this competition depends on the precursor concentration (in line with the high-dilution principle). Thus, the yield of $(MTP\text{-}Ag)_6$ macrocycle can be derived as a function of the precursor deposition rate (*f*) as follows:

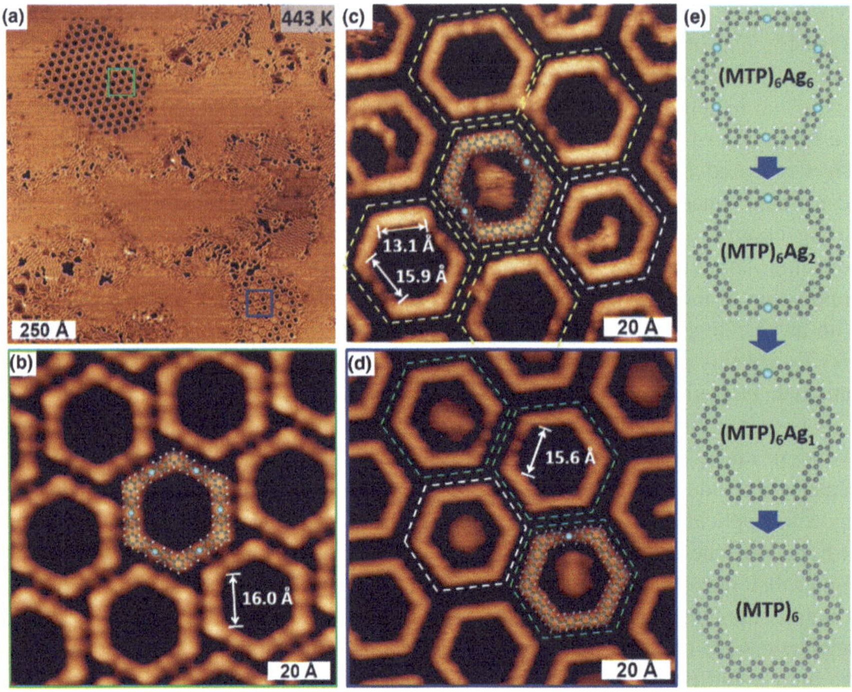

Fig. 7 **a** STM image taken after deposition ($f = 0.5$ ML/h) of 0.5 ML DMTP onto Ag(111) held at 443 K. **b**, **d** Zoom-in of the green and blue framed region in panel (**a**). **c** Section of partially metalated hyperbenzene molecules observed on the sample in panel (**a**). **e** Scheme for the evolution of cyclic $(MTP\text{-}Ag)_6$ to hyperbenzene $(MTP)_6$. Grey spheres represent carbon atoms; white, hydrogen; cyan, silver. Adapted with permission from Ref. [6]

$$\%\ (\mathrm{MTP{-}Ag})_6\ \mathrm{cycle} = \frac{1}{1 + \frac{K_L}{6K_C} f\tau} \times 100 \tag{1}$$

In Eq. 1, τ denotes the reaction time for the C–Ag–C bond formation, while K_C and K_L are the rate constants for cyclisation and growth of the six-membered $(MTP\text{-}Ag)_6$ chains, respectively. According to Eq. 1, the yield of hyperbenzene depends reciprocally on the deposition rate (flux) f. However, this is contrasted by experimental findings for DMTP on Ag(111), which show that the yield of hyperbenzene decreases approximately exponentially with increasing deposition rate f. This is visualized by the red curve in Fig. 8f, which is based on data derived from the STM images in Fig. 8a–e. This curve is inconsistent with the reciprocal type curve (blue, Fig. 8f) according to Eq. 1, especially at high f. The fact that at high fluxes the yield is higher than expected from the J-S model relates to the reversibility of the C–Ag bond formation under the reaction conditions: Organometallic chains $(MTP\text{-}Ag)_n$ that have already exceeded the length for cyclization (i.e., $n > 6$) can dissociate into smaller segments with $n \leq 6$ and

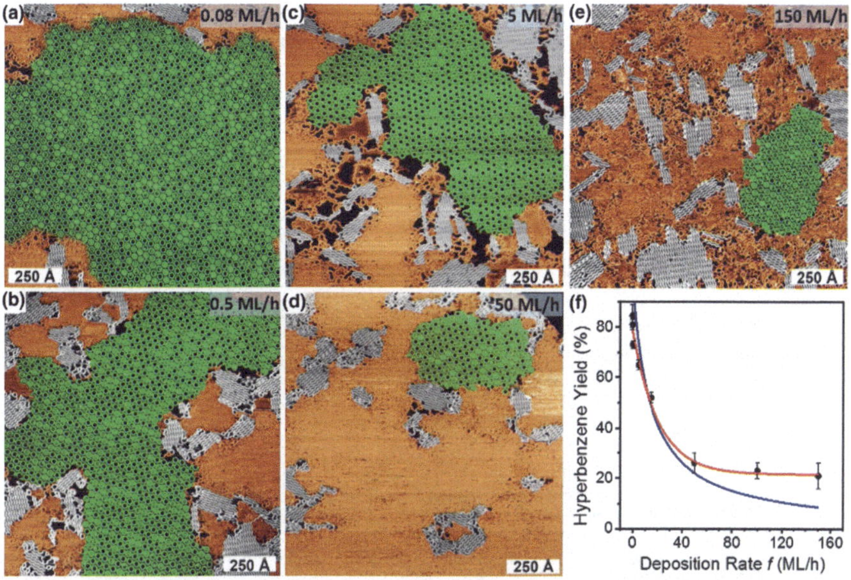

Fig. 8 Overview STM images taken after deposition of 0.5 ML DMTP onto Ag(111) held at 463 K with different deposition rates: **a** 0.08 ML/h, **b** 0.5 ML/h, **c** 5 ML/h, **d** 50 ML/h, and **e** 150 ML/h. Tunneling parameters: **a** $U = 1.2$ V, $I = 0.13$ nA; **b** $U = 1.3$ V, $I = 0.21$ nA; **c** $U = 1.6$ V, $I = 0.12$ nA; **d** $U = 1.5$ V, $I = 0.16$ nA; **e** $U = 1.6$ V, I = 0.13 nA. The islands of hyperbenzene and zigzag oligophenylene chains are indicated by green and grey shading, respectively. **f** Hyperbenzene yield versus DMTP deposition rate *f*. See the text for the meaning of the blue and red curve. Adapted with permission from Ref. [6]

therefore still contribute to the formation of organometallic $(MTP\text{-}Ag)_6$ macrocycles. This is not possible for a chain in which the segments are irreversibly linked by covalent bonds, as is assumed in the J-S model.

In the following, we discuss an alternative mechanism, by which hyperbenzene is formed through cyclization of covalently bonded oligophenylene chains. Cyclization of such a chain is only possible if it has the all-*cis* geometry of the $(MTP)_6$ Chain 6 in Scheme 4. To reach this geometry, an all-*trans* chain (Chain 1 in Scheme 4) needs to undergo several flipping events, which implies transient desorption of parts of the chain. The flipping is necessary because the irreversibility of the C–C bond formation under the reaction conditions makes it impossible to isomerize the covalently linked chains via the detachment and re-attachment of chain segments (as was shown in Scheme 3 for the organometallic chain). Some of the flipping events necessarily involve rather large segments of the chain. For instance, the isomerization from Chains 4 to 5 requires the partial desorption of a quaterphenyl or quinquephenyl unit. The energy barriers for such processes can be estimated from molecular desorption energies.

It has been reported that the desorption temperature for *para*-quaterphenyl on Au (111) is around 550 K [80]. An even higher temperature is to be expected for

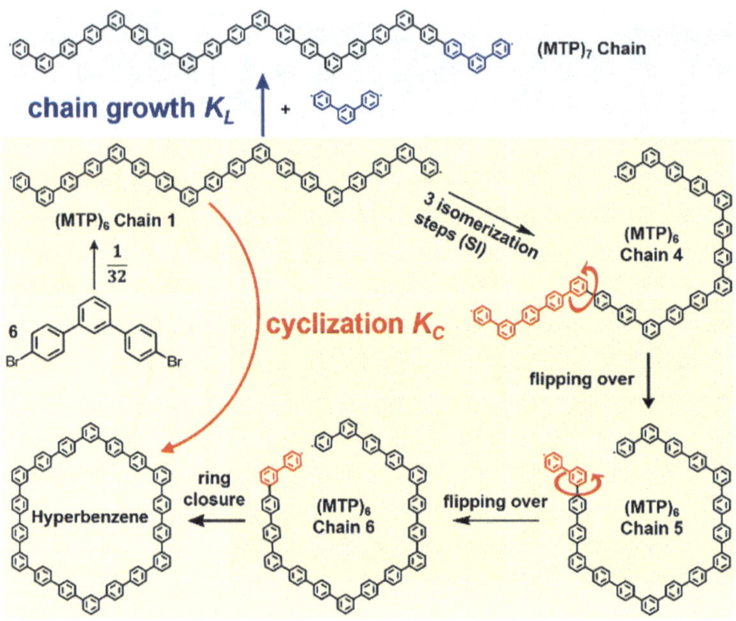

Scheme 4 Chain growth (light blue shaded) and chain cyclization (light brown shaded) of the all-*trans* configured $(MTP)_6$ oligophenylene Chain 1 into an $(MTP)_7$ oligophenylene chain or hyperbenzene. The isomerization process of $(MTP)_6$ chains is necessary for cyclization and is achieved by the flipping of the chain segments (red colored). K_C and K_L denote the rate constants for cyclization and chain growth processes, respectively

desorption from Ag(111) due to the typically larger adsorption energy of phenyl units on Ag(111) than on Au(111) [39, 41]. Therefore, it is unlikely that the transition from Chains 4 to 5 on Ag(111) has a substantial rate at temperatures below 550 K. This means that, except for the $(MTP)_6$ Chains 5 and 6 in Scheme 4, other types of $(MTP)_6$ chains are completely blocked from cyclization under the employed reaction temperatures (483, 503, and 523 K) due to the high barrier for the necessary flipping of chain segments. However, the probability for the direct formation of the $(MTP)_6$ Chains 5 and 6 (Scheme 4) by C–C coupling of six MTP units is 3/32. This is much lower than the experimentally achieved yield of hyperbenzene, which indicates the existence of an additional ring-closure pathway with a higher probability. Thus, these considerations provide additional evidence for a cyclization on the stage of organometallic phase [6].

In view of this, the high-yield formation of hyperbenzene on metal surfaces relies strongly on the ring-closure process on the stage of the organometallic phase. Therefore, any conditions that reduce the lifetime of the organometallic species would undermine the probability for ring-closure, and hence the yield of $(MTP\text{-}Ag)_6$ and of the final hyperbenzene macrocycles. This assumption is evidenced by the decrease of the hyperbenzene yield with the increase of the

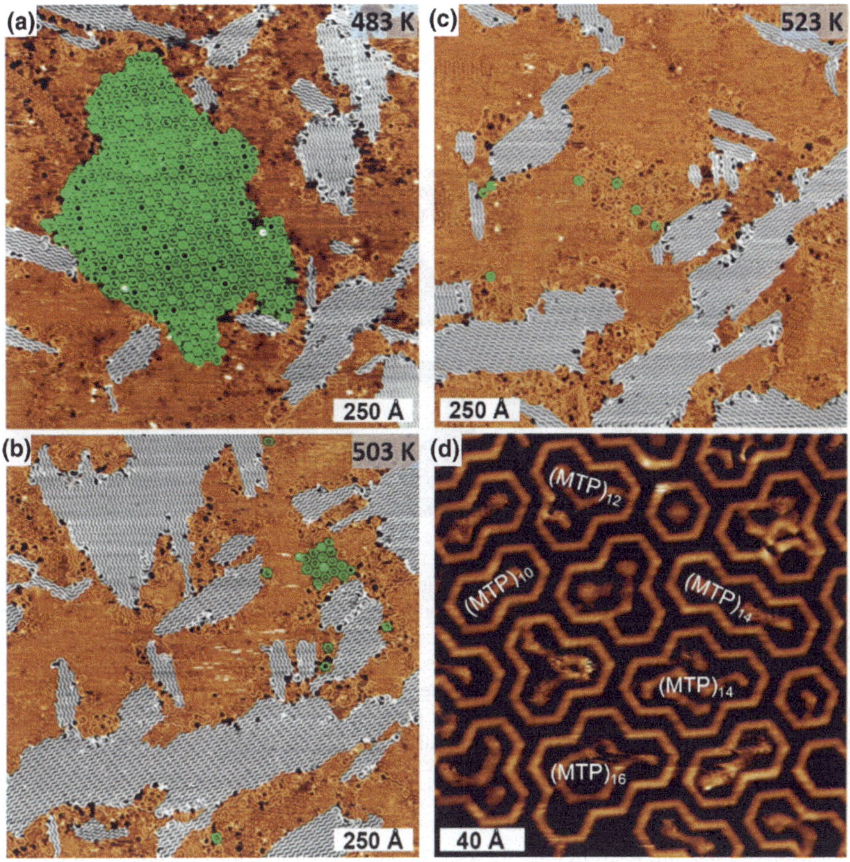

Fig. 9 STM images taken after deposition ($f = 0.5$ ML/h) of 0.5 ML DMTP onto Ag(111) held at **a** 483 K, **b** 503 K, and **c** 523 K. The islands of hyperbenzene and zigzag oligophenylene chains are marked by green and grey shading. **d** Magnified view of the center region in panel (**c**). Adapted with permission from Ref. [6]

temperatures of both Ag(111) and Au(111) surfaces, as shown by Figs. 3d–f and 9a–c, because higher substrate temperature results in reduced lifetimes of the organometallic species. Additional evidence is the considerably lower hyperbenzene yield achieved on Au(111) compared to that on Ag(111) under otherwise identical conditions. This can be explained by the fact that C–Au bonds are less stable and thus have intrinsically shorter lifetimes than C–Ag bonds. Another consequence caused by the reduced lifetime of the organometallic species is the formation of macrocycles that are larger than hyperbenzene, because the cyclization occurs more and more likely on the stage of the covalent species. Since this process is of low probability, it finally leads to a higher relative probability of the competing chain growth and therefore to the formation of larger macrocycles from these longer chains, even though the yields are limited (Fig. 9d).

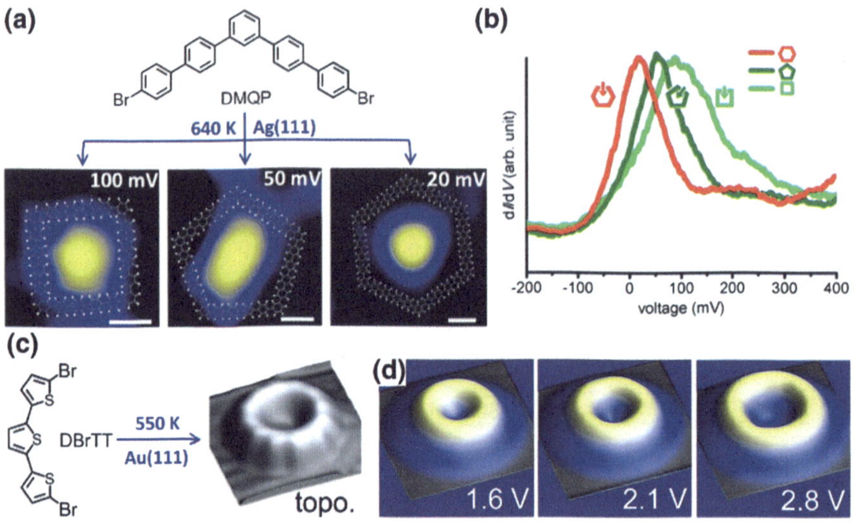

Fig. 10 **a** Experimental dI/dV maps at the indicated bias voltages showing the electron standing-wave pattern inside the central cavities of the square [20]-honeycombene (HC), pentagonal [25]-HC, and hexagonal [30]-HC macrocycles (from the left) formed from DMQP on Ag(111) at 640 K. The white scale bars correspond to 1 nm. **b** Experimental dI/dV spectra measured at the centers of the three macrocycles. **c** Formation of an oligothiophene macrocycle consisting of 12 thiophene rings from DBrTT, the topology of which is shown in the STM image. **d** Constant-height conductance maps acquired at the indicated voltages. Adapted with permission from Refs. [5, 15], © (2017) American Chemical Society, © (2013) American Physical Society

The synthesized macrocycles have various interesting properties. For examples, they function as single-molecule quantum corrals for the confinement of electrons, metal clusters or small organic molecules. Chen et al. reported the on-surface synthesis of conjugated [30]- and [18]-honeycombene (HC) macrocycles, as well as macrocycles with other geometries, which act as molecular quantum corrals resulting in the confinement of surface-state electrons inside their central cavity [5]. Figure 10a shows the electron standing-wave pattern in the center of macrocycles formed from four, five, and six 4,4''''-dibromo-*meta*-quinquephenyl (DMQP) monomers on Ag (111) at 640 K. The energy of the confined surface state increases with the shrinking size of the macrocycles (Fig. 10b). However, due to the relatively low potential barrier imposed by the organic macrocycles, no higher-energy states have been found inside the cavities. Other interesting properties of conjugated macrocycles were reported by Reecht et al. [15], who employed an cyclo-oligothiophene as electron resonators for whispering gallery modes. This conjugated macrocycle forms from four 5,5″-dibromo-2,2′:5′,2″-terthiophene (DBrTT) molecules and consists of 12 thiophene units as shown in Fig. 10c. Its energy-dependent conductance maps are compared to those of the related open-chain oligothiophenes with the same number of thiophene units. While the electronic states of the latter show the typical behavior of a particle in a one-dimensional box, the conductance maps of the macrocycles indicate that these systems must be treated with periodic boundary conditions, i.e., as particle-on-a-ring systems. In particular, the maps show a constant intensity along the ring perimeter and an increasing diameter of the unoccupied orbitals with increasing energy (Fig. 10d).

4 Summary

In this chapter, we have shown the synthesis of hydrocarbon polymers and macrocycles via Ullmann reaction of haloarenes on metal surfaces. The usage of haloarenes containing two C–X groups (X = halogen) on Cu(111) and Ag(111) commonly leads, after C–X bonds scission, to the preliminary formation of organometallic chains, in which the segments are linked by carbon-metal-carbon bonds. The stabilization of these organometallic chains in ordered domains is achieved by hydrogen bonds perpendicular to the chain direction. The covalently bonded polymers form either by post-annealing of the preformed organometallic chains or by direct deposition of the precursor monomers onto hot surfaces. The latter procedure leads to the formation of longer polymers due to the more freely diffusing monomers. High diffusion rates of the monomers also contribute to the increase of the lengths of polymer chain. The electronic properties of the synthesized polymers can be controlled to a large degree by the design of the precursor monomers. A particularly interesting class is that of the graphene nanoribbons, which have a band gap that can be tuned by control of their width and edge topology or by doping with heteroatoms.

The on-surface synthesis of macrocycles typically requires haloarene monomers with bent carbon skeletons such as DMTP. Due to the alternative connections (*cis* or *trans*) between the bent monomers, linear open chains or closed macrocycles are formed. The six-fold symmetric metal (111) surfaces are more favorable for the formation of hexagonal organometallic/covalent macrocycles than a two-fold symmetric (110) surfaces, a finding that is attributed to template effects. Even more pronounced template effects occur on the Cu(110)–(1 × 2)O supergrating surface, where organometallic macrocycles grow on the narrow bare Cu(110) stripes, while they are absent on the pristine Cu(110) surface. The on-surface synthesis of hyperbenzene macrocycles with yields of up to 83% has been achieved on Ag(111) at 463 K by application of the high-dilution principle to the on-surface synthesis reaction. The hyperbenzene macrocycle is proposed to be formed predominantly by demetalation of the $(MTP\text{-}Ag)_6$ organometallic macrocycle preformed via ring-closure of an $(MTP\text{-}Ag)_6$ open chain under pseudo-high dilution conditions. This ring-closure process on the stage of the organometallic phase is effective due to the reversibility of the C–Ag–C bond. In contrast, the direct cyclization of covalent $(MTP)_6$ chains into hyperbenzene is unfavorable due to the high energy barriers for the flipping of chain segments necessary for their ring-closure. Hyberbenzene, honeycombenes and other in situ synthesized conjugated hydrocarbon macrocycles on surfaces can act as effective quantum corrals for the confinement of surface-state electrons or as particle-on-a-ring systems.

References

1. Wöhler, F.: Über künstliche Bildung des Harnstoffs. Ann. Phys. **88**, 253–256 (1828)
2. Held, P.A., Fuchs, H., Studer, A.: Covalent-bond formation *via* on-surface chemistry. Chem. Eur. J. **23**, 5874–5892 (2017)
3. Grave, C., Schlüter, A.D.: Shape-persistent, nano-sized macrocycles. Eur. J. Org. Chem. 3075–3098 (2002)
4. Fan, Q., Wang, C., Han, Y., Zhu, J., Hieringer, W., Kuttner, J., Hilt, G., Gottfried, J.M.: Surface-assisted organic synthesis of hyperbenzene nanotroughs. Angew. Chem. Int. Ed. **52**, 4668–4672 (2013)
5. Chen, M., Shang, J., Wang, Y., Wu, K., Kuttner, J., Hilt, G., Hieringer, W., Gottfried, J.M.: On-surface synthesis and characterization of honeycombene oligophenylene macrocycles. ACS Nano **11**, 134–143 (2017)
6. Fan, Q., Wang, T., Dai, J., Kuttner, J., Hilt, G., Gottfried, J. M., Zhu, J.: On-surface pseudo-high dilution synthesis of macrocycles: principle and mechanism. ACS Nano **11**, 5070–5079 (2017)
7. Fan, Q., Gottfried, J.M., Zhu, J.: Surface-catalyzed C-C covalent coupling strategies toward the synthesis of low-dimensional carbon-based nanostructures. Acc. Chem. Res. **48**, 2484–2494 (2015)
8. Dong, L., Liu, P.N., Lin, N.: Surface-activated coupling reactions confined on a surface. Acc. Chem. Res. **48**, 2765–2774 (2015)
9. Ullmann, F., Meyer, G., Loewenthal, O., Gilli, E.: Symmetric biphenyl derivates. Eur. J. Org. Chem. **332**, 38–81 (1904)
10. Lipton-Duffin, J.A., Ivasenko, O., Perepichka, D.F., Rosei, F.: Synthesis of polyphenylene molecular wires by surface-confined polymerization. Small **5**, 592–597 (2009)
11. Di Giovannantonio, M.E.G.M., Lipton-Duffin, J., Meunier, V., Cardenas, L., Fagot Revurat, Y., Cossaro, A., Verdini, A., Perepichka, D.F., Rosei, F., Contini, G.: Insight into organometallic intermediate and its evolution to covalent bonding in surface-confined Ullmann polymerization. ACS Nano 7, 8190–8198 (2013)
12. Di Giovannantonio, M., Tomellini, M., Lipton-Duffin, J., Galeotti, G., Ebrahimi, M., Cossaro, A., Verdini, A., Kharche, N., Meunier, V., Vasseur, G., Fagot-Revurat, Y., Perepichka, D.F., Rosei, F., Contini, G.: Mechanistic picture and kinetic analysis of surface-confined Ullmann polymerization. J. Am. Chem. Soc. **138**, 16696–16702 (2016)
13. Wang, W.H., Shi, X.Q., Wang, S.Y., Van Hove, M.A., Lin, N.: Single-molecule resolution of an organometallic intermediate in a surface-supported Ullmann coupling reaction. J. Am. Chem. Soc. **133**, 13264–13267 (2011)
14. Basagni, A., Sedona, F., Pignedoli, C.A., Cattelan, M., Nicolas, L., Casarin, M., Sambi, M.: Molecules–oligomers–nanowires–graphene nanoribbons: a bottom-up stepwise on-surface covalent synthesis preserving long-range order. J. Am. Chem. Soc. **137**, 1802–1808 (2015)
15. Reecht, G., Bulou, H., Scheurer, F., Speisser, V., Carriere, B., Mathevet, F., Schull, G.: Oligothiophene nanorings as electron resonators for whispering gallery modes. Phys. Rev. Lett. **110**, 056802 (2013)
16. Talirz, L., Ruffieux, P., Fasel, R.: On-surface synthesis of atomically precise graphene nanoribbons. Adv. Mater. **28**, 6222–6231 (2016)
17. Pham, T.A., Tran, B.V., Nguyen, M.-T., Stöhr, M.: Chiral-selective formation of 1D polymers based on Ullmann-type coupling: The role of the metallic substrate. Small **13**, 1603675 (2017)
18. Ullmann, F., Bielecki, J.: Synthesis in the biphenyl series. Ber. Dtsch. Chem. Ges. **34**, 2174–2185 (1901)
19. Xi, M., Bent, B.E.: Iodobenzene on Cu(111): formation and coupling of adsorbed phenyl groups. Surf. Sci. **278**, 19–32 (1992)
20. Xi, M., Bent, B.E.: Mechanisms of the Ullmann coupling reaction in adsorbed monolayers. J. Am. Chem. Soc. **115**, 7426–7433 (1993)

21. Grill, L., Dyer, M., Lafferentz, L., Persson, M., Peters, M.V., Hecht, S.: Nano-architectures by covalent assembly of molecular building blocks. Nat. Nanotechnol. **2**, 687–691 (2007)
22. Hla, S.W., Bartels, L., Meyer, G., Rieder, K.H.: Inducing all steps of a chemical reaction with the scanning tunneling microscope tip: towards single molecule engineering. Phys. Rev. Lett. **85**, 2777–2780 (2000)
23. Cai, J., Ruffieux, P., Jaafar, R., Bieri, M., Braun, T., Blankenburg, S., Muoth, M., Seitsonen, A.P., Saleh, M., Feng, X., Müllen, K., Fasel, R.: Atomically precise bottom-up fabrication of graphene nanoribbons. Nature **466**, 470–473 (2010)
24. Eichhorn, J., Nieckarz, D., Ochs, O., Samanta, D., Schmittel, M., Szabelski, P.J., Lackinger, M.: On-surface Ullmann coupling: the influence of kinetic reaction parameters on the morphology and quality of covalent networks. ACS Nano **8**, 7880–7889 (2014)
25. Eichhorn, J., Strunskus, T., Rastgoo-Lahrood, A., Samanta, D., Schmittel, M., Lackinger, M.: On-surface Ullmann polymerization *via* intermediate organometallic networks on Ag(111). Chem. Commun. **50**, 7680–7682 (2014)
26. Bieri, M., Nguyen, M.T., Groning, O., Cai, J.M., Treier, M., Ait-Mansour, K., Ruffieux, P., Pignedoli, C.A., Passerone, D., Kastler, M., Müllen, K., Fasel, R.: Two-dimensional polymer formation on surfaces: insight into the roles of precursor mobility and reactivity. J. Am. Chem. Soc. **132**, 16669–16676 (2010)
27. Fan, Q., Wang, C., Liu, L., Han, Y., Zhao, J., Zhu, J., Kuttner, J., Hilt, G., Gottfried, J.M.: Covalent, organometallic, and halogen-bonded nanomeshes from tetrabromo-terphenyl by surface-assisted synthesis on Cu(111). J. Phys. Chem. C **118**, 13018–13025 (2014)
28. Fan, Q., Wang, T., Liu, L., Zhao, J., Zhu, J., Gottfried, J.M.: Tribromobenzene on Cu(111): temperature-dependent formation of halogen-bonded, organometallic, and covalent nanostructures. J. Chem. Phys. **142**, 101906 (2015)
29. Klappenberger, F., Zhang, Y.-Q., Björk, J., Klyatskaya, S., Ruben, M., Barth, J.V.: On-surface synthesis of carbon-based scaffolds and nanomaterials using terminal alkynes. Acc. Chem. Res. **48**, 2140–2150 (2015)
30. Sun, Q., Cai, L., Ding, Y., Xie, L., Zhang, C., Tan, Q., Xu, W.: Dehydrogenative homocoupling of terminal alkenes on copper surfaces: a route to dienes. Angew. Chem. Int. Ed. **54**, 4549–4552 (2015)
31. Sun, Q., Zhang, C., Kong, H., Tan, Q., Xu, W.: On-surface aryl-aryl coupling *via* selective C-H activation. Chem. Commun. **50**, 11825–11828 (2014)
32. Sun, Q., Zhang, C., Cai, L., Xie, L., Tan, Q., Xu, W.: On-surface formation of two-dimensional polymer *via* direct C-H activation of metal phthalocyanine. Chem. Commun. **51**, 2836–2839 (2015)
33. He, Y., Garnica, M., Bischoff, F., Ducke, J., Bocquet, M.-L., Batzill, M., Auwärter, W., Barth, J.V.: Fusing tetrapyrroles to graphene edges by surface-assisted covalent coupling. Nat. Chem. **9**, 33–38 (2017)
34. Zhang, H., Lin, H., Sun, K., Chen, L., Zagranyarski, Y., Aghdassi, N., Duhm, S., Li, Q., Zhong, D., Li, Y., Müllen, K., Fuchs, H., Chi, L.: On-surface synthesis of rylene-type graphene nanoribbons. J. Am. Chem. Soc. **137**, 4022–4025 (2015)
35. Fan, Q., Wang, C., Han, Y., Zhu, J., Kuttner, J., Hilt, G., Gottfried, J.M.: Surface-assisted formation, assembly, and dynamics of planar organometallic macrocycles and zigzag shaped polymer chains with C-Cu-C bonds. ACS Nano **8**, 709–718 (2014)
36. Chung, K.-H., Koo, B.-G., Kim, H., Yoon, J.K., Kim, J.-H., Kwon, Y.-K., Kahng, S.-J.: Electronic structures of one-dimensional metal-molecule hybrid chains studied using scanning tunneling microscopy and density functional theory. Phys. Chem. Chem. Phys. **14**, 7304–7308 (2012)
37. Fan, Q., Dai, J., Wang, T., Kuttner, J., Hilt, G., Gottfried, J.M., Zhu, J.: Confined synthesis of organometallic chains and macrocycles by Cu–O surface templating. ACS Nano **10**, 3747–3754 (2016)
38. Blake, M.M., Nanayakkara, S.U., Claridge, S.A., Fernandez-Torres, L.C., Sykes, E.C.H., Weiss, P.S.: Identifying reactive intermediates in the Ullmann coupling reaction by scanning tunneling microscopy and spectroscopy. J. Phys. Chem. A **113**, 13167–13172 (2009)

39. Chung, K.-H., Kim, H., Jang, W.J., Yoon, J.K., Kahng, S.-J., Lee, J., Han, S.: Molecular multistate systems formed in two-dimensional porous networks on Ag(111). J. Phys. Chem. C **117**, 302–306 (2013)
40. Park, J., Kim, K.Y., Chung, K.-H., Yoon, J.K., Kim, H., Han, S., Kahng, S.-J.: Interchain interactions mediated by Br adsorbates in arrays of metal-organic hybrid chains on Ag(111). J. Phys. Chem. C **115**, 14834–14838 (2011)
41. Huang, H., Tan, Z., He, Y., Liu, J., Sun, J., Zhao, K., Zhou, Z., Tian, G., Wong, S.L., Wee, A.T.S.: Competition between hexagonal and tetragonal hexabromobenzene packing on Au(111). ACS Nano **10**, 3198–3205 (2016)
42. Barton, D., Gao, H.-Y., Held, P.A., Studer, A., Fuchs, H., Doltsinis, N.L., Neugebauer, J.: Formation of organometallic intermediate states in on-surface Ullmann couplings. Chem. Eur. J. **23**, 6190–6197 (2017)
43. Björk, J., Hanke, F., Stafström, S.: Mechanisms of halogen-based covalent self-assembly on metal surfaces. J. Am. Chem. Soc. **135**, 5768–5775 (2013)
44. Basagni, A., Vasseur, G., Pignedoli, C.A., Vilas-Varela, M., Pena, D., Nicolas, L., Vitali, L., Lobo-Checa, J., de Oteyza, D.G., Sedona, F., Casarin, M., Ortega, J.E., Sambi, M.: Tunable band alignment with unperturbed carrier mobility of on-surface synthesized organic semiconducting wires. ACS Nano **10**, 2644–2651 (2016)
45. Nacci, C., Ample, F., Bleger, D., Hecht, S., Joachim, C., Grill, L.: Conductance of a single flexible molecular wire composed of alternating donor and acceptor units. Nat. Commun. **6**, 7397 (2015)
46. Lafferentz, L., Ample, F., Yu, H., Hecht, S., Joachim, C., Grill, L.: Conductance of a single conjugated polymer as a continuous function of its length. Science **323**, 1193–1197 (2009)
47. Liu, J., Li, B.W., Tan, Y.Z., Giannakopoulos, A., Sanchez-Sanchez, C., Beljonne, D., Ruffieux, P., Fasel, R., Feng, X., Müllen, K.: Toward cove-edged low band gap graphene nanoribbons. J. Am. Chem. Soc. **137**, 6097–6103 (2015)
48. Ruffieux, P., Wang, S., Yang, B., Sánchez-Sánchez, C., Liu, J., Dienel, T., Talirz, L., Shinde, P., Pignedoli, C.A., Passerone, D., Dumslaff, T., Feng, X., Müllen, K., Fasel, R.: On-surface synthesis of graphene nanoribbons with zigzag edge topology. Nature **531**, 489–492 (2016)
49. Schulz, F., Jacobse, P.H., Canova, F.F., van der Lit, J., Gao, D.Z., van den Hoogenband, A., Han, P., Klein Gebbink, R.J.M., Moret, M.-E., Joensuu, P.M., Swart, I., Liljeroth, P.: Precursor geometry determines the growth mechanism in graphene nanoribbons. J. Phys. Chem. C **121**, 2896–2904 (2017)
50. Chen, Z., Wang, H.I., Teyssandier, J., Mali, K.S., Dumslaff, T., Ivanov, I., Zhang, W., Ruffieux, P., Fasel, R., Rader, H.J., Turchinovich, D., De Feyter, S., Feng, X., Klaui, M., Narita, A., Bonn, M., Müllen, K.: Chemical vapor deposition synthesis and terahertz photoconductivity of low-bandgap n = 9 armchair graphene nanoribbons. J. Am. Chem. Soc. **139**, 3635–3638 (2017)
51. Talirz, L., Söde, H., Dumslaff, T., Wang, S., Sanchez-Valencia, J.R., Liu, J., Shinde, P., Pignedoli, C.A., Liang, L., Meunier, V., Plumb, N.C., Shi, M., Feng, X., Narita, A., Müllen, K., Fasel, R., Ruffieux, P.: On-surface synthesis and characterization of 9-atom wide armchair graphene nanoribbons. ACS Nano **11**, 1380–1388 (2017)
52. Chen, Y.-C., de Oteyza, D.G., Pedramrazi, Z., Chen, C., Fischer, F.R., Crommie, M.F.: Tuning the band gap of graphene nanoribbons synthesized from molecular precursors. ACS Nano **7**, 6123–6128 (2013)
53. Kawai, S., Saito, S., Osumi, S., Yamaguchi, S., Foster, A.S., Spijker, P., Meyer, E.: Atomically controlled substitutional boron-doping of graphene nanoribbons. Nat. Commun. **6**, 8098 (2015)
54. Cloke, R.R., Marangoni, T., Nguyen, G.D., Joshi, T., Rizzo, D.J., Bronner, C., Cao, T., Louie, S.G., Crommie, M.F., Fischer, F.R.: Site-specific substitutional boron doping of semiconducting armchair graphene nanoribbons. J. Am. Chem. Soc. **137**, 8872–8875 (2015)
55. Cai, J., Pignedoli, C.A., Talirz, L., Ruffieux, P., Sode, H., Liang, L., Meunier, V., Berger, R., Li, R., Feng, X., Müllen, K., Fasel, R.: Graphene nanoribbon heterojunctions. Nat. Nanotechnol. **9**, 896–900 (2014)

56. Dow, A.R., Wittrig, A.M., Kenttämaa, H.I.: Laser-induced acoustic desorption mass spectrometry. Eur. J. Mass Spectrom. **18**, 77–92 (2012)
57. Saywell, A., Magnano, G., Satterley, C.J., Perdigao, L.M., Britton, A.J., Taleb, N., del Carmen Gimenez-Lopez, M., Champness, N.R., O'Shea, J.N., Beton, P.H.: Self-assembled aggregates formed by single-molecule magnets on a gold surface. Nat. Commun. **1**, 75 (2010)
58. Ruggli, P.: A ring with a threefold bond. Eur. J. Org. Chem. **392**, 92–100 (1912)
59. Ziegler, K., Eberle, H., Ohlinger, H.: Polynomial Ring Systems I. The preparative, productive synthesis of polymethyl ketones with more than 6 mmbers of the ring. Eur. J. Org. Chem. **504,** 94–130 (1933)
60. Jia, Z., Monteiro, M.J.: Synthesis of cyclic polymers *via* ring closure. Adv. Polym. Sci. **262**, 295–327 (2013)
61. Melson G.A., Bush D.H.: Cyclic tetramerisation of o-aminobenzaldehyde in presence of metal ions. Proc. chem. Soc. 223 (1963)
62. Laughrey, Z.R., Gibb, B.C.: Macrocycle synthesis through templation. In: Schalley, C.A., Vögtle, F., Dötz, K.H. (eds.) Templates in Chemistry II vol. 249, pp. 67–125 (2005)
63. Hoffmann, M., Wilson, C.J., Odell, B., Anderson, H.L.: Template-directed synthesis of a pi-conjugated porphyrin nanoring. Angew. Chem. Int. Ed. **46**, 3122–3125 (2007)
64. Sprafke, J.K., Kondratuk, D.V., Wykes, M., Thompson, A.L., Hoffmann, M., Drevinskas, R., Chen, W.H., Yong, C.K., Karnbratt, J., Bullock, J.E., Malfois, M., Wasielewski, M.R., Albinsson, B., Herz, L.M., Zigmantas, D., Beljonne, D., Anderson, H.L.: Belt-shaped pi-systems: relating geometry to electronic structure in a six-porphyrin nanoring. J. Am. Chem. Soc. **133**, 17262–17273 (2011)
65. Svatek, S.A., Perdigao, L.M., Stannard, A., Wieland, M.B., Kondratuk, D.V., Anderson, H. L., O'Shea, J.N., Beton, P.H.: Mechanical stiffening of porphyrin nanorings through supramolecular columnar stacking. Nano Lett. **13**, 3391–3395 (2013)
66. O'Sullivan, M.C., Sprafke, J.K., Kondratuk, D.V., Rinfray, C., Claridge, T.D., Saywell, A., Blunt, M.O., O'Shea, J.N., Beton, P.H., Malfois, M., Anderson, H.L.: Vernier templating and synthesis of a 12-porphyrin nano-ring. Nature **469**, 72–75 (2011)
67. Anderson, S., Anderson, H.L.: Templates in organic synthesis: definitions and roles. Templated organic synthesis, pp. 1–38. Wiley-VCH Verlag GmbH (2007)
68. Lin, N., Stepanow, S., Ruben, M., Barth, J.V.: Surface-confined supramolecular coordination chemistry. In: Broekmann, P., Dötz, K.H., Schalley, C.A. (eds.) Templates in Chemistry III, Topics in Current Chemistry, pp. 1–44 (2009)
69. Chen, M., Feng, X., Zhang, L., Ju, H., Xu, Q., Zhu, J., Gottfried, J.M., Ibrahim, K., Qian, H., Wang, J.: Direct synthesis of nickel(II) tetraphenylporphyrin and its interaction with a Au (111) surface: a comprehensive study. J. Phys. Chem. C **114**, 9908–9916 (2010)
70. Gottfried, J.M.: Surface chemistry of porphyrins and phthalocyanines. Sur. Sci. Rep. **70**, 259–379 (2015)
71. Weckesser, J., De Vita, A., Barth, J.V., Cai, C., Kern, K.: Mesoscopic correlation of supramolecular chirality in one-dimensional hydrogen-bonded assemblies. Phys. Rev. Lett. **87**, 096101 (2001)
72. Gambardella, P., Stepanow, S., Dmitriev, A., Honolka, J., de Groot, F.M.F., Lingenfelder, M., Sen Gupta, S., Sarma, D.D., Bencok, P., Stanescu, S., Clair, S., Pons, S., Lin, N., Seitsonen, A.P., Brune, H., Barth, J.V., Kern, K.: Supramolecular control of the magnetic anisotropy in two-dimensional high-spin Fe arrays at a metal interface. Nat. Mater. **8**, 189–193 (2009)
73. Barth, J.V., Costantini, G., Kern, K.: Engineering atomic and molecular nanostructures at surfaces. Nature **437**, 671–679 (2005)
74. Shi, Z., Lin, N.: Porphyrin-based two-dimensional coordination kagome lattice self-assembled on a Au(111) surface. J. Am. Chem. Soc. **131**, 5376–5377 (2009)
75. Shi, Z., Lin, N.: Structural and chemical control in assembly of multicomponent metal-organic coordination networks on a surface. J. Am. Chem. Soc. **132**, 10756–10761 (2010)
76. Lin, T., Shang, X.S., Adisoejoso, J., Liu, P.N., Lin, N.: Steering on-surface polymerization with metal-directed template. J. Am. Chem. Soc. **135**, 3576–3582 (2013)

77. Dai, J., Fan, Q., Wang, T., Kuttner, J., Hilt, G., Gottfried, J.M., Zhu, J.: The role of the substrate structure in the on-surface synthesis of organometallic and covalent oligophenylene chains. Phys. Chem. Chem. Phys. **18**, 20627–20634 (2016)
78. Jacobson, H., Beckmann, C.O., Stockmayer, W.H.: Intramolecular reaction in polycondensations. 2. Ring-chain equilibrium in polydecamethylene adipate. J. Chem. Phys. **18**, 1607–1612 (1950)
79. Jacobson, H., Stockmayer, W.H.: Intramolecular reaction in polycondensations. 1. The theory of linear systems. J. Chem. Phys. **18**, 1600–1606 (1950)
80. Müllegger, S., Salzmann, I., Resel, R., Hlawacek, G., Teichert, C., Winkler, A.: Growth kinetics, structure, and morphology of *para*-quaterphenyl thin films on gold(111). J. Chem. Phys. **121**, 2272–2277 (2004)

Bottom-Up Fabrication of Atomically Precise Graphene Nanoribbons

Martina Corso, Eduard Carbonell-Sanromà and Dimas G. de Oteyza

Abstract Graphene nanoribbons (GNRs) make up an extremely interesting class of materials. On the one hand GNRs share many of the superlative properties of graphene, while on the other hand they display an exceptional degree of tunability of their optoelectronic properties. The presence or absence of correlated low-dimensional magnetism, or of a widely tunable band gap, is determined by the boundary conditions imposed by the width, crystallographic symmetry and edge structure of the nanoribbons. In combination with additional controllable parameters like the presence of heteroatoms, tailored strain, or the formation of heterostructures, the possibilities to shape the electronic properties of GNRs according to our needs are fantastic. However, to really benefit from that tunability and harness the opportunities offered by GNRs, atomic precision is strictly required in their synthesis. This can be achieved through an on-surface synthesis approach, in which one lets appropriately designed precursor molecules to react in a selective way that ends up forming GNRs. In this chapter we review the structure-property relations inherent to GNRs, the synthesis approach and the ways in which the varied properties of the resulting ribbons have been probed, finalizing with selected examples of demonstrated GNR applications.

M. Corso (✉) · D. G. de Oteyza
Centro de Física de Materiales CSIC-UPV/EHU – Materials Physics Center, Paseo Manuel Lardizabal 5, 20018 San Sebastián, Spain
e-mail: martina.corso@ehu.eus

D. G. de Oteyza
e-mail: d_g_oteyza@ehu.eus

E. Carbonell-Sanromà
CIC NanoGUNE, Avenida Tolosa 76, 20018 San Sebastián, Spain

M. Corso · D. G. de Oteyza
Donostia International Physics Center, Paseo Manuel Lardizabal 4, 20018 San Sebastián, Spain

D. G. de Oteyza
Ikerbasque, Basque Foundation for Science, 48013 Bilbao, Spain

D. G. de Oteyza and C. Rogero (eds.), *On-Surface Synthesis II*, Advances in Atom and Single Molecule Machines,
https://doi.org/10.1007/978-3-319-75810-7_6

1 Introduction

Graphene has attracted enormous interest since its first experimental realization through exfoliation of graphite, mainly because of its many superlative properties [1, 2]. By way of example, it is the thinnest, lightest and strongest material known. It is also the material with highest thermal conductivity and with highest electron mobility. The latter causes graphene to be considered as a potentially revolutionary material for future electronic applications [2, 3]. In this respect, however, graphene also faces a drawback for its implementation in conventional electronics: the lack of a band gap to turn the electron conduction on or off [3].

Various strategies have been investigated to overcome this limitation, and one of them is through electron confinement in nanostructured graphene. In particular, graphene nanoribbons, narrow 1D stripes of graphene with widths in the nanometer range, display remarkably varied electronic properties depending on their particular structure at the very atomic level [4].

A visual framework to understand this is analyzing the aromaticity of some representative GNR structures [5, 6]. According to Kekulé, each carbon in polycyclic aromatic hydrocarbons (PAH) like GNRs or graphene itself has its four valence electrons arranged in single, double or triple bonds with electrons from neighboring atoms. The structure of a PAH is then the superposition of all possible Kekulé bond configurations. Within this picture, the delocalization of six π-electrons in a carbon hexagon due to the resonance of two Kekulé configurations with alternating single and double bonds is called a Clar sextet, pictured as a circle within the corresponding hexagon (Fig. 1a). According to Clar's theory [7], the most representative and stable structure of a PAH is that with the highest number of Clar sextets, which is called the "*Clar formula*". Here it is important to keep in mind that the bonds sticking out of a Clar sextet are formally single bonds and thereby impede two neighboring hexagons to be Clar sextets simultaneously.

Graphene can be represented with three equivalent Clar formulas, in each of which one out of every three hexagons is a Clar sextet, arranged in a $(\sqrt{3} \times \sqrt{3})R30°$ superstructure (Fig. 1b–d). Considering a combination of the three possible Clar

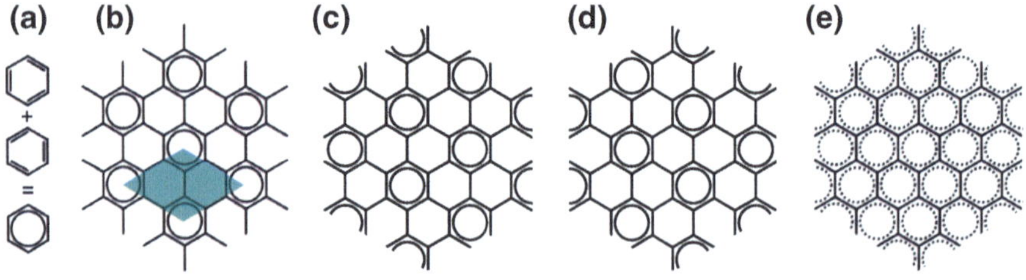

Fig. 1 **a** Graphical explanation of the Clar sextet as being the superposition of two resonant Kekulé configurations. **b–d** Each of the three equivalent Clar formulas of graphene, with its $\sqrt{3} \times \sqrt{3}$ superstructure highlighted in cyan in (**b**). **e** Superposition of all three Clar formulas. Dotted circles correspond to Clar sextets within graphene, although not all simultaneously

formulas, all hexagons in graphene can be Clar sextets and are fully equivalent (Fig. 1e). In addition, bonds within a Clar sextet are equivalent as well. Since bond length alternation (BLA) is a measure of the aromaticity of a system and one of the main causes for the opening of band gaps in conjugated organic materials [8, 9], graphene is an excellent example of a perfectly aromatic system with absent BLA and zero band gap.

Now we place our focus on GNRs. Because GNRs include many different structures, a first classification of GNRs is typically performed depending on their edge orientation. They are called zigzag nanoribbons (zGNR) when the edge runs parallel to one of graphene's lattice vectors, or armchair nanoribbons (aGNR) when the edges run along the high symmetry direction thirty degrees off the lattice vectors (Fig. 2a). GNRs with edges running along any intermediate orientation are called chiral nanoribbons. Their chirality is characterized either by the angle of the edge orientation with respect to a graphene lattice direction, or by the edge unit cell vector in terms of graphene lattice vectors (Fig. 2a). The key in GNRs is that the presence of edges imposes boundary conditions which, among other effects, limit the number of possible Clar sextets.

The width of aGNRs is most commonly given as the number of dimer lines N_a present across the ribbon (Fig. 2b). In the following we assume the most common scenario of purely sp^2 hybridized carbon atoms within the ribbons, consequently with singly hydrogenated edges. Under these conditions, we will see how the boundary conditions divide the aGNRs into three families, depending on the number of Clar formulas they display [5, 6].

For aGNR with $N_a = 3p$, p being an integer, there is only one possible Clar formula maximizing the number of Clar sextets. In this formula, all π-electrons participate in the Clar sextets, consequently displaying no localized double bonds (Fig. 2b). While the π-electrons are delocalized within the Clar sextets, the bonds sticking out of them remain single bonds, causing a substantial variation of the mean bond length within the ribbons that opens a considerable band gap [5, 6]. For aGNRs with $N_a = 3p + 1$ there are two possible Clar formulas, both of which include localized double bonds (Fig. 2b). A combination of the two Clar formulas still leaves hexagons with and without Clar sextets, causing again notable bond length variations within the ribbon. This bond length variation and the electron localization in the structure's double bonds cause this family to have an even larger band gap than the $3p$ family [5, 6]. The scenario is very different with aGNRs of the $N_a = 3p + 2$ family. In this case, there are many different Clar formulas, each of them featuring two localized double bonds (Fig. 2b). However, a linear combination of all the Clar formulas ultimately renders a highly aromatic structure with little bond length variations along the middle part of the ribbon. This implies a high degree of electron delocalization and consequently a low band gap [5, 6].

It is very interesting to analyze zGNRs in this same framework, since they can host only a limited amount of Clar sextets even for infinitely long ribbons (Fig. 2c). Because sextets can be placed anywhere along the longitudinal axis of the ribbon, there are infinite Clar formulas for zGNRs. Their combination ends up being equivalent to a superposition of two fully quinoidal structures, that is, structures

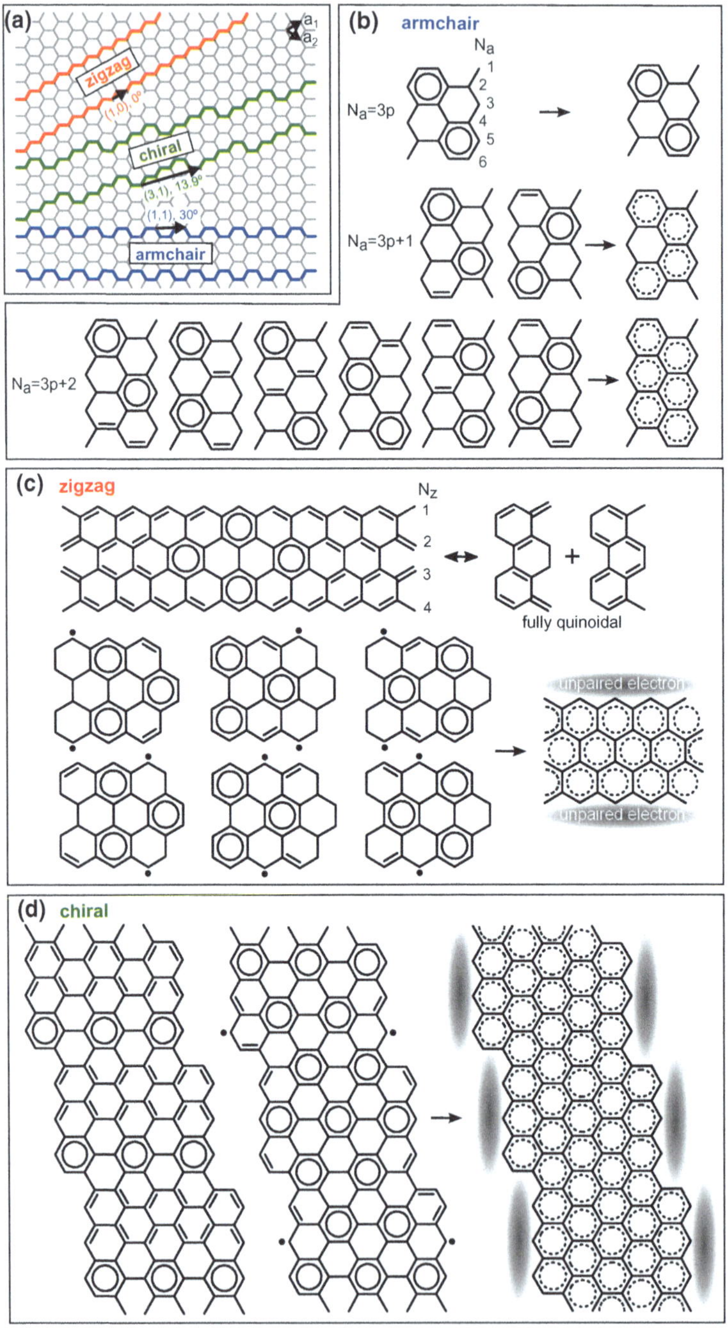
(a)
a1
a2
zigzag
(1,0), 0°
chiral
(3,1), 13.9°
(1,1), 30°
armchair
(b) armchair
Na
Na=3p
Na=3p+1
Na=3p+2
(c) zigzag
Nz
fully quinoidal
unpaired electron
unpaired electron
(d) chiral

◂**Fig. 2** **a** Classification of GNRs by edge orientation into zigzag, chiral or armchair. The respective edge periodicity is given by the unit cell vector in terms of graphene's lattice vectors a_1 and a_2. **b** Clar sexted distribution on aGNR, classified into three families according to its width in terms of dimer lines N_a. The family with $N_a = 3p$ possesses only one formula, exemplified in 6-aGNRs. The family with $N_a = 3p + 1$ possesses two compatible Clar formulas. In each of these formulas there is a localized double bond at the edge of the ribbon, as exemplified with the formulas of 7-aGNRs. On the right, a superposition of both Clar formulas is given, showing all hexagons hosting a Clar sextet (although not all simultaneously). The family with $N_a = 3p + 2$ ribbons have more than two Clar formulas, namely 3p. Some of them are shown for 8-aGNRs. On the right, a superposition of all Clar formulas shows the Clar sextets delocalized along all carbon hexagons. **c** Clar sextet distribution on zGNRs. As exemplified with 4-zGNRs, only a limited number of Clar sextets exists even on infinitely long ribbons. Infinite Clar formulas exist, due to their arbitrary location along the ribbon, and the superposition is equivalent to the two quinoidal structures drawn. Below it is shown how by adding unpaired electrons, Clar sextets can be redistributed along the whole ribbon. Some of the Clar formulas are shown, as well as their superposition evidencing delocalized Clar sextets along the ribbon and unpaired electrons along the edges. **d** Clar sextet distribution on chiral GNRs, evidencing a limited number of Clar sextets per zigzag segment and how addition of unpaired electrons allows creation of additional Clar sextets. One of the multiple Clar formulas is shown for each case, including on the right a superposition of all the possible Clar formulas that evidences the delocalized Clar sextets along the ribbon and unpaired electrons along its edges

with two double bonds per carbon hexagon (Fig. 2c). However, allowing the introduction of unpaired electrons (radicals) into the structure, Clar sextets can be distributed anew along the whole ribbon length (Fig. 2c).

This configuration reduces the overall ribbon energy and readily provides an explanation of some of most intriguing properties of zGNRs. A combination of all possible Clar formulas evidences all hexagons as being Clar sextets, implying a high degree of aromaticity and electron delocalization along zGNRs, as well as an associated low band gap. In addition, because the radicals are always located on the edges, combination of the various Clar formulas evidences unpaired electrons fully delocalized along both ribbon edges. This is the origin of the edge states in zGNR, appearing, as expected from unpaired electrons, at the Fermi level. Finally, since unpaired electrons translate to an uncompensated spin, Clar's formalism also implies the spin polarized nature of the edge states in zGNRs.

Last but not least, chiral GNRs display a combination of alternating zigzag and armchair segments along their edges. The lower the chiral angle the longer the zigzag segments are, and viceversa. As pictured on the (3,1) chiral nanoribbon example in Fig. 2d, there is a limited number of Clar sextets within each zigzag segment. However, as occurs for pure zGNRs, addition of unpaired electrons on the zigzag edges allows formation of additional Clar sextets (Fig. 2d). The combination of the different Clar formulas evidences the high degree of aromaticity and electron delocalization in chiral GNRs that explains the low band gap, as well as the creation of delocalized and spin polarized edge states along the zigzag segments of the ribbon (Fig. 2d). At this point it is important to remind the intricate dependence of these edge states with the detailed GNR structure. In infinitely long zGNRs, addition of unpaired electrons implies changing from a finite number of sextets to an infinite number of sextets, making the edge state creation always favorable. Instead, infinite chiral GNRs readily display infinite Clar sextets (with a limited number of them on each zigzag segment, as mentioned above). Addition of unpaired electrons simply increases the number of Clar sextets per zigzag segment, or in other words the density of Clar sextets per GNR length unit. This has the following implications. On the one hand, there is a width threshold below which the edge states are not present because for narrow ribbons the energy cost of having unpaired electrons may not be compensated by the creation of an insufficient number of new Clar sextets. On the other hand, the chiral angle also plays a major role. The lower the chiral angle (longer zigzag segments), the more new sextets can be created by addition of unpaired electrons, and therefore the lower the width threshold for edge state creation. Lastly, the lower the chiral angle the larger is also the number of unpaired electrons per edge atom and consequently also the net spin moment on each nanoribbon edge.

A similar analysis based on the Clar sextets and the resulting GNR's aromaticity has been also applied to nanoribbons doped with heteroatoms [10]. Depending on the particular atomic species and its bonding site, the Clar sextet distribution and the associated aromaticity can change substantially, having a direct and notable impact on the GNR's electronic properties [10].

All the above shows, from an intuitive and easy to visualize chemical viewpoint, one of the main virtues of GNRs: their amazing variety of electronic properties, ranging from large band gap semiconductors to gap-less structures with spin-polarized edge states. However, it also remarks the stringent need for atomic precision in their synthesis, since minute structural changes at the atomic level can cause unproportioned changes in their electronic properties.

2 Synthesis

Current digital logic devices require excellent switching capabilities and an on-off ratio (I_{on}/I_{off}, corresponding to the current flows under *on* and *off* operation conditions) in the order of 10^4–10^7. The latter in turn requires band gaps of 0.4 eV or more [3]. If graphene nanoribbons are to be integrated into such structures, their width needs to be scaled down to few nanometers [11]. This sets challenging boundary conditions to be solved in the synthesis of graphene nanoribbons, complicated even further by the requisite of atomic precision described in the previous section.

Several groups have readily reported the successful synthesis of GNRs by top-down methods and their subsequent characterization. Amongst the different strategies we find e.g. the unzipping of carbon nanotubes [12], etching of 2D-graphene [13], chemical vapor deposition [14], or scanning probe lithography [15]. However, in spite of the important insight into the properties of GNRs provided in these works, such techniques lack the synthetic reproducibility and/or the atomic precision, the latter being of particular importance to be able to aim at rationally chosen GNR structures (e.g. atomically precise width control) and to avoid the often dominating disorder effects [16, 17].

Instead, Cai and co-workers reported in a seminal work a bottom-up strategy to do just that: selectively grow atomically precise nanoribbons with widths in the nanometer range [18]. That inspiring work has sparked the interest and research efforts in this direction, which basically consists in letting appropriately designed precursor molecules to react in a selective way that ends up forming GNRs [4, 19]. This approach has been followed in solution [19], as well as supported on solid surfaces [4, 19]. Henceforth we will focus on the latter, for which no additional functionalization with side-chains is needed to provide solubility to the GNR products. In a first step, the reactants are deposited on a surface. Thereafter, they are externally activated, triggering polymerization reactions and their subsequent transformation into GNRs. Common to all substrate-supported GNRs successfully synthesized to date, the reactant deposition was by molecular beam deposition, the polymerization based on Ullmann coupling, and the reaction thermally activated. Nevertheless, we would like to remark that there may be useful alternatives to each of these, as are e.g. electrospray deposition [20], other C–C coupling reactions like C–H activation [21] or enediyne cyclizations [22], and photoactivation of the reactions [23, 24]. Such alternatives may help reducing the current constraints on reactants and substrates.

A prototypical reaction process is for example that shown in Fig. 3a [25], which includes polymerization through Ullmann coupling, but also a second cyclo-dehydrogenation step. The whole reaction process has been reviewed extensively [4, 21, 26] and here we only provide a brief description of the main steps. Ullmann coupling itself readily involves various steps [26, 27]. First there is a reactant dehalogenation, resulting in surface or adatom-stabilized radical intermediates. This process is catalyzed by metallic substrates [27], allowing the dehalogenation to occur at temperatures below the temperatures of molecular desorption or of other unwanted side-reactions that could compromise the synthetic selectivity and the overall polymerization process. In a next step, the radical intermediates diffuse along the surface to meet each other and bind covalently. In between the latter two, and depending on the type of substrate, the formation of a metastable metal-organic intermediate is often observed [21, 26]. Finally, cyclo-dehydrogenation leads to the planarization of the polymers, ending up in atomically precise graphene nanoribbons with their structure univocally defined by the design of the reactant. Annealing at higher temperatures can also cause the coupling of neighboring GNRs by inter-ribbon cyclo-dehydrogenation [28–30]. This process, however, results in the fusion of a random number of GNRs, lacking selectivity for the formation of precise widths and generating substantial disorder.

Many of the processes described above and their associated energy barriers depend notably on the substrate material and surface structure, underlining their key role in the overall synthesis. While on the one hand it limits the range of potential substrates to be used, on the other hand their appropriate choice can be used as an additional parameter to control the reaction and thus the resulting products [31, 32], their distribution or alignment [33–36].

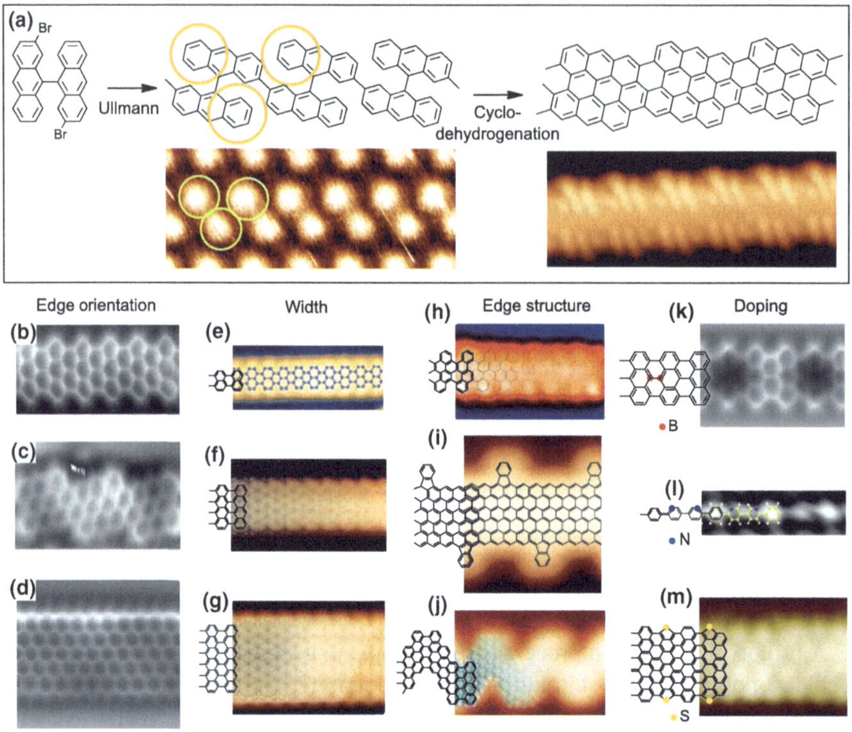

Fig. 3 **a** Schematic representation through wireframe diagrams of an archetypical reaction process. The precursors first undergo Ullmann coupling, resulting in non-planar polymeric structures due to intramolecular steric hindrance. The circles highlight sections that are elevated with respect to the surface. In a second step the planar GNR is formed through cyclodehydrogenation. Associated STM images revealing the non-planar and planar nature of polymer and GNR are added below the wireframe structures. The use of different precursors allows growing different GNRs with varying edge orientation (**b–d**), width (**e–g**), edge morphology (**h–j**) or doping (**k–m**). **a** Reprinted with permission from [25]. **b** Reprinted by permission from Macmillan Publishers Ltd.: Nature Communications [40], copyright (2013), **c** Reprinted with permission from [41]. Copyright (2016) American Chemical Society. **d** and **i** Adapted by permission from Macmillan Publishers Ltd.: Nature [18], copyright (2016). **e** Adapted with permission from [43]. **f** Reprinted with permission from [46]. Copyright (2017) American Chemical Society. **g** Reprinted with permission from [51]. Copyright (2013) American Chemical Society. **h** Reprinted with permission from [45]. **j** Adapted by permission from Macmillan Publishers Ltd.: Nature [18], copyright (2010). **k** Adapted with permission from [48]. **l** Adapted with permission from [36]. **m** Adapted with permission from [49]. Copyright (2016) American Chemical Society

Obviously, the particular reactant used also has a notable effect on the various chemical processes and their barriers. By way of example, one can compare the coupling of 10,10′-dibromo-9,9′-bianthracene [18] and the mono-anthracene equivalent. The latter lies flat on the surface, and the radical coupling is therefore hindered by the steric repulsions between the hydrogens surrounding the carbon radical. Instead, the intramolecular steric hindrance between its two anthracene units makes the former a non-planar precursor. As a result, the steric hindrance between two radical intermediates is reduced, allowing them to get close enough to form the covalent C–C bond [4].

Another beautiful example is the comparison of the same 10,10′-dibromo-9,9′-bianthracene designed to render 7-aGNRs, with 2,2′-dibromo-9,9′-bianthracene, designed to render chiral (3,1)-GNRs [25]. For the latter, the Br atoms lie closer to the surface, lowering the temperature threshold for dehalogenation. However, even more notable is the change in the temperature threshold for the cyclo-dehydrogenation. In this case it rather relates to the completely different strain between the two resulting polymers. With the former, the anthracene units are linked covalently along their short axis by a bond that allows free rotational movement with respect to their neighbors. This freedom results in alternatively tilted anthracene units along the polymer backbone so as to minimize the steric hindrance from opposing H atoms. Instead, with the latter the anthracene units are linked covalently to their neighbors both along their long and short axes (Fig. 3a). Thus, although the anthracene units still display the same alternative tilt to reduce the steric hindrance, the covalent bonds along the long anthracene's axes limit the structure's rotational freedom, resulting in a substantially strained geometry. Sterically induced strain is known to weaken the involved C–H bonds and thereby lower the cyclo-dehydrogenation barriers [37–39] in turn explaining the substantially lower threshold temperature for 2,2′-dibromo-9,9′-bianthracene [25].

Effects like these can be of great importance for potential applications. Focusing on the last example, the lower cyclo-dehydrogenation temperature may on the one hand allow the use of different substrates that could not stand higher temperatures. On the other hand, it may have a strong impact on the resulting products. By way of example, on Au(111) it brings the threshold temperatures for Ullmann coupling and cyclo-dehydrogenation close to each other. Under this scenario, radical quenching by liberated H atoms competes with the radical step growth polymerization, greatly lowering the average length of the resultant GNRs [25].

However, most importantly, the appropriate design of precursors allows growing GNRs with different and tailored structures, all of them with atomic precision. This is shown in Fig. 3, displaying selected GNRs with different edge orientations (Fig. 3b–d) [40–42], different width (Fig. 3e–g) [43–46] different edge structure (Fig. 3h–j) [18, 42, 47] and different doping (Fig. 3k–m) [36, 48, 49], all of them allowing to tune in a controlled way the electronic properties of GNRs.

3 Characterization Tools and Associated Insight

3.1 Electronic Properties Determination

The electronic structure of GNRs and in particular their band gap (E_g) size and the values of bands' effective masses (m_{VB} and m_{CB}), have been addressed locally with scanning tunneling spectroscopy (STS) [34, 40, 42, 43, 45, 50, 51], and with averaging experimental techniques as angle-resolved-, inverse- and two-photon-photoemission [34, 35, 52–54], high-resolution electron energy loss [52, 55] and optical spectroscopy [56].

The most widespread method used so far is STS, which measures the local density of states (LDOS) as a function of tip position and applied sample bias. It allows determining the electronic band dispersion of occupied and empty energy levels for the vary same GNR. Precise quantification of bands' energy level alignments and effective masses is given by Fourier-transformed scanning tunneling spectroscopy (FT-STS) [57]. The presence of edges in GNRs gives rise to standing wave patterns that arise due to scattering of electronic Bloch wave functions against such termini. Recording standing waves for different energies along one GNRs edge, allows determining the *energy versus momentum* dispersion of GNRs' states via Fourier transform processing of STS data. Precise bandgap values of $E_g = 2.37 \pm 0.06$ eV and $E_g = 1.38 \pm 0.03$ eV have been extrapolated from valence band maximum and conduction band minimum for 7-aGNR (Fig. 4c–f) [50] and 9-aGNRs [45] grown on Au(111), respectively.

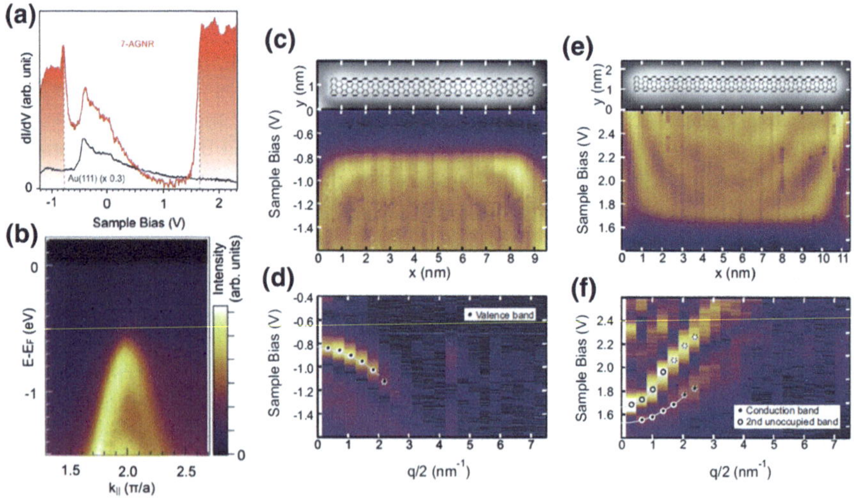

Fig. 4 Electronic structure of 7-aGNRs. **a** dI/dV spectrum (red) of a 7-aGNR grown on Au(111). The onsets of valence and conduction bands are indicated by the shaded areas. Electronic features in the gap are due to the Au surface state (visible in the black spectrum). **b** ARPES intensity plot recorded for 1 ML of 7-aGNRs grown on Au(788) recorded along the ribbon axis. The VB onset is found at −0.7 eV below the Fermi level. **c** STM image and plot consisting on equidistant dI/dV spectra of occupied states recorded along the edge of a 7-aGNR. **d** Line-by-line Fourier transform of (**c**) including a parabolic fit of the VB. The same is done for the CB on a longer ribbon in (**e**, **f**). **a** Reprinted with permission from [51]. Copyright (2017) American Chemical Society. **b** Reprinted with permission from [34]. Copyright (2012) American Chemical Society. **c–f** Reprinted figure with permission from [52]. Copyright (2015) by the American Physical Society

Such precise value could not be simply inferred by single STS spectra (Fig. 4a). On one hand, this is due to the fact that the onset of the GNR's electronic states could be ambiguous due to fingerprints of the electronic structure of the substrate. On the Au(111) surface, for example, the presence of the surface state at −0.45 eV makes it difficult to identify the valence band position in many GNRs. On the other hand the signal from GNRs' valence or conduction bands could be too weak to be detected by STS due to the symmetry of the GNRs' orbitals and the LDOS decay above the GNRs' plane [45, 50].

The occupied frontier band of GNRs have been measured in several cases by angle resolved photoemission (ARPES), which provides access not only to the bands' onset energy, but to the whole *energy versus parallel momentum* dispersion of occupied GNRs states (as displayed in Fig. 4b) [34]. Being an ensemble averaging technique, domains of equally oriented GNRs are needed to get k-resolved ARPES data. This has been managed by growing the GNRs on vicinal surfaces that exhibit a periodic array of steps, which act as templates and drive the uniaxially oriented growth of GNRs along the terraces direction. Slight deviation between the value of the valence band maximum in flat and vicinal surfaces are found due to the different substrate's work functions.

3.2 Chemical Structure

X-ray photoemission spectroscopy (XPS) is the technique of choice to study the chemical composition of a material. In the case of GNRs, it has been used to shade light on the chemical mechanisms underlying the three basic steps in GNRs' formation by on-surface synthesis. After room temperature deposition of precursor molecules on the substrate, the shifts of the core levels (as Br3d and C1s) are measured as a function of increasing sample temperature in different experiments (see Fig. 5 for an example). This allows obtaining invaluable information about the whole reaction process including the step sequence, the halogen desorption from the surface and the associated threshold temperatures [25, 30, 58].

The chemical structure of GNRs has been imaged in UHV with STM and with atomic resolution by *non-contact atomic force microscopy* (nc-AFM). In molecular imaging, STM is sensitive to the density of states near the Fermi level, typically delocalized over the entire molecule. The correlation of bonding structure and STM contrast is thus not straightforward. However, simulations of STM images, reliably performed with DFT, are of great value for the interpretation of the STM data and finally allow identifying the structure and adsorption geometry of GNRs [18, 43, 46].

The AFM technique allows, under specific conditions, sensing the short range forces between the probe and the measured sample, whereby the signal becomes sensitive to the total charge distribution on the adsorbates (which is largest on atomic sites and along chemical bonds). Although not strictly required [59, 60], such imaging is most commonly performed at low temperature (~5 K), in

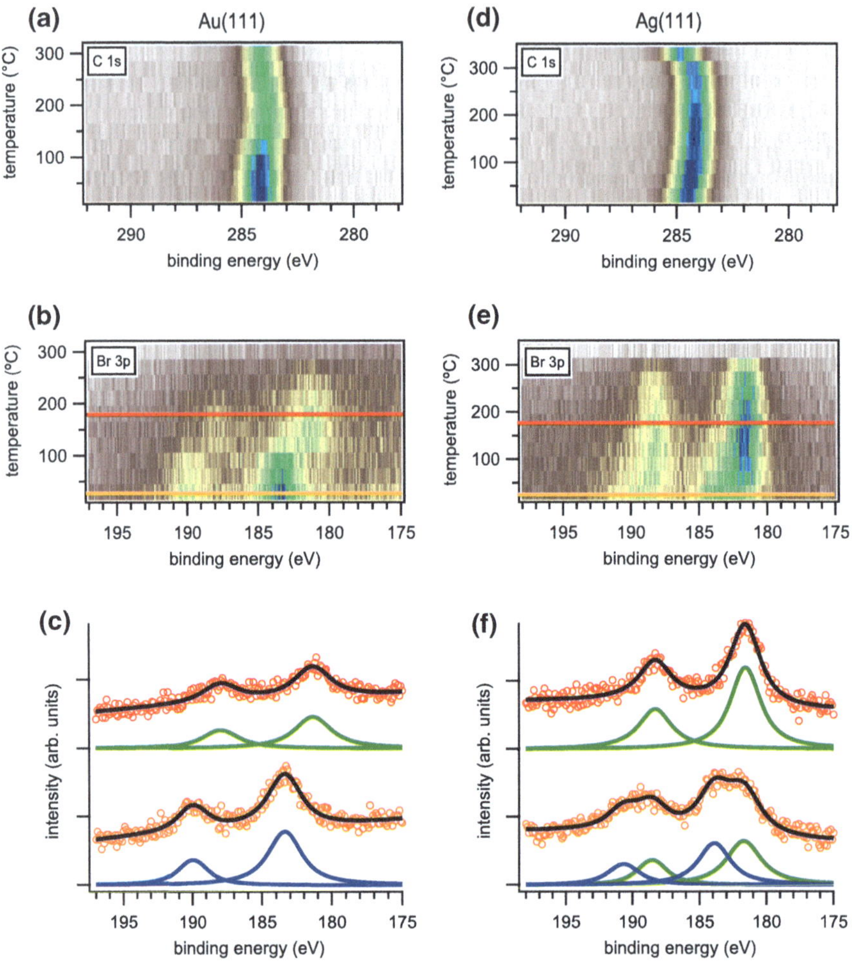

Fig. 5 Photoemission spectra of the C 1s core levels of 2,2′-dibromo-9,9′-bianthracene deposited on (**a**) Au(111) and (**d**) Ag(111) held at room temperature and their evolution as a function of sample annealing temperature. Similar measurements of the Br 3p core levels are shown in panels (**b**) and (**e**). Panels (**c**) and (**f**) depict Br 3p spectra, together with their associated fits (blue and green lines correspond to organic and metal-bound Br components, respectively), of two representative temperatures marked with the colored lines in (**b**) and (**e**), respectively. The spectra are shifted along the intensity axis for better comparison. Reprinted with permission from [25]

frequency modulation and non-contact modes, by keeping an inert probe (typically functionalized with a single CO molecule) close enough to the sample [61]. Doing so and tracking the changes in the frequency shift of the oscillating cantilever force sensor in a plane above the GNR, it has been possible to precisely resolve the bonding structure of pristine and doped GNRs (Fig. 3b–d, k) [40–42, 48], as well as defects [40] and atomically sharp junctions [29].

Besides its imaging capabilities, nc-AFM has been used also to proof fundamental properties of GNRs as their *structural superlubricity* on gold surfaces [62]. GNRs with lengths between 5 and 55 nm could be moved laterally by the force sensor tip with static friction forces between 2 and 200 pN. Such ultralow forces arising in the sliding motion of GNRs pave the way for creating frictionless coatings.

3.3 Detection of Vibrational Modes

The GNRs vibrational structure predicted by theoretical studies, based on first principle methods, is characterized by graphene-like and intrinsic modes [63]. The graphene E_{2g}-like (or G) mode is induced by the relative motion of neighboring atoms and tends to the value of 1580 cm^{-1} of the E_{2g} in graphene as the GNRs' width increases. A localized mode is found at 3000 cm^{-1} which is the typical vibration of the C–H bond. The most prominent peak in the low frequency range (200–800 cm^{-1}) is the *radial breathing mode* (RBLM) due to the outward motion of the nanoribbon's edge atoms by keeping the central atoms at rest (inset in Fig. 6b). This is the most representative vibrational mode of both armchair and zig-zag GNRs and since its frequency is roughly proportional to the inverse square root of the GNR's width, it can allow identifying the types of ribbons present on a surface.

Experimentally, the frequencies related to such vibrational modes of GNRs have been determined by Raman spectroscopy. Raman spectroscopy measures the inelastic scattering of photons by phonons, usually by means of a laser as photon source tuned in a spectral range between the near infrared to the near ultraviolet. Raman spectra measured on a layer of 7- and 9-aGNRs grown on Au(111) identified the different modes predicted by theory, as the RBLM mode at 396 cm^{-1} for 7-aGNRs and 312 cm^{-1} for 9-aGNRs, but also the D peak related to disorder in graphene (Fig. 6 a, b) [18, 45]. Raman peaks showed a strong dependence on the excitation source wavelength so that the RBLM of 9-aGNRs is depleted by incident green laser irradiation (2.33 eV) but enhanced by infrared light (1.58 eV), the opposite occurs for 7-aGNRs. According to tight binding calculations coherent radial-breathing-like phonons in GNRs are excited for photoexcitation near the optical adsorption edge [64]. In fact, in the case of 7-aGNRs, aligned uniaxially on Au(788), the size of the optical gap is 2.1 eV as determined by reflectance difference spectroscopy measurements (RDS) [56]. Such value is then close to the green light excitation while infrared lies within the gap.

Raman is a powerful technique for the characterization of GRNs under several aspects: (i) it does not need ultra-high-vacuum conditions to be operated; (ii) it can be used to detect GNRs vibrational fingerprints also when they are deposited or grown on insulating surfaces; (iii) it allows identifying the orientation of the GNRs edges and the GNRs width [65–68]; (iv) it helps to evaluate the quality of a GNRs

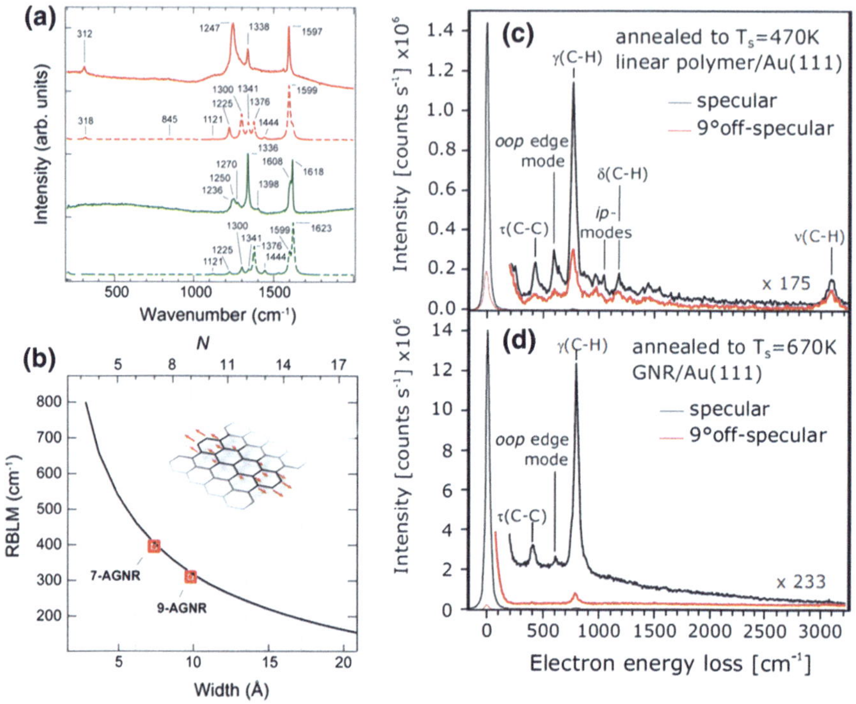

Fig. 6 **a**, **b** Raman characterization of 9-aGNRs grown on Au(111). **a** Experimental Raman spectra taken with two different excitation sources, 785 nm (solid red) and 532 nm (solid green) laser lines. Corresponding DFT based simulations done for photon energies of 1.58 eV (dashed red) and 2.33 eV (dashed green). **b** Theoretical dependence of the wavenumber of the RBML with the aGNRs width and experimental data for 7-aGNRs and 9-aGNRs. **c**, **d** Vibrational HREEL spectra corresponding to (**c**) linear polymer obtained after dehalogenation of DBBA at 470 K on Au(111) and (**d**) 7-aGNRs obtained after cyclodehydrogenation at 670 K. **a**, **b** Reprinted with permission from [45]. Copyright (2017) American Chemical Society. **c**, **d** Reprinted figure with permission from [52]. Copyright (2012) by the American Physical Society

sample by quantifying the defect density via D band [65]. Ideally it will also allow to analyze GNRs length distribution.

The "electron analogue" of Raman spectroscopy, naively, is high-resolution electron energy loss spectroscopy (HREELS). This technique uses the inelastic scattering of low energy electrons to measure the discrete vibrational energies of vibrational modes of adsorbates on a surface. Angle resolved HREELS has been used to follow and characterize the thermally activated steps of GNRs formation [52, 55]. For GNRs well defined changes in the vibrational spectrum from the polymeric phase to the aromatic structure have been measured. The quenching of vibrational modes with dipole moments laying within the phenyl rings plane demonstrated the planarization of all the phenyl rings after cyclodehydrogenation (Fig. 6c, d).

4 Tuning the Electronic Properties

Among the most attractive virtues of graphene nanoribbons is the remarkable dependence of their electronic properties on the detailed GNR structure. As a consequence, even minimal structural changes brought about in a controlled way through the synthesis process, allow tuning the nanoribbon's electronic properties over a wide range. In the following we review some of the different routes that have been proposed and explored to control the atomic structure of graphene nanoribbons, as well as their effect on critical electronic properties as for example the band gap, band dispersion or energy level alignment.

4.1 *Tuning Through Edge Orientation*

As readily discussed in the introduction in the light of Clar's theory, the edge orientation has a critical effect on the nanoribbon's electronic properties. On a more quantitative basis, we will now discuss some of the details in reference to the electronic properties of 2D graphene. Tight-binding calculations have been successfully used to describe with reasonable accuracy the electronic properties of 2D and nanostructured graphene. Electrons involved in gaphene's σ-bonds are more strongly bound and thus the main interest for optoelectronic processes arises from the delocalized π-electrons. Consequently, each atom can be simply characterized by a single $2p_z$ orbital. Applying the tight-binding Hamiltonian at the nearest neighbor level to the two-dimensional honeycomb structure of graphene with its two atoms per unit cell (each forming an equivalent hexagonal sublattice, Fig. 7), the following dispersion relation is obtained [69]:

$$E(k) = st\sqrt{3 + 2\cos\left(\frac{\sqrt{3}k_x a}{2} + \frac{k_y a}{2}\right) + 2\cos\left(\frac{\sqrt{3}k_x a}{2} - \frac{k_y a}{2}\right) + 2\cos(k_y a)} \quad (1)$$

where $s = \pm 1$ (-1 for the valence and +1 for the conduction band), t is the hopping integral and a is graphene's lattice constant. Near the Γ point, both valence and conduction bands depend quadratically on k_x and k_y. At the M points there is a saddle point in the energy dispersion. The most interesting point is around the K points, where the bands show a linear dispersion in so-called Dirac cones that display electron-hole symmetry (Fig. 7c) [69, 70]. That is, this first order tight-binding calculation already predicts the semimetallic, gapless nature of graphene, as well as the linear dispersion around the K points in a massless electron behavior.

Applying the same theoretical analysis to GNRs of arbitrary edge orientation, the following boundary condition is set: because H atoms along the edges do not contribute any π-electron, the GNR's π-electron wave functions on them vanish. In aGNRs, this boundary condition is fulfilled by eigenstates of graphene with

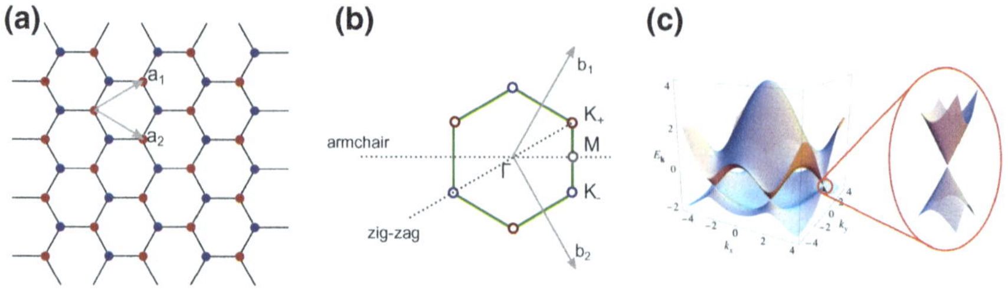

Fig. 7 **a** Graphene lattice and its primitive vectors a_1 and a_2. **b** Reciprocal space of Graphene and its reciprocal primitive vectors b_1 and b_2. The Dirac points are located at $K_{\pm}$. Dashed lines indicate the growth directions of armchair and zig-zag GNRs in the reciprocal space. **c** Band structure of graphene. The close up shows the linear behavior of the bands close to the Dirac points. The slope of these cones correspond to the Fermi velocity v_F. **c** Reprinted figure with permission from [70]. Copyright (2009) by the American Physical Society

perpendicular momenta $k_\perp$ that allow standing waves perpendicular to the aGNR axis. A detailed derivation is provided in reference [69], resulting in the discrete values

$$k_\perp = \frac{r}{N+1}\frac{2\pi}{a}, \quad r = 1, 2, 3, \ldots N \tag{2}$$

The associated aGNR's band structure is thus simply a combination of slices across graphene's band structure at N equidistant $k_\perp$ values. This is shown by way of example in Fig. 8 for 9-aGNRs. The frontier bands stem from the cut closest to

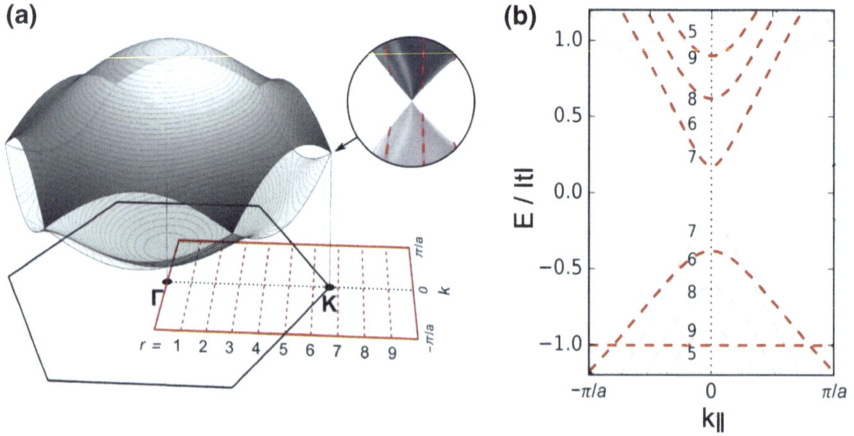

Fig. 8 **a** First Brillouin zone of graphene (black hexagon) and its π-band structure (black/gray surfaces) with inset displaying bands near the Dirac cone. Also shown are the first Brillouin zone (red rectangle) and cutting lines (red dashed lines) for infinite length 9-AGNRs. **b** Energy bands of empty and occupied states corresponding to cutting lines r = 5, 6,..., 9 in (**a**). Reprinted with permission from [45]. Copyright (2017) American Chemical Society

graphene's Dirac point at K. Depending on the nanoribbon's width, the properties can thus vary from a metallic to a semiconducting behavior. The former occurs with aGNRs of the N_a = 3p + 2 family, which always present one cut right through the K point, while the former applies to the other two subfamilies. At this point it is important to note, however, that this tight-binding derivation is based on equivalent carbon atoms throughout the nanoribbons. Taking into account that carbon atoms at the edges are saturated with hydrogen, such equivalence is lifted and the resulting edge distortion actually results in the opening of a band gap also for the 3p + 2 family [11, 71], as obtained also from more refined *ab initio* calculations [11, 71] and as readily proved experimentally with 5-aGNRs [43]. Further experimental examples reporting the electronic properties of different atomically precise aGNRs and confirming the great variety of band gaps as a function of their width will be discussed in the next section.

The case of zGNRs is more complicated. As derived in reference [69], the wave functions satisfying the boundary conditions have transverse momenta displaying a dependence on the parallel momentum such that

$$\sin(k_{\perp}N) + g_k \sin[k_{\perp}(N+1)] = 0, \quad g_k = 2\cos\left(\frac{k_{\parallel}}{2}\right). \tag{3}$$

That is, most of the bands can still be obtained from slices of graphene's band structure, but those slices are now curved, as displayed by way of example with the black solid lines in Fig. 9a for a zGNR with N_z = 4 (4-zGNR). It is immediately

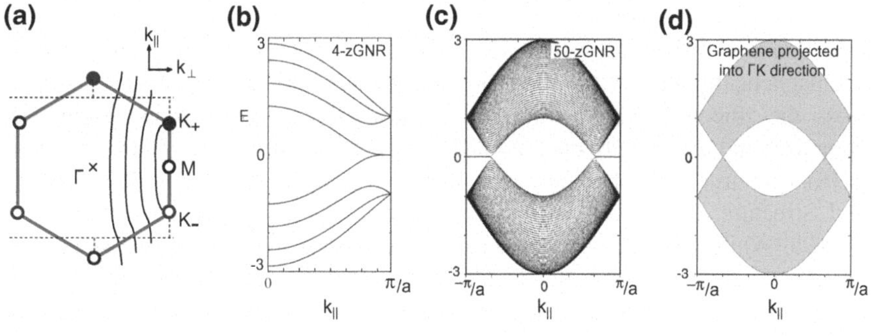

Fig. 9 **a** Reciprocal lattice of graphene showing the discrete values of $k_{\perp}$ for N = 4. The cuts are not linear, in contrast with aGNR. Also note the change in orientation of $k_{\perp}$ and $k_{\parallel}$ with respect to Fig. 8. The red dotted lines mark the regions where the additional edge states appear that cannot be traced back to the band structure of graphene. The reciprocal unit cell of a 4-zGNRs is also shown in black dashed lines. **b** Band structure of a 4-zGNR, revealing the degenerate edge states as $k_{\parallel}$ approaches π/a. **c** Band structure of a 50-zGNR revealing the flat bands of the edge states even more clearly. **d** Graphene's band structure projected into ΓK direction, showing the absence of the flat bands of zGNRs. π/a indicates the boundary of the first 1D Brillouin zone of the GNR (black dotted lines in (**a**)), where the zone-folding technique should be applied. **b** Reprinted figure with permission from [73]. Copyright (1996) by the American Physical Society. **c** and **d** Reprinted figure with permission from [72]. Copyright (1999) by the American Physical Society

obvious that under this tight-binding approximation zGNRs are gapless because one of the slices always goes through graphene's K point. This is also evident from Fig. 9b, which displays the band structure associated with these ribbons. It is interesting to see in Fig. 9b that valence and conduction band join into a degenerate flat band above a critical $k_{\|}$ value k_c. It coincides with the reciprocal space region where only $N - 1$ slices are present instead of N (Fig. 9a), because no $k_{\perp}$ are found as solutions to Eq. (3). In that region, the states giving rise to the flat bands can be understood as imaginary solutions to Eq. (3) at $k_{\perp} = 0$ and $k_{\perp} = \pi/a$, as displayed with dotted red lines in Fig. 9a. The wave functions associated to those solutions are localized at the nanoribbon edges and strongly decay toward the interior of the zGNRs [69, 72, 73]. It is best observed that these edge states cannot be derived from slices in graphene's band structure comparing its projection along the ΓK direction (Fig. 9d) with the band structure of wide z-GNRs (Fig. 9c), which has a large density of slices and could thus be expected to be comparable to that of graphene. Such comparison immediately makes obvious that the flat bands associated to the edge states are not present in 2D-graphene, but that they are a property of zGNRs intrinsic to their zigzag edges.

The strong density of states at the Fermi level caused by the edge states should induce either a lattice distortion via electron-phonon interactions or magnetic polarization via electron-electron interactions [69]. In the case of GNRs the latter is favored and drives the spin polarization of the edge states. Edge sites along a zGNR are made up by atoms of the same sublattice on one side and of the other sublattice on the opposite side. Thus, the edge state being nonzero only on one of the two sublattices at each edge, the magnetic moment selectively increases on that sublattice with a ferromagnetic spin configuration and decays toward the ribbon interior. The opposite edge sites belong to the other sublattice and display magnetization with opposite spin. The edge states are thus not only spin polarized ferromagnetically along each edge, but additionally antiferromagnetically coupled across the ribbon [69]. Notably, the electron-electron interactions driving the magnetization concurrently open a band gap Δ^0 between the edge states, deterring zGNRs from truely being gapless structures. This can be observed in the calculated band structure of zGNRs with and without electron-electron interactions in Fig. 10b. While tight-binding density of states has only one van Hove singularity related to the edge state's flat bands at E = 0, including the electron-electron interaction U term in a mean field Hubbard model results in two pairs of density of states peaks split by Δ^0 and Δ^1 (Fig. 10b). In particular, Δ^0 relates to the antiferromagnetic correlation between the two edges, while the larger splitting Δ^1 relates to the ferromagnetic correlation between the spins of the most strongly localized states at $k_{\|} = \pi/a$ in the same edge. It should be mentioned here that in the nearest-neighbor approximation the electron-hole symmetry remains unchanged, but it is lifted as a next-nearest-neighbor hopping term is included in the Hubbard model [74]. The opening of a bandgap has been confirmed on atomically precise 6-zGNRs synthesized on Au(111) that were subsequently transferred onto NaCl bilayer islands, indeed revealing under such circumstances relatively large splitting values of $\Delta^0 = 1.5$ eV and $\Delta^1 = 1.9$ eV [42]. Although with somewhat disparate

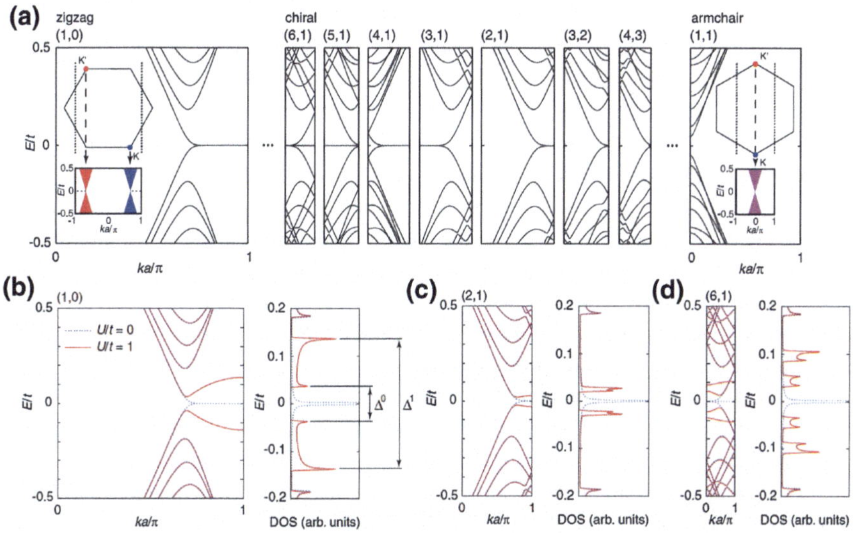

Fig. 10 **a** Band dispersion obtained from tight binding calculations of GNRs of different orientation, from zigzag, through chiral, to armchair. The closer to the armchair direction, the lower the density of states of the flat bands at E = 0, until a band gap opens for armchair ribbons. Similar calculations for zigzag and chiral ribbons are shown in (**b–d**) upon addition of electron-electron interactions, revealing the opening of a bandgap also for these orientations. Reprinted figure with permission from [76]. Copyright (2011) by the American Physical Society

values, similar conclusions have been drawn also from experiments on zGNRs with worse defined and less perfect edges obtained by differen methods like hydrogen etching of 2D-graphene [13] or scanning probe lithography [15].

From a structural point of view, chiral ribbons combine zigzag and armchair segments. From an electronic point of view the zigzag segments are most determining, providing chiral ribbons with low band gaps and spin-polarized edge states. However, these properties vary with the chiral angle (θ, Fig. 2) in a relatively smooth way in between the two limits of zigzag (θ = 0°) and armchair graphene nanoribbons (θ = 30°). In the infinite-width limit, the band structures of chiral GNRs can be obtained from a continuous rotation of the band structure of graphene and the average density of edge states per edge length ρ_0 at the Fermi level is given by [75]

$$\rho_0(\theta) = \frac{2}{3a} cos\left(\theta + \frac{\pi}{3}\right). \tag{4}$$

As discussed above, the large ρ_0 is responsible for the magnetization of GNRs. And as evidenced in Eq. (4), ρ_0 is highest for lowest chiral angles and vanishes as the armchair angle is approached [74, 76]. This has the important implication that the magnetic moment per length unit increases as the chiral angle is reduced. Tight-binding calculations of chiral GNRs of finite width as a function of the chiral angle confirm this picture [76]. For GNRs with edge orientations close to aGNRs the edge states associated to the flat bands at the Fermi level are almost completely suppressed (Fig. 10a). Instead, in ribbons with edge orientation close to a zGNR the flat edge state band extends over the whole 1D Brillouin zone and becomes multiple degenerate due to band folding (Fig. 10a) [76]. As in zGNRs, electron-electron interactions split the degeneracy of the flat bands also in chiral GNRs (Fig. 10c, d). As discussed earlier, each pair of peaks arising in the density of states has a different nature, whereby Δ^0 and Δ^1 relate to the magnetic coupling across the nanoribbon and along its edge, respectively. As can thus be intuitively expected, for a given GNR width Δ^0 hardly changes with the chiral angle. Instead, Δ^1 decreases in a similar way to the net magnetic moment on each edge, as the chiral angle increases and approaches the armchair orientation [76].

Lastly, it should be remarked that the density of edge states as a function of chiral angle depends at the same time on the nanoribbon width. Beyond the visualization through the Clar sextet theory outlined in the introduction, this effect has been computed calculating the number of states at $E \approx 0$, normalized to the total number of states in the GNR [73, 77]. Calculations reveal that there is a critical width below which there is no notable edge state density. Importantly, that critical width decreases with decreasing chiral angle (as the edge orientation approaches that of zGNRs) [73, 77]. In other words, for a given width there is a minimum number of zigzag sites (in relation to armchair sites) along the ribbon's edges to host a notable edge state density. The required ratio of zigzag to armchair sites becomes higher for narrower GNRs.

It is important to observe at this point that the predictions described above for the electronic structure of chiral GNRs still lack experimental proof on atomically precise chiral GNRs. To date, only the precise synthesis of narrow (3,1)-GNRs has been published [25, 31, 32] and only few details of its electronic properties are yet known [65]. There are few other works reporting the electronic structure of chiral graphene nanoribbons with worse defined edges obtained from unzipped carbon nanotubes [12] or chemical vapor deposition [14]. In either case the opening of band gaps through the magnetism-driven splitting of the edge states has been confirmed by scanning tunneling spectroscopy measurements.

4.2 Tuning Through Width Control

Besides the edge orientation, the electronic structure of graphene nanoribbons also displays a marked dependence on their width. It is most straightforward to understand this dependence for armchair graphene nanoribbons, which is furthermore the family with the largest number (albeit still very limited) of successfully synthesized examples on which to experimentally corroborate such behavior. We remind that the band structure of N-aGNRs can be pictured as a combination of 1D-cuts across graphene's band structure at N equidistant $k_{\perp}$ values, namely at

$k_{\perp} = \frac{r}{N+1}\frac{2\pi}{a}$, r being an integer between 1 and N. Keeping in mind that graphene's vanishing band gap and the massless electron behavior occur right at the K-point, the closer those 1D-cuts get to the K-point, the lower the aGNR's band gap and the effective masses of its frontier bands will be. That is, some of the aGNRs' attributes can be readily understood from simple geometric considerations: along the 1D-projection, K and K' are located at $k_{\perp} = \frac{1}{3}\frac{2\pi}{a}$ and $k_{\perp} = \frac{2}{3}\frac{2\pi}{a}$ (Fig. 8a). It is the 3 in the denominator that accounts for the presence of three aGNR families, depending on whether their width in terms of dimer lines *N* is equal to 3p, 3p + 1 or 3p + 2 (p being an integer). There is always a cut going directly through the K-point for the 3p + 2 family. This family thus has the lowest band gap and effective mass, these values being higher on either of the other two families for which no cuts across the K-point are present. However, the wider the GNR, the smaller the spacing between the cuts and the closer they get to the K-point. This determines the inverse proportionality of band gap and effective mass with nanoribbon width within each of the families, although if the width increases atom by atom it would imply changes between families that could result in an increased band gap and effective mass as well.

It is worth remarking here that, under the assumption of equivalent carbon atoms throughout the GNR, tight binding calculations predict the 3p and 3p + 1 families to follow a very similar trend. However, this assumption is inaccurate the closer the atoms are to the edge, most evidently for the hydrogen saturated carbon atoms. Taking into consideration bond length distortions and the associated changes in the hopping constant, important changes appear in the resulting band gap calculations, as are a clear splitting of the 3p and 3p + 1 bandgap trends, as well as the opening of a bandgap for the 3p + 2 family. These findings coincide with more refined *ab initio* calculations revealing the band gaps of the three different families to order according to E_g (3p + 1) > E_g (3p) > E_g (3p + 2) [11, 71].

Besides tight-binding [11, 69, 71], several other methods have been used to quantitatively describe the width-dependent band gap in aGNRs, like for example extended Hückel theory (EHT) [78], density functional theory (DFT) [11, 79], or the GW approximation [79]. At this point, it is important to remind that, to date, all the experimentally reported band gaps of atomically precise GNRs are from surface-supported ribbons. While less important on weakly interacting substrates, if more reactive surfaces are used or reactive functional groups are added to the GNR structure, strong GNR/substrate hybridizations may occur. Under such circumstances the band gap will be substantially affected and the GNR's electronic properties may bear little resemblance with those of free-standing GNRs. In addition, the band gaps of GNRs have been most commonly characterized by ionizing techniques such as STS [34, 43, 45, 46] or a photoemission/inverse photoemission combination [35]. This implies on the one hand that it is not ground states but quasiparticle energies which are probed, and on the other hand that the surrounding dielectric medium (both the GNR itself as well as the substrate) can have a substantial effect on those state's energies through polarization-induced screening. Among the various theoretical methods mentioned above, the intrinsically present

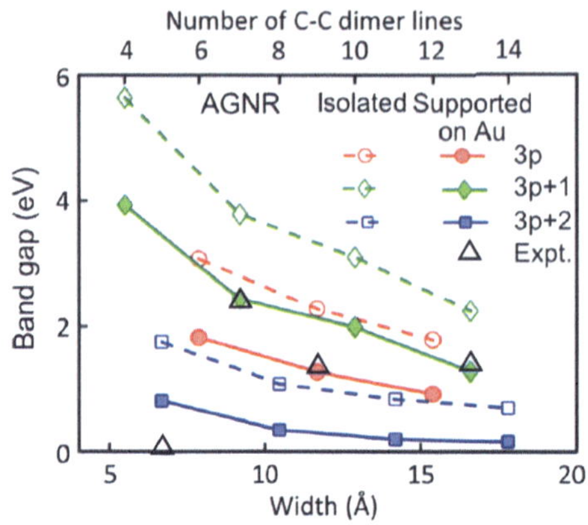

Fig. 11 Calculated bandgap energies for isolated and Au-supported aGNRs of different width for each of the three sub-families (3p, 3p + 1, 3p + 2). Calculations are done with the GW method, further adding substrate screening for the supported ribbons. Superimposed we find the currently available experimental values of Au-supported aGNRs obtained by scanning tunneling spectroscopy, providing good agreement. Adapted with permission from [80]. Copyright (2016) American Chemical Society

electron-electron interactions are best taken into account in the GW approximation. It is, however, computationally very intensive and has thus only been applied for free-standing GNRs. As a result, calculated GW band gaps still overestimate the experimental ones [34]. Best results have been obtained when combining the GW approximation with a semiclassical image-charge model to account for substrate screening [80]. The screening-derived band gap renormalization varies with the substrate and the GNR, being in the 1 eV range for aGNRs on Au (Fig. 11) and slightly lower for aGNRs on a NaCl bilayer on Au [80]. Its combination with the calculated GW band gaps ultimately provides a remarkably good agreement with currently available experimental values for 5 [43], 7 [50], 9 [45] and 13-aGNRS [46] (Fig. 11).

The effective mass is another important figure whose relevance stems from its inverse relation to the charge carrier mobility [81], a critical parameter for potential applications. However, only few studies have characterized effective masses in GNRs. From the experimental side, there are measurements for 7-aGNRs and 9-aGNRs. For the valence band of the former, very disparate values have been reported, ranging from ~0.21 m_e (measured by ARPES [34, 82]) to 0.41 ± 0.08 m_e (measured by STS [50]) and even 1.37 m_e (measured by two photon photoemission [53]). For the valence band of the latter, better matching values of 0.09 ± 0.02 m_e and 0.12 ± 0.03 m_e have been reported from ARPES and STS measurements, respectively [45]. Interestingly, there have been different works proposing analytical relations between the effective mass and other GNR parameters. For example, Arora et al. proposed the effective mass to be linearly dependent on the GNR band gap in a similar way as that claimed for carbon nanotubes:

Table 1 Summary of the calculated and experimentally obtained values for the effective mass of aGNR's valence bands given in electron mass (m_e) units

	Arora et al. [83]		Raza et al. [78]	Experiment		
Width (N)	From calc. E_g	From exp. E_g		ARPES	STS	2PPE
7	0.154	0.21	0.21	0.21 [34] 0.23 [82]	0.41 ± 0.08 [50]	1.37 [53]
9	0.074	0.12	0.09	0.09 ± 0.02 [45]	0.12 ± 0.03 [45]	–

Experimental E_g values for the m* calculation with Arora's relation have been taken from Refs. [45, 50] and for 7-aGNR and 9-aGNR, respectively.

$m^* = (E_g/11.37)\ m_e$, with E_g given in eV [83]. In turn, Raza et al. related it to the aGNR width W through $m^* = (0.091/W)\ m_e$ for the 3p family, $m^* = (0.160/W)\ m_e$ for the 3p + 1 family and $m^* = (0.005/W)\ m_e$ for the 3p − 1 family (W being the distance, in nanometers, between the C atoms on either edge of the ribbon, based on a C–C bond length of 1.44 Å) [78]. In the following we provide Table 1 comparing the experimentally reported effective masses to the values predicted by either of the models mentioned above. It can be seen that there is an excellent agreement between predictions and experiments, best of all when taking the experimental values obtained from ARPES measurements.

The width dependence of the electronic properties of GNRs does not only affect armchair oriented ribbons, but also ribbons with other orientations. As described in the previous section, the edge state magnetization in zigzag and chiral GNRs drives the opening of a band gap Δ^0 and an additionally split resonance Δ^1 related to the inter-edge and intra-edge magnetic coupling, respectively (Fig. 10) [69, 74, 76]. Thus, as expected from their respective nature, Δ^1 hardly varies with the ribbon's width. Instead, Δ^0 decreases with growing width, although in the zGNR case in a monotonic way and faster than for aGNRs [74, 75] [76]. This qualitative behavior of Δ^0 and Δ^1 is at least what results from calculations on free standing zGNRs. However, while the same trend holds for Δ^0 also in calculations of gold-supported ribbons, in that case Δ^1 shows a surprising increase with ribbon width [80]. Experimental confirmation on atomically precise zGNRs of different width is still missing, since only 6-zGNRs have been synthesized to date with atomic precision [42]. Nevertheless, closely related measurements have been performed on the edge states localized at the zigzag ends of 7-aGNRs. For increasing 7-aGNR length, which increases the distance between their ends (in analogy to a width increase in zGNRs), the gap between edge states hardly changes. Because the minimum separation measured in that study is ~3 nm, this observation is in agreement with a rapidly converging band gap at even smaller length scales, further supported with calculations [84]. There are also closely related measurement on zGNRs obtained from scanning probe lithography and hydrogen etching. The latter shows band gaps

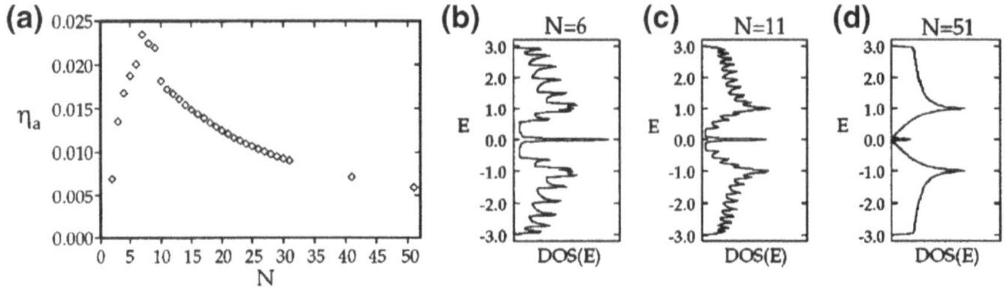

Fig. 12 **a** Flatness index η_a, defined as the ratio between the number of states with E $\approx$ 0 and the total number of states in the ribbon, calculated as a function of zGNR width N_z. Density of states calculated for zGNR widths of N_z = 6 (**b**), N_z = 11 (**c**) and N_z = 51 (**d**), evidencing the decreasing significance of the edge states in the overall density of states for increasingly wide ribbons. Reprinted figure with permission from [73]. Copyright (1996) by the American Physical Society

monotonously decreasing with increasing zGNR width until saturating at widths around 4 nm, in reasonable agreement with the predictions described above [13]. The former shows a smoothly decreasing band gap with increasing width until a threshold value ($\sim$7 nm) above which the gap suddenly drops to zero, proposed to be related to a sharp antiferromagnetic semiconductor to ferromagnetic metal transition [15]. Nevertheless, both studies are based on zGNRs with poorly controlled and defined structures as compared to the atomically precise nanoribbons that are the focus of this chapter, and thus possibly affected by the additional disorder.

Beyond the gap, also the edge state magnetization, which is one of the most coveted electronic properties of zGNRs, varies with GNR width. In particular, it increases as the width grows, rapidly reaching its saturation value at widths around N_z = 10 [69, 74]. However, it is important to consider also the relative weight of those magnetic edge states on the total density of states of the nanoribbon. Initially they increase with increasing N_z, but after a peak around N_z = 7 it ends up decreasing at a rate approximately proportional to 1/N_z (Fig. 12a) [73]. This effect shows how the significance of the special edge state disappears as the ribbon grows infinitely wide and becomes 2D graphene (Fig. 12b–d).

Chiral ribbons also display a dependence of their electronic properties with their width. By way of example, and as readily discussed briefly in the previous section, the edge state density in chiral GNRs is absent for widths under a critical value [73, 77]. That critical width depends on the chiral angle and decreases as the edge orientation gets closer to the zigzag direction. Regarding the band gap, it does not decrease monotonically with increasing width as in zGNRs, but displays a periodic modulation on top of its overall reduction, reminiscent of aGNRs [77, 85]. However, it is worth mentioning that this overall band gap reduction is faster in chiral ribbons than in aGNRs. Describing the band gap dependence as $E_g \propto 1/N^{\beta}$, β increases from 1 for aGNRs, to larger values as the chiral angle is reduced (e.g. β = 1.431 for θ = 23.41°; β = 2.908 for θ = 16.1°; or β = 3.758 for θ = 12.73°) [77]. In contrast, the overimposed band gap modulation becomes less pronounced for lower chiral angles [77, 85].

Lastly we would like to remind that the discussion above has been made considering quasi-infinite ribbons on which only the width is changed. However, atomically precise GNRs obtained by on-surface synthesis are finite, typically with lengths in the range from few nanometers to some tens of nanometers [25, 35]. As for other conjugated molecules or polymers, their band gap scales inversely proportional to their length, saturating towards the "infinite" limit [9, 84]. Thus, although for most applications the growth of GNRs with maximized length is desirable, the growth of monodisperse GNRs of limited length could be seen as an alternative way to tune and control their corresponding band gaps.

4.3 Tuning Through Doping

To understand which effect might doping bring to semiconducting GNRs one can look to the effect of impurities on semiconductors [86, 87]. Three scenarios might be naively distinguished in: (i) the formation of impurity levels in the semiconductor's bandgap; (ii) the appearance of an impurity band; (iii) a rigid shift of the whole band structure or modification of the bandgap. In the first case, at *light* impurities concentrations ($\approx 10^{16}$ atoms cm^{-3}), n-type (or donor) impurities give rise to energy levels below the semiconductor conduction band while p-type impurities yield electron acceptor levels above the valence band. Doping influences the position of the Fermi level, that for n-type moves linearly towards the minimum of the conduction band with exponential increase of the doping concentration. In the second case at *moderate* impurity concentrations ($\approx 10^{18}$ atoms cm^{-3} for Si), neighboring impurities are so close that their wave functions overlap sufficiently to give rise to an impurity band. In *heavily* doped semiconductors ($\approx 2 \times 10^{22}$ cm^{-3} i.e. 3% for Si) the ionization energy of the impurity atoms falls to zero, i.e. the impurity band merges with the conduction (for donors) or valence band (for acceptors) and only one band is formed. In other cases of heavy doping, as in disordered alloys, the band gap can be changed, for example linearly narrowed by adding Al atoms in $Ga_{1-x}Al_xAs$.

For carbon-based materials two doping schemes have been used: (i) *electrical doping* which does not alter the lattice structure or chemical composition of the material, as by adsorption of a gas or a metal; (ii) *chemical doping* which can be achieved by substitution of carbon (C) atoms with heteroatoms like nitrogen (N), boron (B), silicon or sulfur (S). The first type of doping has been reported for Li-doped 7-aGNRs [82]. Heavy doping of 7-aGNRs with a monolayer of Li lowers the valence band by 1.33 eV and brings the conduction band onset below the Fermi level, thereby causing a semiconductor-to-metal transition of the ribbons. Highest doping levels (reached upon saturation at around two monolayers of Li) reveal a transfer of up to 0.05 electrons per carbon atom. Importantly, such doping is accompanied by a quasiparticle bandgap renormalization from 2.4 to 2.07 eV, as well as by a dramatic increase of the conduction band's effective mass.

The second type of doping has been successfully demonstrated on atomically precise GNRs by means of chemically substituted molecular precursors. The inclusion of N, S or B atoms in the GNR matrix corresponds to a heavy doping level of several at.% but the resulting electronic structure of GNRs cannot be directly related to dopant concentration [88]. In fact, extra-electrons may remain in molecular orbitals localized around the dopant heteroatom instead of extending into the GNR π orbitals [10]. The appropriate design of reactants has allowed to precisely introduce dopant heteroatoms either on GNRs edges [36, 55, 89, 90] or within their backbone [48, 91]. Ultimately it is the precise combination of heteroatomic species and their localization which defines their effect on electronic properties like band gap [10] or energy level alignment.

In the first case two different situations have been realized. The substitution of C–H groups with N atoms in the edges of chevron type GNRs leads only to a rigid downshift of the conduction and valence band energies by 0.1–0.13 eV per N atom per precursor molecule unit, thus making it a stronger electron acceptor (n-doping of the ribbon) [55, 89, 90]. The decrease of the position of the valence band by increasing the number of N dopants in the precursor molecule has been also characterized with ARPES, STS and DFT for 3-aGNRs grown on stepped Au(788) as shown in Fig. 13a–c [36]. The electron lone pair of those nitrogen atoms being not in conjugation with the GNR π-system does not further affect the density of states of the frontier bands, leaving critical parameters like the band gap or the band's effective mass unchanged.

On the contrary the substitution of $(C\text{–}H)_2$ groups by S atoms in 13-aGNRs' edges allows hybridization of one of the heteroatom's lone pairs with the delocalized π-bands. Thus, as predicted by ab initio simulation for free standing ribbons, the energy gap decreases 140 meV with respect to undoped ribbons [49]. However, dramatic changes on the energy level alignment are not induced by S atoms because their electronegativity difference with respect to C is smaller than in the case of N atoms.

Finally, B impurities have been successfully introduced in the 7-aGNRs' backbone, allowing a full conjugation of empty B p_z orbitals with the ribbon's π-system [48, 91]. According to calculations, B-doping induces the formation of a new acceptor band lying only 0.8 eV above the valence band. As a result, the band gap is strongly decreased with respect to that of pristine 7-aGNR (displayed in Fig. 13d, e). Experimentally, the acceptor band and its distribution along the B-doped 7-aGNR backbone has been measured by STS, showing good agreement with calculations (Fig. 13f, g). However, the position of the valence band (and consequently the experimentally determined band gap) has not been reported yet.

4.4 Tuning Through Strain

The dependence of the electronic structure of GNRs on the atomic structure of the ribbon orientation, the ribbon width and doping have been experimentally

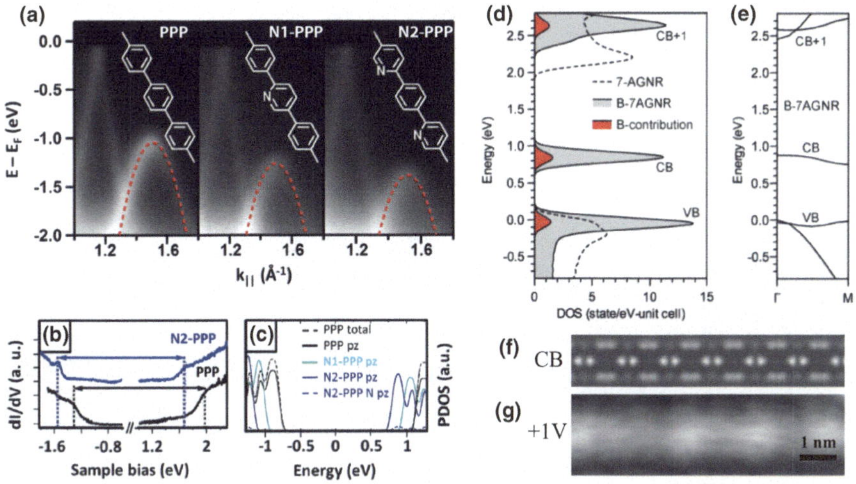

Fig. 13 **a** ARPES spectra of pristine and nitrogen doped poly-para-phenylene (PPP) polymers taken at $k_{||}$ close to the valence band maximum. Red lines correspond to parabolic fits of the GNR's bands revealing an unchanged effective mass of 0.19 m_0. **b** Differential conductance spectra performed on PPP (black) and N2-PPP (blue), displaying the rigid shift of valence and conduction bands with N-doping. **c** Computed projected densities of states (PDOS) of the different polymers, confirming again the rigid shift of valence and conduction bands upon N-doping and underlining the low contribution of the N atoms to the p_z states forming the frontier bands. **d** Calculated total density of states (DOS) for pristine 7-aGNRs (dotted line), for B-doped 7-aGNRs (gray) and contribution from B-atoms to the DOS (red) using the GW approximation and including Au(111) substrate screening. **e** Calculated quasiparticle band structure of B-7aGNRs. **f** Calculated local density of states (LDOS) map of states at the conduction band edge 4 Å above the borylated ribbon. **g** Differential conductance map of B-7aGNRs taken at 1 V. **a–c** Adapted with permission from [36]. **d–g** Adapted with permission from [91]. Copyright (2015) American Chemical Society

demonstrated by on surface synthesized GNRs. Strain engineering in semiconducting two dimensional materials offers the possibility to control their optical and electronic properties [92]. Thus another route to tune the band-gap of GNRs is to exploit strain. To understand the effect of strain on GNRs, it is necessary first to revise what occurs on strained graphene. Two types of strain are usually considered: uniaxial and shear (Fig. 14a, b). When graphene is stretched in one direction, it will shrink in the perpendicular one [93]. Tight binding and first principle calculations demonstrated that the effect of a uniform strain applied to the atomic lattice of graphene is to drive in the reciprocal space the position of the Dirac cone crossing (E_D) away from the K (K') points while maintaining the cone-like energy dispersion. This E_D will follow a path that is different depending on the type of strain applied (shear or uniaxial along armachair or zigzag directions).

For aGNR shear strain modifies only slightly the band structure but uniaxial strain has a relevant effect. The one dimensional Brillouin zone of aGNRs is defined by electronic states with allowed k values that lie on parallel lines ($k = r\pi/N + 1$

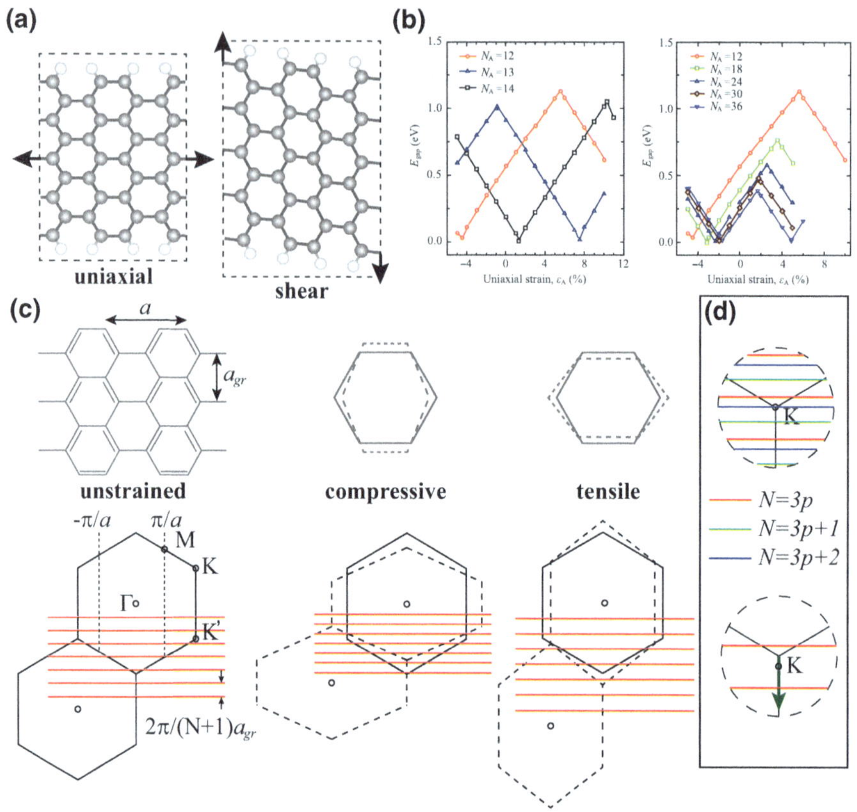

Fig. 14 **a** Schematic view of an aGNR under uniaxial and shear strain. **b** Calculated energy gap of aGNRs as a function of an uniaxial strain ε_A on the left, for three ribbons belonging to the three families $N = 3p$, $3p + 1$ and $3p + 2$ and on the right, for different ribbons of the $N = 3p$ type. **c** Schematic illustration of the effects of strain that induce (i) the deformation of the graphene Brillouin zone and motion of the K and K' points, (ii) a change of the distance between k-lines for allowed aGNRs electronic states and (iii) a different crossing of k-lines with the K and K' points. **d** Behavior of the allowed k-lines near the K point for the three aGNRs families. The green arrow indicates the shift of the Fermi point under uniaxial strain. Adapted with permission from [93]

with $r = 1, 2, \ldots, N + 1$) (Fig. 14c). The three families of aGNRs characterized by their different width ($N = 3p$, $3p + 1$, $3p + 2$; p being an integer) have different conditions for the crossing of their allowed k-lines with the K (K') points of the Brillouin zone (Fig. 14d). When uniaxial strain is applied E_D moves away from K (K') along the strained Brillouin zone in a direction perpendicular to the k-lines (Fig. 14c). In this way when E_D arrives in the middle of two lines, the energy gap will be maximum; when E_D coincides with a k-line, then the gap will close. This finding explains why the energy gap of aGNR is modified in a periodic way with a zigzag pattern which is different for each family (Fig. 14b). Moreover, within each family, the wider the ribbon is, the closer the k-lines are. Thus, the maximum energy gap value will diminish.

For the case of zigzag GNRs, strain can slightly modify the band structure only if spin-polarization is taken into account. In this case uniaxial strain will increase the bandgap but shear strain will not have much influence.

To conclude, as observed in the case of graphene, GNR growth on substrates which are not lattice-matched might induce strain that eventually is released by buckling, fringes, ripples or blisters formation. Interface strain induced on GNRs synthesized on a surface has not been observed experimentally yet, but it might occur by increasing the interaction of the GNR with its support.

4.5 Tuning Through Heterostructure Formation

Heterostructure formation has been successfully used with inorganic semiconductors to modulate their electronic properties, whereby sophisticated devices have been realized [94, 95]. In analogy, the same concept has been proposed to be feasible with GNRs [96, 97], for which one may combine segments with different properties like those described above (edge orientation, width, doping, strain), or also other parameters like edge terminations or edge structures. Experimentally, some examples have been readily demonstrated, creating atomically sharp and precise junctions by combination of different reactants (Fig. 15) [89, 98]. An alternative way to create randomly distributed GNR heterostructures is by the use of only one type of reactant that can, however, react in two different ways [99].

First to be realized was a type II heterojunction in which pure hydrocarbon reactants were combined with doped reactants that included nitrogen heteroatoms (Fig. 15b) [89]. As described in Sect. 4.3 and graphically displayed in Fig. 15b, the substitution along the GNR edges of C–H by N has little impact on the electronic band gap. In turn, nitrogen's more attractive core potential lowers the onset energies of valence and conduction band and turns the N-doped GNRs into n-type semiconductors. On the other hand, due to the particular GNR and substrate combination, undoped GNRs on Au are p-doped, displaying their valence band close to the Fermi level. Combination of pristine and N-doped segments into the same nanoribbon thus creates sharp p-n heterojunctions, with a band offset of around 0.5 eV for the heterojunction displayed in Fig. 15b. Because the band bending occurs over a distance in the order of 2 nm, the resulting electric field at the interface is extremely high (2×10^8 V/m), making these heterostructures highly promising for electronic devices based on p-n junctions [89].

Type I heterojunctions have been synthesized combining precursors that lead to differently wide GNRs [98]. The resulting structure is thus a width-modulated GNR as shown in Fig. 15a, with segments of 7-aGNRs and of 13-aGNRs. The latter has a larger band gap than the former, resulting in the straddling gap evolution that characterizes type I heterojunctions. Because the 7-aGNR segments serve as energy barrier for charge carriers localized along the 13-aGNR segment, the formation of

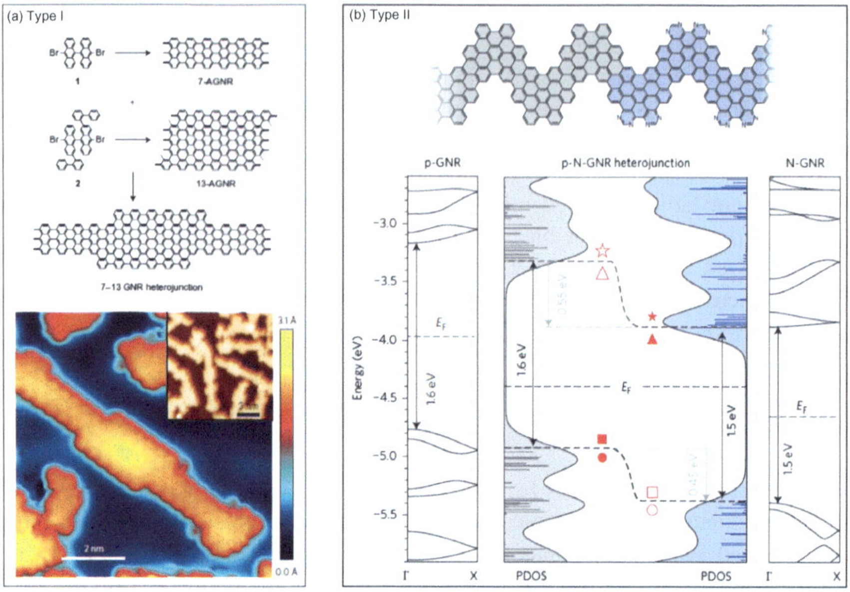

Fig. 15 **a** Schematic representation of the precursors leading to differently wide GNRs and an associated heterostructure, whose staggered bandgap can be classified as a type I heterojunction. Below, constant current STM images display the experimental realization of those heterostructures. **b** Schematic representation of the heterostructure resulting from a combination of pure hydrocarbon and N-doped precursors. Below the calculated electronic properties are displayed: band structure of pure p-GNRs (undoped precursors), of n-GNR (N-doped precursors), and of the density of states of the heterojunction displaying the roughly unchanged bandgap across the interface and the type II heterojunction energy alignment. **a** Reprinted with permission from [98]. **b** Reprinted by permission from Macmillan Publishers Ltd.: Nature Nanotechnology [121], copyright (2014)

quantum well states can be seen for short 13-aGNR segments surrounded by the narrower 7-aGNR, establishing an ideal framework for potential band gap engineering [97]. Furthermore, quantum dots could potentially be used as active components inside GNRs.

An alternative way for the creation of quantum dots embedded inside 7-aGNRs has been demonstrated by mixing pristine 7-aGNR with a small amount of boron doped 7-aGNRs precursors (Fig. 16a) [100]. The pristine regions in these hybrid ribbons preserve the electronic structure of 7-aGNR, while the borylated segments lack an energy level aligned with the pristine VB. As a result, the VB electrons on the pristine segments become confined by the boron atom pairs, since the VB ends abruptly over the borylated regions (Fig. 16b).

The calculated transmission function (Fig. 16c) for these free-standing hybrid 7-aGNR, that is, the transmission of electrons from one side of the ribbon to the other, shows that the boron pairs are very efficient reflectors and that the VB-1 acts as a transmission channel even though the VB is confined (Fig. 16d). This actually

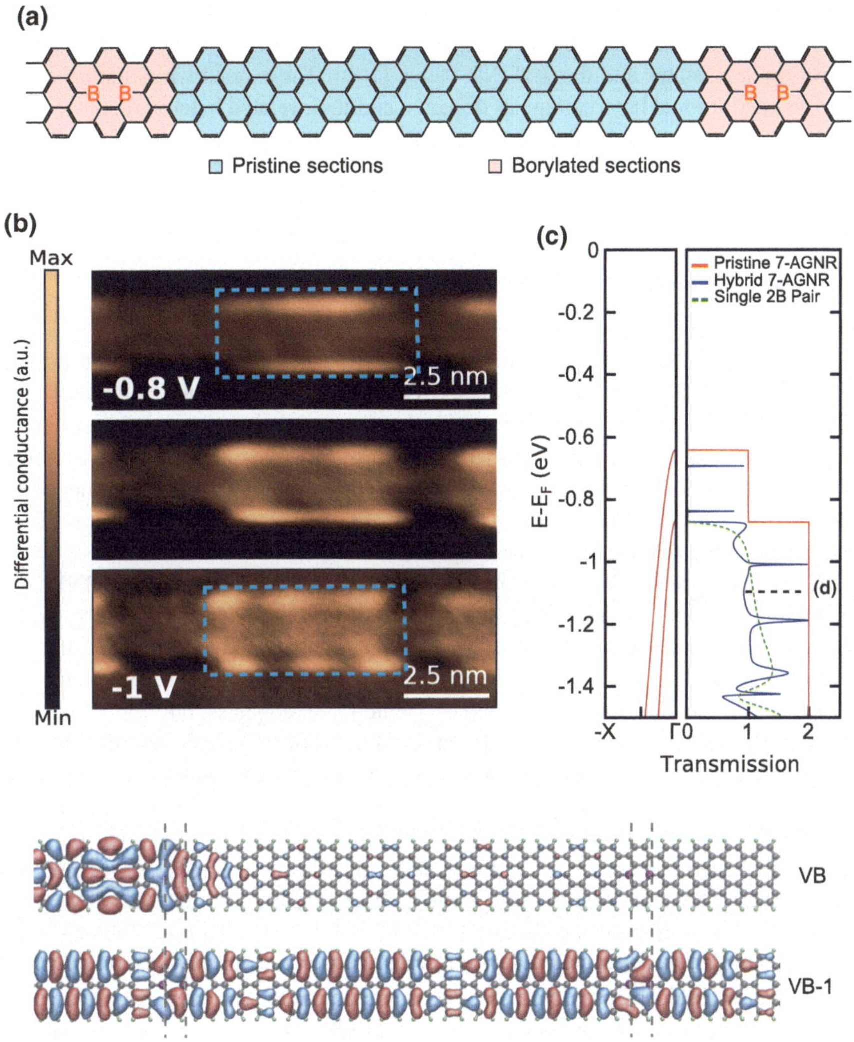

Fig. 16 **a** Schematic representation of the hybrid 7-aGNR, consisting in extended pristine regions and disperse borylated segments. **b** dI/dV maps of a pristine region (dashed blue square) enclosed between two borylated segments. The maps show an increasing number of modes in the conductance with increasing negative bias, fingerprint of the VB confinement. **c** Band structure of a pristine 7-aGNR (left) and transmission function (right) of a 7-aGNR (red) and a hybrid 7-aGNR (blue). The sharp peaks of transmission are related to the quantum well levels caused by the VB confinement. The step-like increase of transmission above the VB-1 onset reveals that the VB-1 is not confined. **d** Eigenchannel wave functions of the bands at the energy indicated in panel c. The VB electrons are strongly scattered, while the VB-1 electrons transmit almost freely. Adapted with permission from [100]. Copyright (2017) American Chemical Society

implies that boron atoms selectively confine VB electrons while leaving the VB-1 unaffected. The reason behind this selectivity stems from the symmetry between the boron induced states and the 7-aGNR bands [100]. These results highlight that the use of substitutional heteroatoms as dopants, and their related heterostructures, goes beyond a simple charge doping model and can strongly modify the transport properties of GNRs.

5 Applications

Graphene nanoribbons have been used in many types of applications as energy storage [101, 102], thin films coatings [103–105] or in composite materials [106–108], but thanks to their attractive physical properties it is in electronic and opto-electronic devices applications where they arouse most attention. As a result, research efforts have pushed towards the integration of GNRs as active components in electronic devices like sensors [109, 110], photodetectors [111–113] and field effect transistors (FETs) [114–119].

A few experiments have already demonstrated the great potential of GNRs for technological applications. For example, nanometer-wide GNRs epitaxially grown on silicon carbide were found to be single-channel ballistic conductors at room-temperature over distances up to 16 micrometers, which is similar in performance to metallic carbon nanotubes (which are edgeless but for which the chirality determines the electronic structure) [120]. For thinner GNRs, the electron transport characteristics have been addressed locally in ultra-high-vacuum by STM with single ribbon precision [121]. Transport occurs in a tunneling regime and the tunneling decay length through a 7-aGNR measured at different bias voltages reveals the dependence of conductance with the ribbon electronic states, that is, energy level alignment and band gap value (Fig. 17a, b).

Graphene nanoribbons technology is still in its infancy and a few problems need still to be solved to speed up the use of GNRs in current technology [4]. On-surface chemistry, as outlined in the previous pages, could overcome the problems of high quality fabrication and functionalization of GNRs. In fact, it allows producing graphene nanoribbons with atomically precise control of widths and edges, and thus with well-defined electronic structure. It allows to dope GNRs selectively and reproducibly, and eventually to increase their functionality by adding active components as photoactive, magnetic or switchable elements. For their operation in devices, as in gated multiterminal devices, GNRs have to be transferred to insulators. The direct growth on insulating substrates remains a challenge. Promising strategies may include the synthesis through photoactivation or through new reactions beyond the Ullmann coupling—cyclodehydrogenation combination. Nevertheless, different routes can be explored to transfer GNRs from a metal to another material as silicon dioxide, as it has been successfully done for two-dimensional materials: exfoliation with or without a solution [118, 119], intercalation assisted exfoliation [51], or chemical vapor deposition [122].

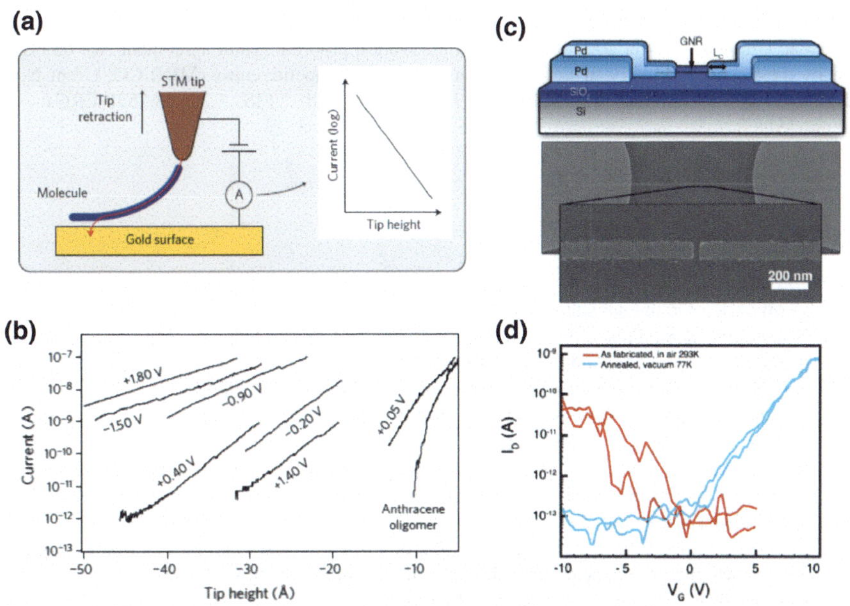

Fig. 17 **a** Schematic of the STM pulling experiment (arrow indicates tunnelling current). A characteristic current signal during the pulling sequence is shown in the right panel. **b** Current as a function of tip height for different experiments at different bias, evidencing the bias dependence of the conductance. **c** Schematic illustration of the device geometry and scanning electron micrograph of the device. **d** Electrical characterization of a typical device at $V_{SD} = 1$ V in both air and under vacuum at 77 K. **a–b** Reprinted by permission from Macmillan Publishers Ltd.: Nature Nanotechnology [121], copyright (2012). **c**, **d** Reprinted from [118], with the permission of AIP Publishing

The last problem of GNR technology is the final fabrication and processing of the device. A few prototypes of field effect transistors have been realized exploiting on-surface synthesized GNRs (Fig. 17c, d) [118, 119]. The interest on FETs stems on the extreme thinness of GNRs which is expected to enable fabricating FETs with very short channels. This should significantly increase the device speed, while avoiding the unfavorable short-channel effects of standard CMOS technologies [123]. In GNR-FETs the driving current is limited by the contact resistance between a single GNR and the metallic contacts caused by the Schottky barrier at the interface. The height of the barrier depends linearly with the GNR band gap. In fact, FETs with 7-aGNR have shown low driving currents of 1 nA per 1 V drain bias [118] while FETs with GNRs displaying lower band gap like 9-aGNR and 13-aGNR exhibited on-currents up to 100 nA per −1 V bias and on/off ratios of 10^2–10^5 [119]. To further improve GNR-FETs performance wider ribbons with smaller band gaps could be used, or materials for metallic contacts could be tailored to the electronic structure of the particular GNR used. In addition, longer ribbons (lengths >30 nm) would assure a better contact overlap with source and drain.

Acknowledgements Financial support by the European Research Council (ERC) under the European Union's Horizon 2020 research and innovation program (grant agreement No. 635919) and by the Spanish Ministry of Economy, Industry and Competitiveness (MINECO, Grant Nos. MAT2016-78293-C6-1-R, MAT2016-78293-C6-5-R and FIS 2015-62538-ERC) is acknowledged.

References

1. Geim, A.K.: Graphene: status and Prospects. Science **324**, 1530–1534 (2009)
2. Novoselov, K.S., Falko, V.I., Colombo, L., Gellert, P.R., Schwab, M.G., Kim, K.: A roadmap for graphene. Nature **490**, 192–200 (2012)
3. Schwiertz, F.: Graphene transistors. Nat. Nanotechnol. **5**, 487–496 (2010)
4. Talirz, L., Ruffieux, P., Fasel, R.: On-surface synthesis of atomically precise graphene nanoribbons. Adv. Mater. **28**, 6222–6231 (2016)
5. Wassmann, T., Seitsonen, A.P., Saitta, A.M., Lazzeri, M., Mauri, F.: Clar's theory, π-electron distribution, and geometry of graphene nanoribbons. J. Am. Chem. Soc. **132**, 3440–3451 (2010)
6. Martín-Martínez, F.J., Fias, S., van Lier, G., de Proft, F., Geerlings, P.: Electronic structure and aromaticity of graphene nanoribbons. Chem. Eur. J. **18**, 6183–6194 (2012)
7. Clar, E.: The Aromatic Sextet. Wiley, London (1972)
8. Kertesz, M., Choi, C.H., Yang, S.: Conjugated polymers and aromaticity. Chem. Rev. **105**, 3448–3481 (2005)
9. Yang, S., Olishevski, P., Kertesz, M.: Bandgap calculations for conjugated polymers. Synth. Met. **141**, 171–177 (2004)
10. Martín-Martínez, F.J., Fias, S., van Lier, G., de Proft, F., Geerlings, P.: Tuning aromaticity patterns and electronic properties of armchair graphene nanoribbons with chemical edge functionalization. Phys. Chem. Chem. Phys. **15**, 12637–12647 (2013)
11. Son, Y.-W., Cohen, M., Louie, S.G.: Energy gaps in graphene nanoribbons. Phys. Rev. Lett. **97**, 216803-1–4 (2006)
12. Tao, C., Jiao, L., Yazyev, O.V., Chen, Y.-C., Feng, J., Zhang, X., Capaz, R.B., Tour, J.M., Zettll, A., Louie, S.G., Dai, H., Crommie, M.F.: Spatially resolving edge states of chiral graphene nanoribbons. Nat. Phys. **7**, 616–620 (2011)
13. Li, Y.Y., Chen, M.X., Weinert, M., Li, L.: Direct experimental determination of onset of electron–electron interactions in gap opening of zigzag graphene nanoribbons. Nat. Commun. **5**, 4311-1–8 (2014)
14. Pan, M., Girao, E.C., Jia, X., Bhaviripudi, S., Li, Q., Kong, J., Meunier, V., Dresselhaus, M. S.: Topographic and Spectroscopic characterization of electronic edge states in CVD grown graphene nanoribbons. Nano Lett. **12**, 1928–1933 (2012)
15. Magda, G.Z., Jin, X., Hagymasi, I., Vancso, P., Osvath, Z., Nemes-Incze, P., Hwang, C., Biro, L.P., Tapaszto, L.: Room-temperature magnetic order on zigzag edges of narrow graphene nanoribbons. Nature **514**, 608–611 (2014)
16. Stampfer, C., Güttinger, J., Hellmüller, S., Molitor, F., Ensslin, K., Ihn, T.: Energy gaps in etched graphene nanoribbons. Phys. Rev. Lett. **102**, 056403-1–4 (2009)
17. Querlioz, D., Apertet, Y., Valentin, A., Huet, K., Bournel, A., galdin-Retailleau, S., Dollfus, P.: Suppression of the orientation effects on bandgap in graphene nanoribbons in the presence of edge disorder. Appl. Phys. Lett. **92**, 042108-1–3 (2008)
18. Cai, J., Ruffieux, P., Jaafar, R., Bieri, M., Braun, T., Blankenburg, S., Muoth, M., Seitsonen, A.P., Saleh, M., Feng, X., Müllen, K., Fasel, R.: Atomically precise bottom-up fabrication of graphene nanoribbons. Nature **466**, 470–473 (2010)
19. Narita, A., Wang, X.-Y., Feng, X., Müllen, K.: New advances in nanographene chemistry. Chem. Soc. Rev. **44**, 6616–6643 (2015)

20. Kahle, S., Deng, Z., Malinowski, N., Tonnoir, C., Forment-Aliaga, A., Thontasen, N., Rinke, G., Le, D., Turkowski, V., Rahman, T.S., Rauschenbach, S., Ternes, M., Kern, K.: The quantum magnetism of individual manganese-12-acetate molecular magnets anchored at surfaces. Nano Lett. **12**, 518–521 (2012)
21. Held, P.A., Fuchs, H., Studer, S.: Covalent-bond formation via on-surface chemistry. Chem. Eur. J. **23**, 5874–5892 (2017) https://doi.org/10.1002/chem.201604047
22. de Oteyza, D.G.: Cyclization chemistry under ultra-high-vacuum. In: Gourdon, A. (ed.) On-Surface Synthesis. Springer International Publishing AG, Switzerland (2016)
23. Palma, C.-A., Diller, K., Berger, R., Welle, A., Björk, J., Cabellos, J.L., Mowbray, D.J., Papageorgiou, A.C., Ivleva, N.P., Matich, S., Margapoti, E., Niessner, R., Menges, B., Reichert, J., Feng, X., Räder, H.J., Klappenberger, F., Rubio, A., Müllen, K., Barth, J.H.: Photoinduced C–C reactions on insulators toward photolithography of graphene nanoarchitectures. J. Am. Chem. Soc. **136**, 4651–4658 (2014)
24. Basagni, A., Ferrighi, L., Cattelan, M., Nicolas, L., Handrup, K., Vaghi, L., Papagni, A., Sedona, F., di Valentin, C., Agnoli, S., Sambi, M.: On-surface photo-dissociation of C–Br bonds: towards room temperature Ullmann coupling. Chem. Commun. **51**, 12593–12596 (2015)
25. de Oteyza, D.G., García-Lekue, A., Vilas-Varela, M., Merino-Diez, N., Carbonell-Sanromà, E., Corso, M., Vasseur, G., Rogero, C., Guitián, E., Pascual, J.I., Ortega, J.E., Wakayama, Y., Peña, D.: Substrate-independent growth of atomically precise chiral graphene nanoribbons. ACS Nano **10**, 9000–9008 (2016)
26. Dong, L., Liu, P.N., Lin, N.: Surface-activated coupling reactions confined to a surface. Acc. Chem. Res. **48**, 2765–2774 (2015)
27. Björk, J., Hanke, F., Stafström, S.: Mechanisms of halogen-based covalent self-assembly on metal surfaces. J. Am. Chem. Soc. **135**, 5768–5775 (2013)
28. Huang, H., Wei, D., Sun, J., Wong, S.L., Feng, Y.P., Castro Neto, A.H., Wee, A.T.S.: Spatially resolved electronic structures of atomically precise armchair graphene nanoribbons. Sci. Rep. **2**, 983-1–7 (2012)
29. Dienel, T., Kawai, S., Söde, H., Feng, X., Müllen, K., Ruffieux, P., Fasel, R., Gröning, O.: Resolving atomic connectivity in graphene nanostructure junctions. Nano Lett. **15**, 5185–5190 (2015)
30. Basagni, A., Sedona, F., Pignedoli, C.A., Cattelan, M., Nicolas, L., Casarin, M., Sambi, M.: Molecules–oligomers–nanowires–graphene nanoribbons: a bottom-up stepwise on-surface covalent synthesis preserving long-range order. J. Am. Chem. Soc. **137**, 1802–1808 (2015)
31. Han, P., Akagi, K., Canova, F.F., Mutoh, H., Shiraki, S., Iwaya, K., Weiss, P.S., Asao, N., Hitosugi, T.: Bottom-up graphene-nanoribbon fabrication reveals chiral edges and enantioselectivity. ACS Nano **8**, 9181–9187 (2014)
32. Schulz, F., Jacobse, P.H., Canova, F.F., van der Lit, J., Gao, D.Z., van den Hoogenband, A., Han, P., Gebbink, R.J.M.K., Moret, M.-E., Joensuu, P.M., Swart, I., Liljeroth, P.: Precursor geometry determines the growth mechanism in graphene nanoribbons. J. Phys. Chem. C **121**, 2896–2904 (2017)
33. Saywell, A., Schwartz, J., Hecht, S., Grill, L.: Polymerization on stepped surfaces: alignment of polymers and identification of catalytic sites. Angew. Chem. Int. Ed. **51**, 5096–5100 (2012)
34. Ruffieux, P., Cai, J., Plumb, N., Patthey, L., Prezzi, D., Ferretti, A., Molonari, E., Feng, X., Müllen, K., Pignedoli, C., Fasel, R.: Electronic structure of atomically precise graphene nanoribbons. ACS Nano **6**, 6930–6935 (2012)
35. Linden, S., Zhong, D., Timmer, A., Aghdassi, N., Franke, J. H., Zacharias, H.: Electronic structure of spatially aligned graphene nanoribbons on Au(788). Phys. Rev. Lett. **108**, 216801-1–5 (2012)
36. Basagni, A., Vasseur, G., Pignedoli, C.A., Vilas-Varela, M., Peña, D., Nicolas, L., Vitali, L., Lobo-Checa, J., de Oteyza, D.G., Sedona, F., Casarin, M., Ortega, J.E., Sambi, M.: Tunable band alignment with unperturbed carrier mobility of on-surface synthesized organic semiconducting wires. ACS Nano **10**, 2644–2651 (2016)

37. Treier, M., Pignedoli, C.A., Laino, T., Rieger, R., Müllen, K., Passerone, D., Fasel, R.: Surface-assisted cyclodehydrogenation provides a synthetic route towards easily processable and chemically tailored nanographenes. Nat. Chem. **3**, 61–67 (2011)
38. Björk, J., Stafström, S., Hanke, F.: Zipping up: cooperativity drives the synthesis of graphene nanoribbons. J. Am. Chem. Soc. **133**, 14884–14887 (2011)
39. Blankenburg, S., Cai, J., Ruffieux, P., Jaafar, R., Passerone, D., Feng, X., Müllen, K., Fasel, R., Pignedoli, C.A.: Intraribbon heterojunction formation in ultranarrow graphene nanoribbons. ACS Nano **6**, 2020–2025 (2012)
40. van der Lit, J., Bonenschanscher, M.P., Vanmaekelbergh, D., Ijas, M., Uppstu, A., Ervasti, M., Harju, A., Liljeroth, P., Swart, I.: Suppression of electron–vibron coupling in graphene nanoribbons contacted via a single atom. Nat. Comm. **4**, 2023-1–6 (2013)
41. Sánchez-Sánchez, C., Dienel, T., Deniz, O., Ruffieux, P., Berger, R., Feng, X., Müllen, K., Fasel, R.: Purely armchair or partially chiral: noncontact atomic force microscopy characterization of dibromo-bianthryl-based graphene nanoribbons grown on Cu(111). ACS Nano **10**, 8006–8011 (2016)
42. Ruffieux, P., Wang, S., Yang, B., Sánchez-Sánchez, C., Liu, J., Dienel, T., Talirz, L., Shinde, P., Pignedoli, C.A., Passerone, D., Dumslaff, T., Feng, X., Müllen, K., Fasel, R.: On-surface synthesis of graphene nanoribbons with zigzag edge topology. Nature **531**, 489–532 (2016)
43. Kimouche, A., Ervasti, M.M., Drost, R., Halonen, S., Harju, A., Joensuu, P.M., Sainio, J., Liljeroth, P.: Ultra-narrow metallic armchair graphene nanoribbons. Nat. Comm. **6**, 10177-1–6 (2015)
44. Zhang, H., Lin, H., Sun, K., Chen, L., Zagranyarski, Y., Aghdassi, N., Duhm, S., Li, Q., Zhong, D., Li, Y., Müllen, K., Fuchs, H., Chi, L.: On-surface synthesis of rylene-type graphene nanoribbons. J. Am. Chem. Soc. **137**, 4022–4025 (2015)
45. Talirz, L., Söde, H., Dumslaff, T., Wang, S., Sanchez-Valencia, J.R., Liu, J., Shinde, P., Pignedoli, C.A., Liang, L., Meunier, V., Plumb, N.C., Shi, M., Feng, X., Narita, A., Müllen, K., Fasel, R., Ruffieux, P.: On-surface synthesis and characterization of 9-atom wide armchair graphene nanoribbons. ACS Nano **11**, 1380–1388 (2017)
46. Chen, Y.-C., de Oteyza, D.G., Pedramrazi, Z., Chen, C., Fischer, F.R., Crommie, M.F.: Tuning the band gap of graphene nanoribbons synthesized from molecular precursors. ACS Nano **7**, 6123–6128 (2013)
47. Liu, J., Li, B.-W., Tan, Y.-Z., Giannakopoulos, A., Sánchez-Sánchez, C., Beljonne, D., Ruffieux, P., Fasel, R., Feng, X., Müllen, K.: Toward cove-edged low band gap graphene nanoribbons. J. Am. Chem. Soc. **137**, 6097–6103 (2015)
48. Kawai, S., Saito, S., Osumi, S., Yamaguchi, S., Foster, A. S., Spijker, P., Meyer, E.: Atomically controlled substitutional boron-doping of graphene nanoribbons. Nat. Commun. **6**, 8098-1–6 (2015)
49. Nguyen, G.D., Toma, F.M., Cao, T., Pedramrazi, Z., Chen, C., Rizzo, D.J., Joshi, T., Bronner, C., Chen, Y.-C., Favaro, M., Louie, S.G., Fischer, F.R., Crommie, M.F.: Bottom-up synthesis of N = 13 sulfur-doped graphene nanoribbons. J. Phys. Chem. C **120**, 2684–2687 (2016)
50. Söde, H., Talirz, L., Gröning, O., Pignedoli, C. A., Berger, R., Feng, X., Müllen, K., Fasel, R., Ruffieux, P.: Electronic band dispersion of graphene nanoribbons via Fourier-transformed scanning tunneling spectroscopy. Phys. Rev. B **91**, 045429-1–6 (2015)
51. Deniz, O., Sánchez-Sánchez, C., Dumslaff, T., Feng, X., Narita, A., Müllen, K., Kharche, N. K., Meunier, V., Fasel, R., Ruffieux, P.: Revealing the electronic structure of silicon intercalated armchair graphene nanoribbons by scanning tunneling spectroscopy. Nano Lett. **17**, 2197–2203 (2017)
52. Bronner, C., Leyssner, F., Stremlau, S., Utecht, M., Saalfrank, P., Klamroth, T., Tegerder, P.: Electronic structure of subnanometer wide bottom-up fabricated graphene nanoribbon: end states, band gap, and dispersion. Phys. Rev. B **86**, 085444 (2012)

53. Bronner, C., Utecht, M., Haase, A., Saalfrank, P., Klamroth, T., Tegeder, P.: Electronic structure changes during the surface-assisted formation of a graphene nanoribbon. J. Chem. Phys. **140**, 024701-1–7 (2014)
54. Bronner, C., Gerbert, D., Broska, A., Tegeder, P.: Excitonic states in narrow armchair graphene nanoribbons on gold surfaces. J. Phys. Chem. C **120**, 26168–26172 (2016)
55. Bronner, C., Stremlau, S., Gille, M., Brauße, F., Haase, A., Hecht, S., Tegeder, P.: Aligning the band gap of graphene nanoribbons by monomer doping. Angew. Chem. Int. Ed. **52**, 4422–4425 (2013)
56. Denk, R., Hohage, M., Zeppenfeld, P., Cai, J., Pignedoli, C.A., Söde, H., Fasel, R., Feng, X., Müllen, K., Wang, S., Prezzi, D., Ferretti, A., Ruini, A., Molinari, E., Ruffieux, P.: Exciton-dominated optical response of ultra-narrow graphene nanoribbons. Nat. Comm. **5**, 4253-1–7 (2014)
57. Bergvall, A., Löfwander, T.: Spectral footprints of impurity scattering in graphene nanoribbons. Phys. Rev. B **87**, 205431-1–14 (2013)
58. Batra, A., Cvetko, D., Kladnik, G., Adak, O., Cardoso, C., Ferretti, A., Prezzi, D., Molinari, E., Morgante, A., Venkataraman, L.: Probing the mechanism for graphene nanoribbon formation on gold surface through X-ray spectroscopy. Chem. Sci. **5**, 4419–4423 (2014)
59. Moreno, C., Stetsovych, O., Shimizu, T.K., Custance, O.: Imaging three-dimensional surface objects with submolecular resolution by atomic force microscopy. Nano Lett. **15**, 2257–2262 (2015)
60. Huber, F., Matencio, S., Weymouth, A.J., Ocal, C., Barrena, E., Giessibl, F.J.: Intramolecular force contrast and dynamic current-distance measurements at room temperature. Phys. Rev. Lett. **115**, 066101-1–4 (2015)
61. Gross, L., Mohn, F., Moll, N., Liljeroth, P., Meyer, G.: The chemical structure of a molecule resolved by atomic force microscopy. Science **325**, 1110–1114 (2009)
62. Kawai, S., Benassi, A., Gnecco, E., Söde, H., Pawlak, R., Feng, X., Müllen, K., Passerone, D., Pignedoli, C.A., Ruffieux, P., Fasel, R., Meyer, E.: Superlubricity of graphene nanoribbons. Science **351**, 957–961 (2016)
63. Zhou, J., Dong, J.: Vibrational property and Raman spectrum of carbon nanoribbon. Appl. Phys. Lett. **91**, 173108–173111 (2007)
64. Sanders, G.D., Nugraha, A.R.T., Saito, R., Stanton, C.J.: Coherent radial-breathing-like phonons in graphene nanoribbons. Phys. Rev. B **85**, 205401 (2012)
65. Han, P., Akagi, K., Canova, F.F., Shimizu, R., Oguchi, H., Shiraki, S., Weiss, P.S., Asao, N., Hitosugi, T.: Self-assembly strategy for fabricating connected graphene nanoribbons. ACS Nano **9**, 12035–12044 (2015)
66. Saito, R., Furukawa, M., Dresselhaus, G., Dresselhaus, M.S.: Raman spectra of graphene ribbons. J. Phys. Cond. Mat. **22**, 334203-1–6 (2010)
67. Sasaki, K., Yamamoto, M., Murakami, S., Saito, R., Dresselhaus, M.S., Takai, K., Mori, T., Enoki, T., Wakabayashi, K.: Kohn anomalies in graphene nanoribbons. Phys. Rev. B **80**, 155450 (2009)
68. Yang, R., Shi, Z., Zhang, L., Shi, D., Zhang, G.: Observation of Raman G-peak split for graphene nanoribbons with hydrogen-terminated zig-zag edges. Nano Lett. **11**, 4083–4088 (2011)
69. Wakabayashi, K., Sasaki, K.-I., Nakanishi, T., Enoki, T.: Electronic states of graphene nanoribbons and analytical solutions. Sci. Technol. Adv. Mater. **11**, 054504-1–18 (2010)
70. Castro Neto, A.H., Guinea, F., Peres, N.M.R., Novoselov, K.S., Geim, A.K.: The electronic properties of graphene. Rev. Mod. Phys. **81**, 109–162 (2009)
71. Gunlycke, D., White, C.T.: Tight-binding energy dispersions of armchair-edge graphene nanostrips. Phys. Rev. B **77**, 115116-1–6 (2008)
72. Wakabayashi, K., Fujita, M., Ajiki, H., Sigrist, M.: Electronic and magnetic properties of nanographite ribbons. Phys. Rev. B **59**, 8271–8282 (1999)
73. Nakada, K., Fujita, M., Dresselhaus, G., Dresselhaus, M.S.: Edge state in graphene ribbons: nanometer size effect and edge shape dependence. Phys. Rev. B **54**, 17954–17961 (1996)

74. Carvalho, A.R., Warnes, J.H., Lewenkopf, C.H.: Edge magnetization and local density of states in chiral graphene nanoribbons. Phys. Rev. B **89**, 245444-1–8 (2014)
75. Akhmerov, A.R., Beenakker, C.W.J.: Boundary conditions for Dirac fermions on a terminated honeycomb lattice. Phys. Rev. B **77**, 085423-1–10 (2008)
76. Yazyev, O.V., Capaz, R.B., Louie, S.G.: Theory of magnetic edge states in chiral graphene nanoribbons. Phys. Rev. B **84**, 115406-1–5 (2011)
77. Jiang, Z., Song, Y.: Band gap oscillation and novel transport property in ultrathin chiral graphene nanoribbons. Phys. Rev. B **464**, 61–67 (2015)
78. Raza, H., Kan, E.C.: Armchair graphene nanoribbons: electronic structure and electric-field modulation. Phys. Rev. B **77**, 245434-1–5 (2008)
79. Yang, L., Park, C.-H., Son, Y.-W., Cohen, M.L., Louie, S.G.: Quasiparticle energies and band gaps in graphene nanoribbons. Phys. Rev. Lett. **99**, 186801-1–4 (2007)
80. Kharche, N., Meunier, V.: Width and crystal orientation dependent band gap renormalization in substrate-supported graphene nanoribbons. J. Phys. Chem. Lett. **7**, 1526–1533 (2016)
81. Karl, N.: Charge carrier transport in organic semiconductors. Synth. Met. **133–134**, 649–657 (2003)
82. Senkovskiy, B.V., Fedorov, A.V., Haberer, D., Farjam, M., Simonov, K.A., Preobrajenski, A.B., Martensson, N., Atodiresei, N., Caciuc, V., Blügel, S., Rosch, A., Verbitskiy, N.I., Hell, M., Evtushinsky, D.V., German, R., Marangoni, T., van Loosdrecht, P.H.H., Fischer, F.R., Grüneis, A.: Semiconductor-to-metal transition and quasiparticle renormalization in doped graphene nanoribbons. Adv. Electron. Mater. 1600490-1–8 (2017)
83. Arora, V.K., Bhattacharyya, A.: Unified bandgap engineering of graphene nanoribbons. Phys. Stat. Sol. B **251**, 2257–2264 (2014)
84. Wang, S., Talirz, L., Pignedoli, C.A., Feng, X., Müllen, K., Fasel, R., Ruffieux, P.: Giant edge state splitting at atomically precise graphene zigzag edges. Nat. Comm. **7**, 11507-1–6 (2016)
85. Barone, V., Hod, O., Scuseria, G.E.: Electronic structure and stability of semiconducting graphene nanoribbons. Nano Lett. **6**, 2748–2754 (2006)
86. Balkanski, M., Fisher Wallis, R.: Semiconductor Physics and Applications. Oxford University Press, Oxford (2000)
87. Fistul', V.I.: Heavily Doped Semiconductors, Monographs in Semiconductor Physics, vol. 1. Springer, Boston (1969)
88. Sheng Yu, S., Zheng, W.T., Jiang, Q.: IEEE Trans. Nanotechnol. Electronic properties of nitrogen-/boron-doped graphene nanoribbons with armchair edges. **9**, 78–81 (2010)
89. Cai, J., Pignedoli, C.A., Talirz, L., Ruffieux, P., Söde, H., Liang, L., Meunier, V., Berger, R., Li, R., Feng, X., Müllen, K., Fasel, R.: Graphene nanoribbon heterojunctions. Nat. Nanotech. **9**, 896–900 (2014)
90. Zhang, Y., Zhang, Y., Li, G., Lu, J., Lin, X., Du, S., Berger, R., Feng, X., Müllen, K., Gao, H.-J.: Direct visualization of atomically precise nitrogen-doped graphene nanoribbons. Appl. Phys. Lett. **105**, 023101 (2014)
91. Cloke, R.R., Marangoni, T., Nguyen, G.D., Joshi, T., Rizzo, D.J., Bronner, C., Cao, T., Louie, S.G., Crommie, M.F., Fischer, F.R.: Site specific substitutional boron doping of semiconducting armchair graphene nanoribbons. J. Am. Chem. Soc. **137**, 8872–8875 (2015)
92. Roldan, R., Castellanos-Gomez, A., Cappelluti, E., Guinea, F.: Strain engineering in semiconducting two-dimensional crystals. J. Phys. Condens. Matter **27**, 313201–1–18 (2015)
93. Li, Y., Jiang, X., Liu, Z., Liu, Z.: Strain effects in graphene and graphene nanoribbons: the underlying mechanism. Nano Res. **3**, 545–556 (2010)
94. Zwanengurg, F.A., Dzurak, A.S., Morello, A., Simmons, M.Y., Hollenberg, L.C.L., Klimeck, G., Rogge, S., Coppersmith, S.N., Eriksson, M.A.: Silicon quantum electronics. Rev. Mod. Phys. **85**, 961–1019 (2013)
95. Bastard, G.: Wave Mechanics Applied to Semiconductor Heterostructures. Les Editions de Physique, Les Ullis, France (1988)

96. Sevincli, H., Topsakal, M., Ciraci, S.: Superlattice structures of graphene-based armchair nanoribbons. Phys. Rev. B **78**, 245402-1–8 (2008)
97. Fan, B., Chang, S.: Confined state energies in AGNR semiconductor–semiconductor heterostructure. Phys. Lett. A **381**, 309–322 (2017)
98. Chen, Y.-C., Cao, T., Chen, C., Pedramrazi, Z., Haberer, D., de Oteyza, D.G., Fischer, F.R., Louie, S.G., Crommie, M.F.: Molecular bandgap engineering of bottom-up synthesized graphene nanoribbon heterojunctions. Nat. Nanotech. **10**, 156–160 (2015)
99. Marangoni, T., Haberer, D., Rizzo, D.J., Cloke, R.R., Fischer, F.R.: Heterostructures through divergent edge reconstruction in nitrogen-doped segmented graphene nanoribbons. Chem. Eur. J. **22**, 13037–13040 (2016)
100. Carbonell-Sanromà, E., Brandimarte, P., Balog, R., Corso, M., Kawai, S., García-Lekue, A., Saito, S., Yamagushi, S., Meyer, E., Sánchez-Portal, D., Pascual, J.I.: Quantum dots embedded in graphene nanoribbons by chemical substitution. Nano Lett. **17**, 50–56 (2017)
101. Yang, Y., Li, L., Fei, H., Peng, Z., Ruan, G., Tour, J.M.: Graphene nanoribbon/V_2O_5 cathodes in lithium–ion batteries. ACS Appl. Mater. Interfaces **6**, 9590–9594 (2014)
102. Li, L., Raji, A.R.O., Tour, J.M.: Graphene-wrapped MnO_2-graphene nanoribbons as anode materials for high-performance lithium ion batteries. Adv. Mater. **25**, 6298–6302 (2013)
103. Zhu, Y., Tour, J.M.: Graphene nanoribbon thin films using layer-by-layer assembly. Nano Lett. **10**, 4356–4362 (2010)
104. Volman, V., Zhu, Y., Raji, A.-R.O., Genorio, B., Lu, W., Xiang, C., Kittrell, C., Tour, J.M.: Radio-frequency-transparent, electrically conductive graphene nanoribbons thin films as deicing heating layers. ACS Appl. Mater. Interfaces. **6**, 298–304 (2014)
105. Raji, A.R., Salters, S., Samuel, E.L., Zhu, Y., Volman, V., Tour, J.M.: Functionalized graphene nanoribbon films as radiofrequency and optically transparent material. ACS Appl. Mater. Interfaces. **6**, 16661–16668 (2014)
106. Lin, J., Raji, A.R.O., Nan, K., Peng, Z., Yan, Z., Samuel, E.L.G., Natelson, D., Tour, J.M.: Iron oxide nanoparticle and Graphene nanoribbon composite as an anode material for high-performance Li-ion batteries. Adv. Funct. Mater. **24**, 2044–2048 (2014)
107. Jian, L., Peng, Z., Xiang, C., Ruan, G., Yan, Z., Natelson, D., Tour, J.M.: Graphene nanoribbon and nanostructured SnO_2 composite anodes for lithium ion batteries. ACS Nano **7**, 6001–6006 (2013)
108. Li, L., Ruan, G., Peng, Z., Yang, Y., Fei, H., Raji, A.R., Samuel, E.L., Tour, J.M.: Enhanced cycling stability of lithium sulfur batteries using sulfur-polyaniline-graphene nanoribbon composite cathodes. ACS Appl. Mater. Interfaces **6**, 15033–15039 (2014)
109. Johnson, J.L., Behnam, A., Pearton, S.J., Ural, A.: Hydrogen sensing using Pd-functionalized multi-layer graphene nanoribbon networks. Adv. Mater. **22**, 4877–4880 (2010)
110. Traversi, F., Raillon, C., Benameur, S.M., Liu, K., Khlybov, S., Tosun, M., Krasnozhon, D., Kis, A., Radenovis, A.: Detecting the translocation of DNA through a nanopore using graphene nanoribbons. Nat. Nanotechnol. **8**, 939–945 (2013)
111. Wei, D., Xie, L., Lee, K., Hu, Z., Tan, S., Chen, W., Sow, C.H., Chen, K., Liu, Y., Wee, A.: Controllable unzipping for intramolecular junctions of graphene nanoribbons and single-walled carbon nanotubes. Nat. Comm. **4**, 1374 (2013)
112. Freitag, M., Low, T., Zhu, W., Yan, H., Xia, F., Avouris P.: Photocurrent in graphene harnessed by tunable intrinsic plasmons. Nat. Comm. **4**, 1951-1–8 (2013)
113. Ryzhii, V., Ryabova, N., Ryzhii, M., Baryshnikov, N., Karasik, V., Mitin, V., Otsuji, T.: Terahertz and infrared photodetectors based on multiple graphene layer and nanoribbon structures. Opto-Electron. Rev. **20**, 15–25 (2012)
114. Schwiertz, F.: Graphene transistors: status, prospects, and problems. IEE **101**, 1557–1566 (2013)

115. Fantuzzi, P., Martini, L., Candini, A., Corradini, V., del Pennino, U., Hu, Y., Feng, X., Müllen, K., Narita, A., Affronte, M.: Fabrication of three terminal devices by electrospray deposition of graphene nanoribbons. Carbon **104**, 112–118 (2016)
116. Hwang, W.S., Zhao, P., Tahy, K., Nyakiti, L., Wheeler, V., Myers-Ward, R., Eddy Jr., C., Gaskill, D., Robinson, J., Haensch, W., Xing, H., Seabaugh, A., Jena, D.: Graphene nanoribbon field-effect transistors on wafer-scale epitaxial graphene on SiC substrates. APL Mater. **3**, 011101-1–9 (2015)
117. Wang, X., Ouyang, Y., Li, X., Wang, H., Guo, J., Dai, H.: Room-temperature all-semiconducting sub-10-nm graphene nanoribbon field-effect transistors. Phys. Rev. Lett. **100**, 206803-1–4 (2008)
118. Bennet, P., Pedramrazi, Z., Madani, A., Chen, Y.-C., de Oteyza, D., Chen, C., Fischer, F., Crommie, M., Bokor, J.: Bottom-up graphene nanoribbon field-effect transistors. App. Phys. Lett. **103**, 253114-1–4 (2013)
119. Linas, J.P., Fairbrother, A., Barin, G.B., Shi, W., Lee, K., Wu, S., Choi, B.Y., Braganza, R., Lear, J., Kau, N., Choi, W., Chen, C., Pedramrazi, Z., Dumslaff, T., Narita, A., Feng, X., Müllen, K., Fischer, F., Zettl, A., Ruffieux, P., Yablonovitch, E., Crommie, M., Fasel, R., Bokor, J.: Short-channel field effect transistor with 9-atom and 13-atom wide graphene nanoribbons. Nat. Comm. **8**, 633-1-6 (2017)
120. Baringhaus, J., Ruan, M., Edler, F., Tejeda, A., Sicot, M., Taleb-Ibrahimi, A., Li, A.-P., Jiang, Z., Conrad, E.H., Berger, C., Tegenkamp, C., de Heer, W.A.: Exceptional ballistic transport in epitaxial graphene nanoribbons. Nature **506**, 349–354 (2014)
121. Koch, M., Ample, F., Joachim, C., Grill, L.: Voltage-dependent conductance of a single graphene nanoribbon. Nat. Nanotechnol. **7**, 713–717 (2012)
122. Sakaguchi, H., Kawagoe, Y., Hirano, Y., Iruka, T., Yano, M., Nakae, T.: Width-controlled sub-nanometer graphene nanoribbon films synthesized by radical-polymerized chemical vapor deposition. Adv. Mater. **26**, 4134–4138 (2014)
123. Kreupl, F.: Advancing CMOS with carbon electronics. In: Proceedings of DATE'14, vol. 237 (2014)

Aryl–Aryl Covalent Coupling on Rutile TiO_2 Surfaces

Marek Kolmer and Jakub S. Prauzner-Bechcicki

Abstract In recent years, enormous progress has been made in developing bottom-up strategies based on the polymerization of specially designed building blocks directly on a supporting surface. So far, selected noble metals have been mostly used as substrates for such on-surface chemical reactions. For the sake of practical applications the semiconductor surfaces clearly represent much more attractive platforms. Especially transition metal oxides exhibiting advantageous optical as well as photo- and electrochemical properties seem to be particularly interesting. In this chapter we describe the strategies for thermally triggered on-surface covalent coupling of aryl halides performed directly on rutile titanium dioxide surfaces. We focus our work on important parameters that need to be considered for understanding and optimization of the polymerization reactions on this model transition metal oxide system.

1 Introduction

On-surface synthesis is a very promising approach to construct desired and seemingly complex molecular nanostructures in a precisely controlled way. In this approach, first, molecular precursors equipped with judiciously chosen linking groups are designed. Then, such precursors are deposited onto selected surface. It is envisaged that linking groups once activated (by means of sample heating, light illumination, etc.) will react with each other leading to formation of covalent bond/s between precursors. The position, number and type of linking groups will determine shape, size and dimensionality of fabricated nanostructures. Eventually, construction of various 1D and 2D structures becomes feasible [1–7]. The on-surface synthesis gives an unprecedented opportunity to increase control over formation of

M. Kolmer (✉) · J. S. Prauzner-Bechcicki
Faculty of Physics, Astronomy and Applied Computer Science,
Centre for Nanometer-Scale Science and Advanced Materials,
NANOSAM, Jagiellonian University, Lojasiewicza 11, 30-348 Krakow, Poland
e-mail: marek.kolmer@uj.edu.pl

D. G. de Oteyza and C. Rogero (eds.), *On-Surface Synthesis II*,
Advances in Atom and Single Molecule Machines,
https://doi.org/10.1007/978-3-319-75810-7_7

organic nanostructures while keeping non-local character of the whole process. The method may be tough considered as a Golden mean between self-assembly based processes and scanning tip-induced procedures, as it possesses non-locality and easiness of self-assembly and yet it is atomically precise and controllable as tip-induced reactions.

One of the most extensively studied examples of on-surface synthesis is an aryl halide homo-coupling on metal substrates, considered as Ullmann-like reaction [8–12]. In the reaction on metal surfaces aryl halide species are usually thermally activated. Resultantly, a carbon–halogen bond is cleaved which yields a radical/ion form of molecular precursor stabilized by a metal substrate [13]. In subsequent steps the radicals recombine to form covalently bonded oligomeric structures. Although details of reaction pathway on metal surfaces are still debated, it is unquestionable that extension of the same approach to other technologically relevant types of substrates is important albeit quite challenging [6, 14]. Several attempts to perform the aryl halide C–C coupling on semiconducting [15–20] and insulating [21, 22] substrates have been reported recently.

Needless to say, details of reaction pathway are depended on the type of substrate onto which precursors are deposited. Regardless of the substrate, one of the fundamental and universal parameters controlling the output of polymerization reaction of aryl halides is mobility of molecular precursors. In case of metal surfaces this fact has already been addressed by Bieri et al. [23]. In case of more reactive, semiconducting surfaces, the mobility of precursors plays a pivotal role. For example on a pristine Ge(001) surface the lack of precursors' mobility was responsible for failure of an attempt to form covalently bonded structures [15]. This problem was then overcome by introduction of an atomic hydrogen monolayer, which effectively decoupled adsorbed molecules from the substrate and thus increased their mobility [24, 25]. As a result dimers and longer structures were observed, however, a total reaction yield was very low [20]. Entirely different picture is observed for rutile titania surfaces in terms of on-surface mobility of molecular adsorbates. Many experimental reports justify a conclusion that generally molecules are mobile enough to form various ordered arrangements on rutile titania surfaces [26–37]. The comparison between the two types of semiconductor surfaces, i.e. Ge(001) and TiO_2, addressing mainly precursors' mobility has been made elsewhere [38].

In the following we would like to move beyond the view which is focused on mobility of molecular adsorbates and discuss other aspects of the aryl halide C–C coupling mechanism on rutile titania surfaces. Available experimental reports focus on two most stable rutile faces, i.e. (110) and (011). Structure and chemical activity of these surfaces are considerably different, what may lead to different reaction pathways. In the subsequent sections, we pinpoint important parameters that need to be considered in the aryl halide C–C coupling on rutile titania surfaces.

2 TiO_2 Properties

In the realm of on-surface synthesis, it is evident that an essential role is played by the substrate itself. Obviously knowledge and understanding of physical and chemical properties of the surface are then of uttermost importance. Several extensive and valuable review papers on structural characterisation and consequent properties of titanium dioxide surfaces, and their interactions with organic and non-organic adsorbates are available in the literature [37, 39–50]. We shall not reproduce these reports here in extenso, instead we recall the most important general features accompanied by brief comparison of the two most often studied faces of rutile titania, i.e. (110)-(1 × 1) and (011)-(2 × 1), and, finally, focus on the issue of crystal reduction. The more interested reader is referred to the above mentioned reviews for further details. Let us just mention in this place, that both surfaces discussed in this chapter are prepared under ultra-high vacuum (UHV) conditions in a similar way by repeated cycles of ion bombardment and annealing.

2.1 General

Surfaces of rutile titanium dioxide, especially the (110) face, may be considered as the model metal oxide surfaces. Already back in 2003, they were the most investigated systems in the surface science of metal oxides [39]. Titanium dioxide is used in many applications, e.g. as a white pigment in paints, personal care and food products, as a photocatalyst, in self-cleaning exterior building materials, as a coating (e.g. optical coatings, corrosion-protective coatings), in photovoltaics, in ceramics, in electronics, in biomedical implants, and many other fields. Functionalization of its surface by on-grown molecular layers and nanostructures is widely explored in the context of almost all above mentioned technologies [37, 39–50].

As pointed by Pang, Lindsay and Thornton in their review addressing chemical reactions on rutile $TiO_2(110)$ [40], inaccurate characterization of a clean surface structure may at worst lead to incorrect mechanistic interpretation of reactions taking place on it. This observations can be generalized to all faces of rutile titania. Thus in our pursuit for reaction pathway for the aryl halide C–C coupling on rutile titania surfaces we will begin with description of the (110)-(1 × 1) and (011)-(2 × 1) faces. Fortunately, model of the (110)-(1 × 1) face is well established and has been cross-checked with various experimental techniques [39]. The (110)-(1 × 1) surface contains two different kinds of titanium atoms, i.e. a six-fold coordinated (Ti_{6c}) and a five-fold coordinated (Ti_{5c}), and two different kinds of oxygen atoms, i.e. a three-fold coordinated (O_{3c}) and a two-fold coordinated (O_{2c}) (Fig. 1a). Both kinds of titanium atoms and O_{3c} atoms constitute the main surface plane, whereas O_{2c} atoms form rows protruding from the main surface and running along the [001] direction. Therefore these atoms are also known as bridging oxygen atoms. Here it is worth noticing that Ti_{5c} and O_{2c} atoms are less coordinated than Ti and O atoms in the bulk. The five-fold coordinated titanium atoms expose their dangling bonds perpendicular to the surface and, thus, become anchor points for molecular adsorbates. On the other hand, the coordinative undersaturation of bridging oxygen atoms is responsible for easiness in formation of oxygen vacancies by thermal annealing during surface preparation procedure [39]. If special sample treatment is not performed during preparation of the surface (i.e. annealing in O_2 partial pressure), the oxygen vacancies are inevitable and, as such, they are very important surface point defects and affect the surface chemistry [39, 40, 42, 44]. Surface density of oxygen vacancies depends on crystal reduction state and typically is about 0.05–0.1 monolayer (ML, 1 ML refers to one vacancy per surface unit cell) [44]. The bridging oxygen vacancies are also partially responsible for the formation of an occupied band gap state at 1 eV below the Fermi level [40, 42, 44]. Residual water quite easily dissociates in these oxygen vacancies giving rise to another typical, and the most visible, surface defect, i.e. hydroxyl groups, $–O_{2c}H$ [44]. The appearance of the (110)-(1 $\times$ 1) face in scanning probe microscopies (SPM) strongly depends not only on position of atomic species on the surface but also on their electronic properties. Mostly, images obtained in scanning tunnelling microscope (STM) present empty states of the sample and, thus, show bright rows of Ti_{5c} atoms running along the [001] surface direction, whereas bridging oxygen atom rows, running in the same direction, are imaged as dark (Fig. 1b). The oxygen vacancies and hydroxyl groups are both seen as bright protrusions. Contrarily, in a typical non-contact atomic force microscopy (ncAFM) image rows of O_{2c} atoms dominate and are observed as bright, as they are closer to the scanning tip. In this case, oxygen vacancies are dark depressions on bright rows. However, as the tip termination is vulnerable to changes during scanning, the contrast inversion (with bright Ti_{5c} rows and dark O_{2c} rows) is possible in ncAFM images [44].

There is an utterly different situation when it comes to settling a model for the (011) surface of rutile titania. In equilibrium rutile crystal the (011) face is the second most abundant surface [51]. Debate on the structural model of the (011) surface of a crystal after routine preparation in ultrahigh vacuum conditions is still vivid, supported by new results from both experimental and theoretical research [52–59]. At the present stage, to the best of our knowledge, two types of models of the (011) surface are opposed to each other, i.e. the 'brookite (001)-like' model [52, 53] and other derived from it [55] are confronted with the microfacet model [56, 57]. Such a situation makes the quest for resolving the aryl halide C–C coupling mechanism on the (011) surface of rutile titanita a more difficult task. Nevertheless, some general features are common to both discussed models. In both models the top-most atoms are two-fold coordinated oxygen atoms, O_{2c} (Fig. 1c). Typical point defects are hydrogen atoms adsorbed on these oxygen atoms, referred to as hydroxyl groups. Another important observation is made with respect to Ti_{5c} atoms, these are also present on the surface, however, are sterically hardly accessible for adsorbed molecules. Unquestionable is appearance of the (011) surface in scanning probe experiments. Both in STM (Fig. 1d) and high-resolution ncAFM

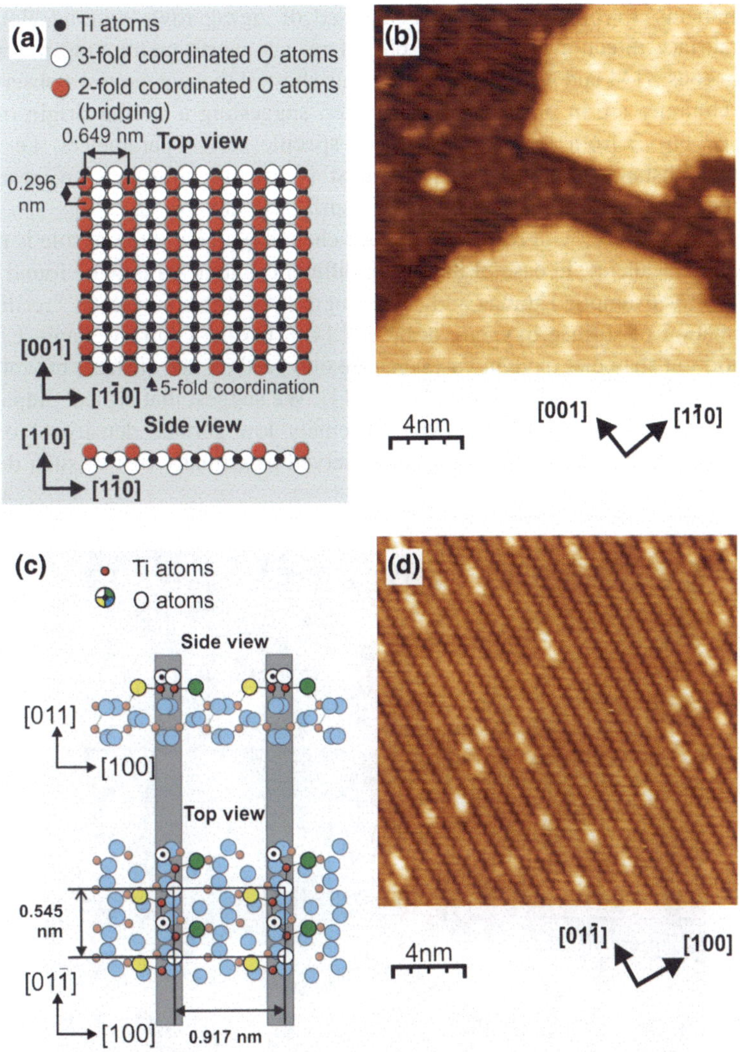

Fig. 1 The models and empty states STM images of the rutile titania surfaces. **a** and **b** Refer to $TiO_2(110)$-(1 × 1) surface. The red, white and black balls in (**a**) represent O_{2c}, O_{3c} and Ti atoms respectively. Five-fold coordinated titanium atoms (Ti_{5c}) are located between O_{2c} rows, what is pointed by a black arrow. **c** and **d** Refer to $TiO_2(011)$-(2 × 1) surface. The model in (**c**) bases on structure proposed in [53]. White, yellow and green balls correspond to two-fold coordinated oxygen atoms (O_{2c}), whereas red balls are Ti_{5c} atoms. Light red and light blue balls are bulk-like coordinated atoms: Ti_{6c} and O_{3c}, respectively. STM scan areas in (**b**) and (**d**) are in both cases 20 × 20 nm^2. Tunneling currents: 100 pA (**b**) and 2pA (**d**); bias voltages: +1.3 V (**b**) and +2 V (**d**). Reprinted from [30], with permission from Elsevier

measurements a typical contrast is comprised of zigzag rows running along the [0-11] surface direction [52, 54, 58, 60]. Here it is worth mentioning that recent simultaneous STM and ncAFM measurements showed that zigzag rows observed in both techniques are in phase with each other, suggesting a similar origin of both [54] (see Fig. 2 from the reference). For specific set of parameters, i.e. small tip-sample distances, another STM contrast is usually observed, i.e. a row of bean-like feature running along the [0-11] surface direction [52, 58].

On the (110) face, with respect to surface chemistry, the important role is played by oxygen vacancies. In case of the (011), different statements may be found in the literature, for example Thomas and Syres suggest, that this surface is "resilient to the formation of surface oxygen vacancies" [42]. On the contrary, Pang, Lindsay and Thornton state that oxygen vacancies are observed on that surface, but rather as pairs of vacancies instead of isolated ones [44]. We suggest that typical preparation procedure leaves the (011) face with extremely low surface density of oxygen vacancies and, thus, the surface may be observed as virtually free of such defects.

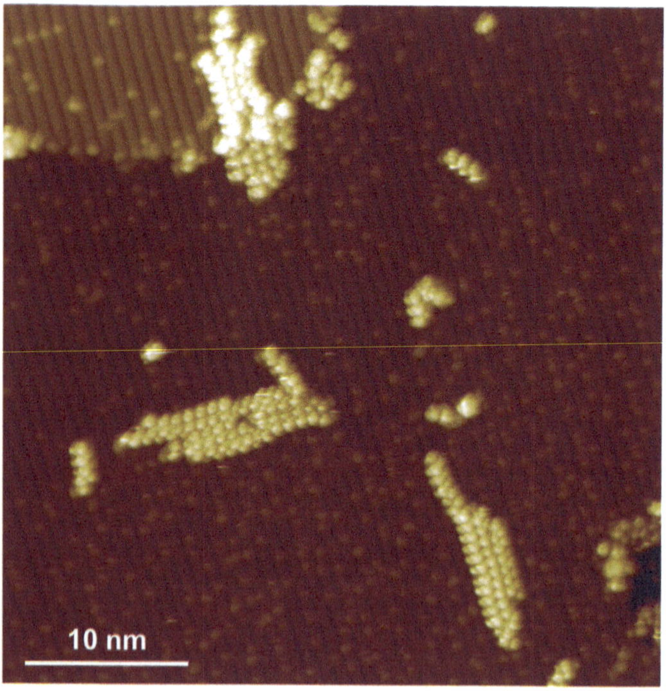

Fig. 2 STM image of polyanthrylene chains formed on the $TiO_2(011)$ surface. Image taken at LNT, 10 pA, +2 V, size 50 nm × 50 nm. Reprinted from [17], with permission from John Wiley and Sons

2.2 Comparison Between Rutile Faces

Compiling the above characterization of the (110) and (011) faces of rutile titania we would like to point to similarities and dissimilarities of both surfaces with respect to on-surface aryl–aryl C–C coupling. First, both surfaces are anisotropic. In each case, that aspect is revealed by presence of surface reconstruction rows that run in one specific surface direction, the [001] on the (110) face and the [0-11] on the (011) face, respectively. The anisotropy is also present in energy profiles for adsorption of molecules, see for example the case of PTCDA on TiO_2(110) [29]. Furthermore, it reveals itself in anisotropic diffusion [61]. The anisotropy of the surface structure and surface diffusion may be an advantage in particulars designs, for instance, when ordered molecular structures like oligomers aligned in a specific directions are targeted.

Second, both surfaces constitute of Ti_{5c} and O_{2c} atoms. The (110) face has also Ti_{6c} and O_{3c} atoms in its structure. Here comes, however, a prominent difference. On the (110) face Ti_{5c} atoms are well exposed and serve as anchor points for molecular species. On the (011) face, though, Ti_{5c} atoms are sterically hardly accessible and binding molecular adsorbates to them is hindered.

Finally, at least to some extent, the two surfaces have variant defects. For the (110) face, the main point defects are bridging oxygen vacancies. Their presence on surface is associated with formation of other important surface point defects, i.e. hydroxyl groups. In the case of the (011) face and slightly reduced crystals hydroxyl groups are dominant type of defects and oxygen vacancies are rarely observed [54, 62].

2.3 Crystal Reduction and Surface Defects

We would like to address one more issue which is, in our opinion, important in view of the aryl halide C–C coupling on rutile titania surfaces. Both surfaces in question are prepared in a similar way, by repeated cycles of ion bombardment and annealing. The preparation procedure leads to inevitable bulk crystal reduction [63]. On one hand, the reduction of the crystalline sample allows its imaging with use of STM due to lowered bandgap. Stoichiometric rutile is an insulator with an approx. 3 eV band gap, that is turned into a n-type semiconductor upon preparation by adding excess electrons [64]. On the other hand, the level of reduction and the presence of excess electrons, especially in a form of subsurface polarons [64], may be of importance for various on-surface chemical reactions.

Deskins et al. [65] basing on a systematic study using density functional theory (DFT) suggest, that all three major rutile defects, i.e. oxygen vacancies, hydroxyl groups and titanium interstitials, donate electrons in a similar manner. In a result these three defects lead to similar surface reduction states. Therefore, the same redox chemistry on surface is independent of the source of excess electrons. Finally, these authors show a semi-quantitative correlation between the number of excess electrons, and the strength of adsorption at Ti_{5c} sites. Thereby, the number of defects cannot be ignored in analysis of molecular species on the rutile surfaces.

Setvin et al. [64] characterized excess electrons in rutile $TiO_2(110)$ and showed that they can localize at any lattice Ti atom forming small polarons. Such polarons hop rapidly among Ti lattice sites changing their localization between surface and subsurface atoms. In 75% cases they are staying in the first subsurface. However, in 25% cases the polarons move to the surface Ti_{5c} sites [66]. What could be of importance for the aryl halide C–C coupling is the fact that such surface and subsurface polarons should facilitate an efficient charge transfer to adsorbed species [45]. An open question remains, whether different surface structure of the (011) face of rutile has an influence on behaviour of such small polarons in subsurface and surface regions.

As already discussed above, the excess electrons are donated in a similar manner by oxygen vacancies, hydroxyl groups and titanium interstitials [65]. Consequently, presence of hydroxyl groups influences adsorption of organic species on either of two surfaces. It is related to the fact, that electrons of adsorbed hydrogen occupy states at the conduction band minimum. Then, if a subsequently adsorbed molecule creates free states that lay lower in the energy, these states will become populated by electrons donated by hydrogen. In a consequence the heat of adsorption will increase [67]. Because carboxylic acids do not offer free states below the conduction band no increase in adsorption energy is expected. Indeed, at least with respect to theoretical studies within DFT framework, preadsorbed hydrogen leads to slight lowering of the binding energy of acetic acid [67]. Such an interplay between hydroxyl groups and adsorbed molecular precursors should be addressed in more detailed studies of the aryl–aryl C–C coupling to define its influence on the reaction pathway.

There are also other theoretical approaches that study various consequences of surface hydroxylation [68, 69]. For the (011)-(2 $\times$ 1) face [68] these calculations, suggest a roughly 10–15% increase of the band gap energy upon hydrogen adsorption. That increase is accompanied by a shift of the Fermi level from the top of the valence band to the bottom of the conduction band. Additionally, a well-defined peak appears below the valence band. Similar behaviour with respect to the band gap energy and the shift of the Fermi energy level is observed for the isolated oxygen vacancies introduced on the clean surface on the top of zigzag rows and in the valleys. Only vacancies on the side of zigzag rows lead to lowering of the band gap energy [68]. For the (110)-(1 $\times$ 1) face [69] DFT calculations with the Hubbard U correction predict that hydroxylation leads to negligible changes of the band gap. Yet, the valence and conduction band edges experience alike energy shifts towards the vacuum level. The authors explain this effect by presence of the electric dipoles arising from the hydroxyl groups on the surface [69].

In the context of influence of hydroxyl groups on on-surface reactions, especially interesting are new experimental results reported by Zhang and Yates [70]. Their work is focused on hydrogen atoms adsorbed onto $TiO_2(110)$-(1 $\times$ 1) surface. The authors point that there is a significant difference between surface hydrogen atoms produced by H_2O dissociation on bridging oxygen vacancies and those adsorbed by dosing atomic hydrogen. The former H species are called α–H, the latter are called β–H. A systematic study showed that β–H are less thermally stable and extremely sensitive to electron and UV excitations when compared to α–H [70]. Finally, Zhang and Yates Jr. suggest that these β–H are formed as subsurface hydroxyl species via diffusion of atomic hydrogen from the surface to the bulk. Sources of atomic hydrogen are experimentally readily accessible and the control over β–H species surface density is reachable, thus it would be intriguing to see whether that group plays a role in aryl–aryl coupling.

Diffusion of atomic hydrogen from the surface to the bulk has also been reported for the (011)-(2 × 1) face [62]. In their experiment, Tao et al. obtained the surface hydrogen atoms by dosing atomic hydrogen to the sample. However, the process of subsurface diffusion did not happen spontaneously. Moreover, it turned out to be reversible. Tao et al. report that the subsurface diffusion and its reversal are thermally driven, i.e. the adsorbed hydrogen disappears from the surface at temperatures ~500 K, but then reappears upon cooling the sample. Yet, UV light irradiation leads to desorption via hydrogen excitation [62].

Although all discussed important defects lead to similar surface reduction states, the control over their density is not equally straightforward [71]. For example, the creation of bridging oxygen vacancies via ion bombardment and annealing at UHV conditions inevitably produces subsurface defects. Even though oxygen vacancies might be healed by annealing in oxygen partial pressure, the subsurface defects will remain. What is more, they are difficult to quantitatively assessed. Thus, in studies on influence of crystal reduction state on on-surface chemistry special care has to be taken with respect to sample preparation. Mao et al. [71] even suggest that in real studies and applications the band gap states introduced by surface defects, especially hydroxyl groups, are of major importance.

Furthermore, quite recently it has been shown that oxygen vacancies on reduced $TiO_2(110)$ surface are key players in another C–C coupling reaction. Yang et al. [72] examined reductive C–C coupling of acetaldehyde with controlled oxygen vacancy coverage by means of temperature-programmed desorption. Their studies showed that with increasing vacancy coverage the reaction yield goes up. The authors point out that the reactions of acetaldehyde on oxidized and reduced $TiO_2(001)$ surfaces [73] (it is the second, with respect to energy, stable face of rutile titania) differ significantly from those obtained on the (110) face. In the former case formation of crotonaldehyde and crotyl alcohol via aldol condensation of CH_3CHO is favoured, in the latter formation of 2-butanone and butane is dominant. Although we do not analyse dissimilarities between the (001) and (110) faces here, one important deduction can be made, i.e. the surface chemistry of rutile is not only vulnerable to reduction state of the bulk, but also to the kind of surface terminating the crystal—both aspects have to be taken into account simultaneously.

Let us add one more piece to the mosaic. Benz et al. reported a thorough study of reductive coupling of benzaldehyde on $TiO_2(110)$ surface [74, 75]. In contrast to Yang et al. [72] result discussed above, they show that Ti interstitials govern the analysed reaction. The titanium interstitials migrate to the surface from subsurface regions and form a reactive intermediate by coupling two benzaldehyde molecules on a surface. The intermediate is then reduced to gaseous stilbene upon heating to ~400 K. As a result two oxygen atoms from the molecular species that reacted with Ti interstitial are left behind on the surface as a nanoisland [75].

The image that emerges from the analysis of the cited research proves to be very complex. Adsorbed hydrogen atoms, oxygen vacancies and titanium interstitials are expected to lead to similar surface reductions states. However, their role in on-surface reactions can be very different. The number of parameters that must be considered in resolving aryl halide C–C coupling on rutile titania surfaces is quite large. Among others, the list includes the differences in topography and its influence on behaviour of small polarons in subsurface regions of different faces of rutile titania, dissimilarity in surface defects (e.g. the presence of oxygen vacancies on the (011) face after typical preparation procedure is still elusive), the source of defects (see discussion on α–H and β–H species on the (110) face). In addition, it turns out that the degree of reduction of the surface and the bulk need to be treated individually to some extent. As Cremer, Jensen and Friend studies show [76], the oxidized $TiO_2(110)$ surface is much more efficient in formate production in photo-oxidation of formaldehyde than a highly reduced one, provided that bulk reduction is similar in both cases.

The following sections are meant to shed some more light on aryl–aryl coupling on rutile titania surfaces. We start with recalling the first successful realization of that reaction. Next, an effort that has been made in determining reaction pathway is reviewed. Eventually, we summarize and give an outlook for future research directions.

3 Aryl Halide Coupling on the Rutile (011) Surface

3.1 First Successful C–C Coupling on a Rutile Surface

The first successful aryl halide C–C coupling on rutile titania surface has been realized with 10,10′-dibromo-9,9′-bianthryl (DBBA) molecules deposited onto the (011)-(2 × 1) face [17]. DBBA molecular precursors were used earlier to obtain extended polyanthrylenes directly on the Au(111) surface through thermally initiated Ullmann-type reaction [6, 77]. Further annealing of the polyanthrylenes on the gold substrate resulted in formation of extended graphene nanoribbons of width N = 7. The triggering temperature for the first step was around 200 °C, for the second step the triggering temperature was around 400 °C.

As already discussed in Sect. 1, one of the critical requirements for oligomer formation through on-surface coupling reaction is considerable mobility of the adsorbed precursors. Basing on previous experience [26–37] rather high mobility of DBBA molecules on the $TiO_2(011)$ surface was anticipated. Indeed, when molecules deposited on the surface were imaged at room temperature stable STM imaging conditions were hardly achievable due to substantial mobility of adsorbates. Lowering temperature during imaging to liquid nitrogen temperatures allowed for observation of single molecules. Intact precursors were found at step-edges and randomly distributed on terraces. None of polymers, short oligomers or even dimers were found. First short oligomers were observed, when sample with molecular precursors was annealed at 300 °C [17]. Further annealing of the same sample at the same temperature neither induce elongation of the oligomers nor ordering of them. The authors report that more efficient coupling is obtained when precursors are deposited onto substrate kept at elevated temperature [17]. They have found that already temperature of 270 °C is high enough to obtain oligomers (Fig. 2).

Comparison of STM images of polyanthrylene obtained on rutile titania [17] and gold surfaces [6] reveals an excellent agreement between the two. In both cases, the oligomers are seen as composed of protrusions appearing alternately on both sides with periodicity of 0.86 nm which is in line with the periodicity of the bianthryl core. Furthermore, the width of observed structures is the same no matter, which kind of the substrate is taken into account. The above discussed agreement was a strong evidence that observed molecular objects are indeed polyanthrylene oligomers. Additional prove was provided by means of STM tip-induced manipulation. Kolmer et al. [17] applied rather harsh scanning conditions to observed objects showing that they may be moved to another location in an intact form. If the molecular precursors have not been bound by covalent bonds in the molecular chain they would have surely fallen apart in such conditions. It was not the case, demonstrating that precursors are bonded by a strong covalent bond.

Having achieved formation of polyanthrylene oligomers Kolmer et al. [17] tried to transform these into graphene nanoribbons in a similar way as reported by Cai et al. for polyanthrylene on gold substrate [6]. In the latter case annealing sample with oligomers at 400 °C was enough to induce a cyclodehydrogenation reaction and formation of fully aromatic system. It appeared that for polyanthrylene on $TiO_2(011)$-(2×1) surface even annealing at 450 °C failed to trigger the second step and none of the nanoribbons were found on the surface. Meanwhile it became evident that the oligomers are quite robust to thermal treatment.

Finally, Kolmer and co-authors explored possible reaction pathways for the aryl halide C–C coupling on the (011)-(2×1) surface of rutile titania. On metal surfaces it is accepted that the thermally triggered coupling reaction involves: (1) radicals formation, which are stabilized by the substrate and (2) radicals recombination. Completely different mechanism is proposed for the (011) face of rutile titania. Basing on DFT screening, Kolmer et al. suggest that C–C coupling between DBBA monomers is realized via a multistep proton assisted process. Surface hydroxyl groups serve as a reservoir of protons. Proton from a surface hydroxyl group is transferred to a DBBA molecule and migrates on the aromatic framework of the latter to eventually attach to a Br-bearing carbon atom. Once the additional proton is attached to the Br-bearing atom the C–Br bond is significantly weakened. The molecular precursor become activated. The highest energy barrier in the activation stage is expected for proton transfer from the surface to the molecule, the migration is anticipated to be a rather easy process. Next, the activated DBBA molecule diffuse on the surface and may encounter inactivated species. Such a meeting of activated and inactivated precursor enables C–C bond formation and release of by products, the most probably Br_2 or HBr (this issue will be discussed in the next section). The proton which activated the linking site may now migrate to another Br-bearing carbon atom in the newly formed dimer and the whole process may repeat. It may be envisaged that the density of surface hydroxyl groups is an important feature in the proposed scenario and may serve as a control parameter for the reaction. That aspect will be discussed in the next section.

Furthermore, the proton transfer from the surface to the molecule should be the limiting factor for the C–C coupling, thus the reaction outcome should not depend strongly on a type of halogen atom used. Particularly it is expected that the whole process should be triggered at comparable temperatures for Br and I atoms used at linking sites. The latter observation is in a striking contrast with respect to what is observed for Ullmann-type coupling on metal surfaces. There, the C–I bonds are cleaved at lower temperatures in comparison to bond scission temperatures for the C–Br bonds. That difference is a foundation of hierarchical growth of extended structures via the aryl halide C–C coupling as proposed by Lafferentz et al. [7].

Lastly, the reaction pathway for the other titania face, i.e. the (110)-(1 $\times$ 1) surface, not necessarily needs to be similar. However, in view of the discussed properties of the surface hydrogen atoms (see previous section) it could be anticipated that some role might be played by the protons in this case too. For example, it has been shown that capture and release of proton is responsible for switching between mobile and immobile states of catechol molecules on a $TiO_2(110)$ surface [78]. Influence of the surface hydrogen atoms on aryl–aryl coupling becomes even more apparent when the difference between α–H and β–H species is taken into account [70].

3.2 Insight into Aryl Halide Coupling on the Rutile (011) Surface (Adapted from Ref. [18] with Permission from the Royal Chemical Society)

The reaction pathway proposed by Kolmer et al. [17] still deserved a further experimental and theoretically investigations. In the following work Kolmer et al. [18] used the diiodoterfluorene (DITF) and dibromoterfluorene (DBTF) molecules as monomers to study the polymerization reaction outcomes on $TiO_2(011)$ surfaces. The latter monomer bearing two terminal bromine atoms had been previously reported to form π-conjugated polymer chains on gold substrates [5, 79, 80] and investigated with regard to their single molecule charge transport properties. In the first part of their experiment the authors followed the behavior of polymerization reaction byproducts as a function of the substrate temperature for both studied monomers. Kolmer et al. combined STM measurements with an in situ quadrupole mass spectrometer (QMS) analysis. Firstly, a full monolayer of monomers was deposited on $TiO_2(011)$ kept at RT. The evaporation of the complete layer was chosen to achieve the sufficient number of reacting molecules that would enable detection of the reaction byproducts in QMS. Afterwards, the QMS signals attributed to the HI (HBr) and I_2 (Br_2) molecules were subsequently recorded during the controlled sample heating up to 400 °C. The QMS signals of both analyzed species for DITF monomer showed the appearance of the pronounced peaks starting at around 260 °C (see Fig. 3a dashed black line). Complementary STM measurements showed that during such thermal treatment an initially unordered layer of precursors was transferred into oligomers (Fig. 3b, c). Presented QMS analysis explains the reduced number, or even complete disappearance of surface defects related to the polymerization byproducts for the DITF precursors deposited at 260 °C (see next paragraphs in this section). At this temperature the iodine atoms are effectively desorbed from the surface in the form of the HI and I_2 molecules. In contrary to that, a similar QMS experiment for DBTF had not yielded any substantial signal coming from the reaction byproducts (Br, HBr and Br_2) for temperatures up to 300 °C. In agreement with the QMS experiment the authors showed that after DBTF deposition on $TiO_2(011)$ kept at 260 °C polymerization byproducts stay on the surface. Moreover, Kolmer et al. showed that the polymerization reaction of the DITF and DBTF precursors takes place also at lower temperatures than 260 °C, i.e. at 200 °C, but in this case the reaction byproducts are present on the surface, which agrees with QMS analysis. Importantly, despite the byproducts behavior Kolmer et al. have not found any qualitative difference in reaction outcome for both studied monomers for temperatures higher than 200 °C.

In the main part of their work Kolmer et al. studied the role of the surface hydroxyl groups in the on-surface polymerization [18]. They prepared a reduced *r*-$TiO_2(011)$-(2×1) surface containing hydroxyl groups with the less than 0.5% coverage (Fig. 4a), where the coverage is relative to the maximum number of the available two-fold coordinated terminal oxygen atoms. The *r*-$TiO_2(011)$ substrate was then used to prepare two other types of well-defined surfaces under UHV conditions. First, the sample was exposed to atomic hydrogen and produced the hydroxylated surfaces. For the purpose of this study, the authors used a reduced rutile surface with the moderate coverage of $5 \pm 1\%$ (5%H-$TiO_2(011)$), shown in Fig. 4b, and with a saturation coverage of $20 \pm 1\%$ (20%H-$TiO_2(011)$, Fig. 4c). The oxidized *o*-$TiO_2(011)$ surface, in turn, was obtained by annealing the hydroxylated *H*-$TiO_2(011)$ surface in an O_2 atmosphere. This process led to formation of the well-defined surface with low density of the surface hydroxyl groups amounting again to less than 0.5%. As a result, such protocols allowed for an effective control over the density of the surface hydroxyl groups and the oxidation state of the substrate as well.

In this part of the work DITF is used as a monomer. The choice of DITF instead of the DBTF was though motivated by desorption of iodine-related reaction byproducts at elevated temperatures. Submonolayer deposition of DITF on the 5% H-$TiO_2(011)$ substrate kept at room temperature gives rise to randomly distributed monomers still preserving their terminal iodine atoms after deposition on the rutile

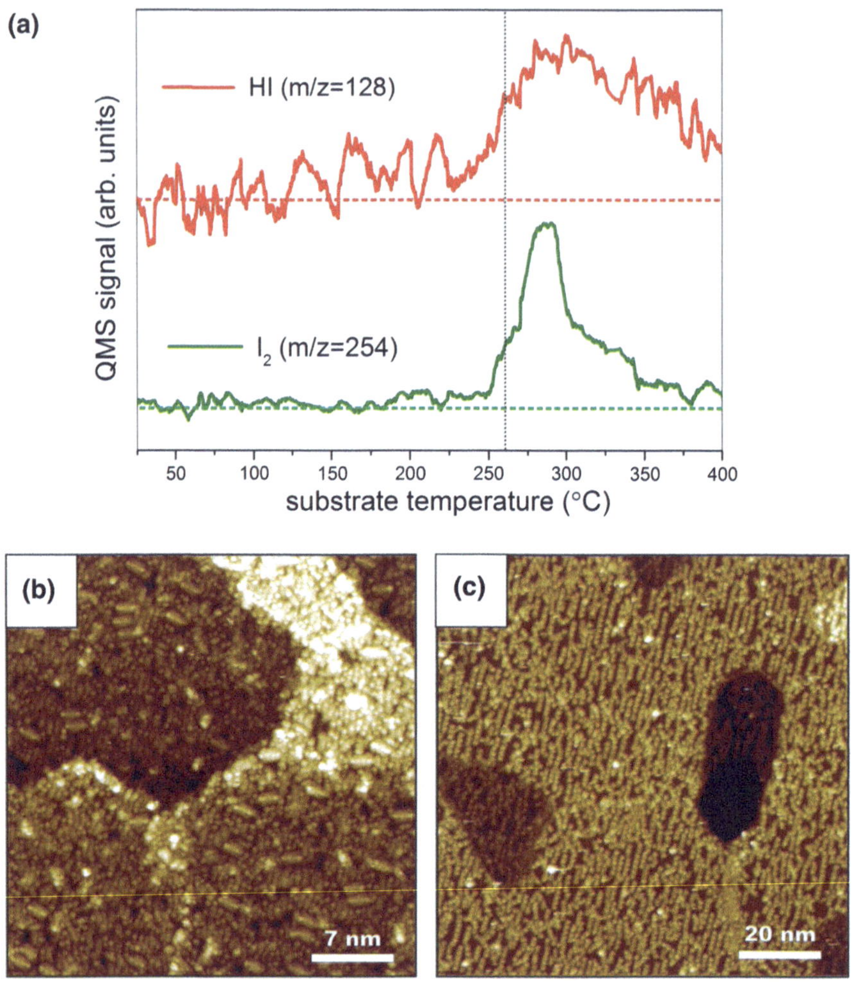

Fig. 3 **a** QMS spectra for m/z = 128 (HI) and m/z = 254 (I_2) obtained during controlled heating of the 5%H-TiO_2(011) covered by one monolayer of DITF molecules deposited at RT. STM images of DITF molecules deposited on the 5%H-TiO_2(011) substrate kept at RT obtained **b** before and **c** after controlled heating of the sample to 350 °C. The STM parameters are +2 V, 5 pA in (**b**) and +1.2 V, 5 pA in (**c**). Adapted from Ref. [18] with permission from The Royal Society of Chemistry

substrate. To study the influence of the density of the surface hydroxyl groups on the polymerization process, the DITF monomers were evaporated on the substrates of different hydroxylation extent, all kept at 260 °C, as at this temperature the reaction byproducts are effectively desorbed from the surface. The STM images of the resulting molecular structures reveal that neither on the *r*-TiO_2(011) surface (Fig. 4d), nor on the *o*-TiO_2(011) surface oligomers are formed. Both substrates

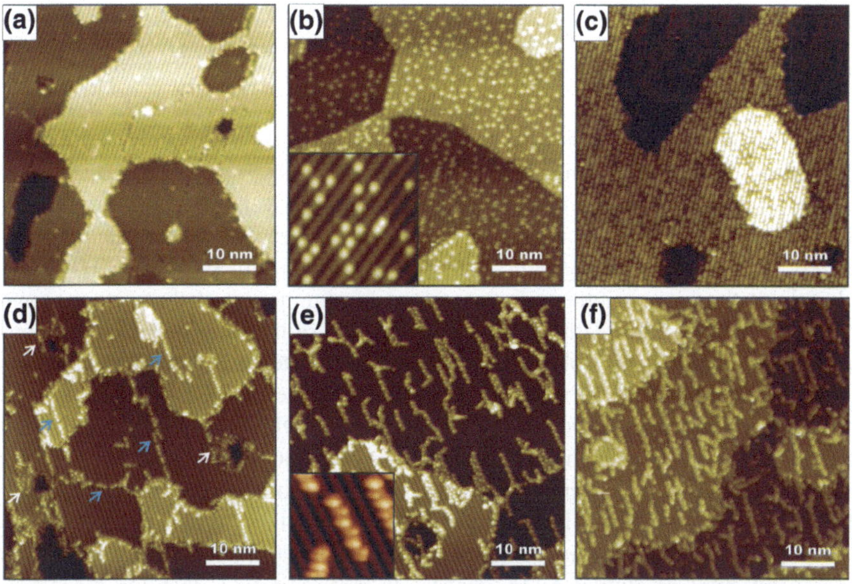

Fig. 4 **a–c** Aryl halide C–C coupling controlled by density of surface hydroxyl groups on rutile $TiO_2(011)$. STM images of (**a**) reduced r-$TiO_2(011)$, (**b**) moderately hydroxylated 5% H-$TiO_2(011)$, and (**c**) fully hydroxylated 20%H-$TiO_2(011)$. The inset in (**b**): high resolution STM image showing surface hydroxyl groups (+1.3 V, 10 pA, 7.5 × 7.5 nm^2). For all large scale STM images, the parameters are +2 V and 10 pA. **d–f** Polymerization of the DITF monomers on rutile surfaces with various hydroxyl groups coverages. **d** STM image of DITF molecules deposited on reduced r-$TiO_2(011)$ kept at 260 °C. **e** STM image of DITF molecules deposited on moderately hydroxylated 5%H-$TiO_2(011)$ kept at 260 °C. Inset: high resolution STM image showing a dimer (7 × 7 nm^2). **f** STM image of DITF molecules deposited on fully hydroxylated 20%H-$TiO_2(011)$ kept at 260 °C. For all STM images the parameters are +2 V and 10 pA. Arrows in (**d**) point to DITF monomers trapped on domain boundaries (blue) and DITF forming supramolecular aggregates (white). Adapted from Ref. [18] with permission from The Royal Society of Chemistry

have almost no surface hydroxyl groups, and in such a case the DITF monomers are found only on the terrace edges and domain boundaries, which are chemically more active than the bare terraces. In particular, the molecules trapped on the domain boundaries are clearly intact and rather well separated (blue arrows in Fig. 4d), proving that under these conditions the polymerization reaction did not occur. The remaining admolecules form supramolecular aggregates, which are also clearly seen in STM images (white arrows in Fig. 4d). In dramatic variance to these results, evaporation of the DITF monomers on the hydroxylated 5%H-$TiO_2(011)$ and 20% H-$TiO_2(011)$ substrates leads to formation of well-developed oligomeric species (Fig. 4e, f), which are preferentially oriented along the surface reconstruction rows. Analysis of the intramolecular STM contrast of the dimer structure (inset in Fig. 4e) proves clearly a covalent coupling between the monomer building blocks. The alternating protrusions along the [01-1] direction, which correspond to the differently oriented dimethyl groups in the polyfluorene chain, all occur with the same separation distance of 0.84 ± 0.01 nm. This value is in agreement with the expected length of a single fluorene unit (0.845 nm).

In the STM images (Fig. 4e, f) one can also observe a decrease of the density of the surface hydroxyl groups as compared to the substrate state prior to the polymerization. This observation is more evident for the 5%H-TiO_2(011) surface, where almost no hydroxyl groups remain on the surface, whereas in the case of the 20%H-TiO_2(011) surface their number is largely reduced, yet they do not vanish completely. It is worth noticing here, that simple annealing of the rutile TiO_2(011)-(2 × 1) at these temperatures does not lead to elimination of the surface-bound hydrogen [62]. Therefore, the spontaneous removal of the surface hydroxyl groups during the course of the polymerization clearly indicates that the hydrogen atoms directly take part in this process. For example, hydrogen could be involved in the formation of hydrogen iodide as a byproduct of the DITF deiodination activating step. Moreover, the oligomers grown on the 5%H-TiO_2(011) substrate (Fig. 4e) exhibit increased chain lengths as compared to those grown on the 20%H-TiO_2(011) substrate (Fig. 4f). This qualitative observation is substantiated by statistical analysis of the polymerization reaction products from the detailed inspection of several high resolution STM images. It proves indeed, that for the 5%H-TiO_2(011) substrate the oligomers are longer than for 20%H-TiO_2(011), where the mere dimer structures dominate. Then, the efficiency of the DIFT coupling reaction can be described by a conversion parameter, expressed as the ratio (percentage) of the number of monomer connections (C–C bonds) successfully made during the polymerization to the total number of reactive groups (C–I bonds) initially present. The conversion (evaluated using the total number of counted DITF monomers $N = 901$) drops from 55% for 5%H-TiO_2(011) to 40% for 20%H-TiO_2(011) ($N = 521$). The lower efficiency of the reaction leading to the shorter oligomers that take place on the highly hydroxylated substrate, strongly points to the surface hydroxyl groups serving as the key species controlling the polymerization process.

All experimental findings showed by Kolmer et al. [18] support proton-mediated polymerization reaction mechanism described in the previous section. It nicely explains the observed optimal surface hydroxyl group coverage for the best polymerization outcome, since for coupling to occur the activated, protonated monomers (and growing oligomers) may only react with the intact monomers (due to charge repulsion between the protonated species). Thus, the surface hydroxyl coverage, and hence the proton availability, controls the ratio of the activated to non-activated monomers, and thereby the entire polymerization process.

The results of DFT molecular modeling presented by Kolmer et al. confirm the kinetic feasibility of the postulated reaction steps for DITF homocoupling on TiO_2(011). The energetic barrier for the initial proton transfer is quite low (0.92 eV), and the subsequent coupling proceeds spontaneously with an associated activation energy of 0.98 eV. The activation energies for the analogous coupling of two intact (DITF–DITF) and two protonated ($DITFH^+$–$DITFH^+$) molecules are much higher (2.45 and 1.95 eV, respectively), accounting well for the experimental observations. The high barrier for the DITF–DITF dimerization results from the fact that two inactivated C–I bonds have to be strongly distorted before the formation of C–C bond. In the case of the protonated monomers the acquired charge give rise to their already mentioned strong electrostatic repulsion enhancing the reaction barrier.

4 Aryl Halide Coupling on the Rutile (110) Surface

Experiments described in last two sections were performed on the (011) face of rutile titania. Due to different physio-chemical properties of (110) face discussed in Sect. 2, which are directly related to their chemical activity it was not straightforward to assume that aryl halide coupling is also feasible on the (110) face of rutile TiO_2. Particularly, the surface reconstruction structure with easily accessible five-fold coordinated titanium atoms causes lower diffusion rates for adsorbed organic molecules, which could be a fundamental obstacle against efficient reaction conditions. Moreover, the five-fold coordinated Ti might tend to bind activated molecular precursors and block formation of new C–C bonds between them.

However, in a recent work by Vasseur and co-authors [19] it was shown that clever choice of molecular precursor, in this case 4,4″-dibromoterphenyl (DBTP), results in a similarly successful polymerization outcome on the (110) face of rutile as on the previously reported reactions on the (011) surface. DBTP precursor structure is tailored to the reconstruction of the (110)-(1×1) surface, which is reflected in a well-ordered supramolecular face obtained after deposition of molecules on substrate kept at RT. The molecules are oriented along the [001] crystallographic direction of the substrate, that is along the (1×1) reconstruction rows. The supramolecular layer structure is confirmed by RT-STM and low-energy electron diffraction (LEED) studies, which both show periodicity of 0.65 nm in the [1–10] direction, that is across the reconstruction rows. This distance matches exactly to periodicity of a bare rutile (110)-(1×1), which proves the closely packed molecular arrangement with one molecule per reconstruction row in the [1–10] direction of the substrate. Due to limited STM resolution the authors did not described in detail the adsorption geometry, however, it is expected that in this highly ordered molecular face neighbouring DBTP precursors on the same row of the substrate are oriented with adjacent Br atoms. This is in a high contrast to the previous studies on the (011) face [17, 18], where non-activated precursors did not arrange in such an ordered structures. The fact of ordering molecular precursors on the (110) face prior to the coupling reaction seems to be a pivotal element for the successful reaction outcome.

As in the previous studies on the (011) faces Vasseur et al. used two strategies to activate the monomers. In the first type of experiments the RT supramolecular face was progressively annealed and system was monitored in situ by X-ray photoemission spectroscopy (XPS). Thermal activation of the RT supramolecular face occurs at about 200 °C. It is reflected by shifts of both C 1s and Br 3d core levels observed in XPS (Fig. 5b), which are considered as fingerprints of the

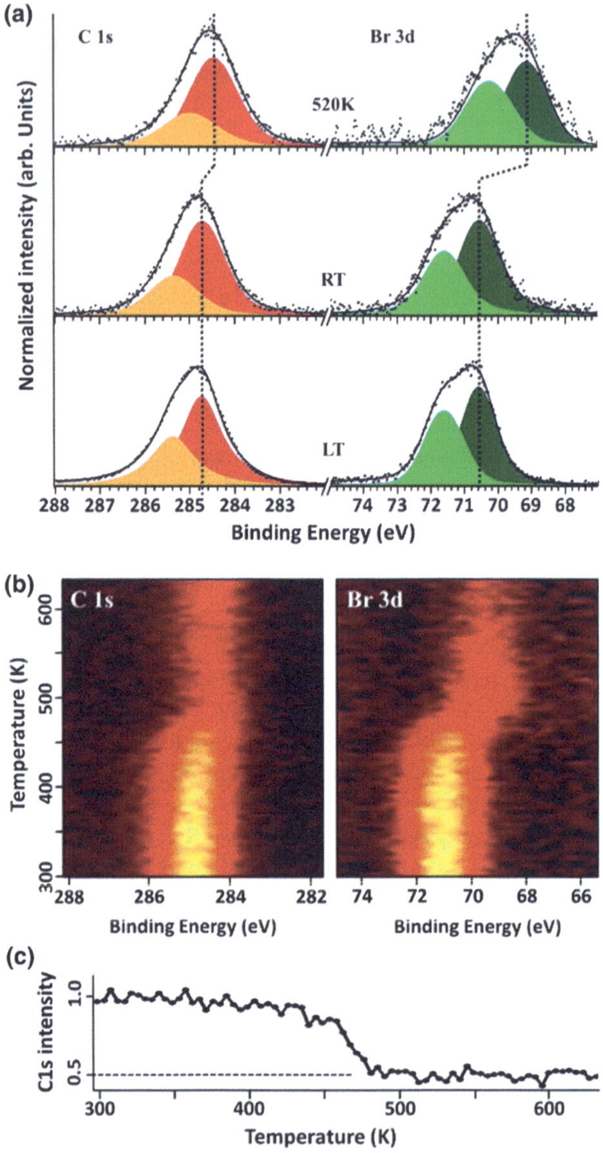

Fig. 5 X-ray photoemission spectroscopy: **a** C 1s and Br 3d core level spectra measured after a multilayer deposition of DBTP with the sample kept at 80 K (bottom), after 1 ML deposition at RT (middle), and after 1 ML deposition at 520 K (top); **b** Evolution of C 1s and Br 3d core levels as a function of temperature, measured after deposition of 1 ML at RT; **c** Evolution of the integrated C 1s signal as a function of temperature, extracted from (**b**). Reprinted from Ref. [19] with permission from The American Chemical Society

dehalogenation of precursors [81]. The corresponding shifts are accompanied with a general decrease in intensity of XPS signals assigned to a significant desorption of molecules upon thermal treatment (see Fig. 5b, c). The temperature coincidence of both processes leads the authors to a conclusion of comparable activation energy for C–Br scission and the binding energy of DBPT molecules in the RT phase. The change in supramolecular phase is also reflected in LEED experiment, which did not show initially present supramolecular ordering pattern after post-annealing of the sample to about 220 °C. RT-STM studies of a sample post-annealed to about 250 °C show short chain-like structures aligned with the substrate [001] direction. The chains correspond to poly(p-phenylene) (PPP), which are however surrounded by many non-assigned cluster-like defects resulting in an overall low-quality of surface morphology.

In order to improve the quality of formed PPP chains Vasseur et al. decided to follow the other strategy and deposit DBTP precursors directly on a substrate kept at elevated temperature of about 250 °C. The resulting XPS analysis of corresponding C 1s and Br 3d core level peak intensities match qualitatively the one obtained by the first strategy. Interestingly, in these optimal conditions, which are close to the one reported previously by Kolmer et al. [17, 18], the PPP chains revealed by RT-STM analysis are effectively longer than after post-annealing of the RT molecular phase. The high resolution STM images and LEED studies confirm the formation of polymers as obtained lattice parameter along the [001] direction is about 0.42 nm, which matches the theoretically predicted inter-phenyl distance obtained by DFT for a PPP chain, as well as, the value measured experimentally for a PPP chain on Cu(110) [81].

The formation of PPP chains on rutile $TiO_2(110)$ surface is unambiguously proven by angle resolved photoelectron spectroscopy (ARPES) studies, which were performed by Vasseur et al. to directly probe the electronic structure of different molecular phases. The authors mapped the electron dispersion with ARPES along the [001] direction, parallel to the reconstruction rows with incident photons of 21.22 eV energy (Fig. 6). The formation of supramolecular face of DBTP at RT results in a well-defined molecular states appear just over valence band of the bare $TiO_2(110)$ substrate (see Fig. 6a, b, d) at energy around −2.9 eV and are attributed to the highest occupied molecular orbital (HOMO) states, the π states. Importantly, the ARPES data obtained for PPP chains formed by deposition of DBTP at about 300 °C (Fig. 6c) present a strongly dispersive band with its top shifted to about −2 eV. This about 0.9 eV shift toward the Fermi energy is in agreement with the HOMO/LUMO (lowest unoccupied molecular orbital) gap reduction related to formation of C–C bonds between the precursors. Moreover, by the direct parabolic

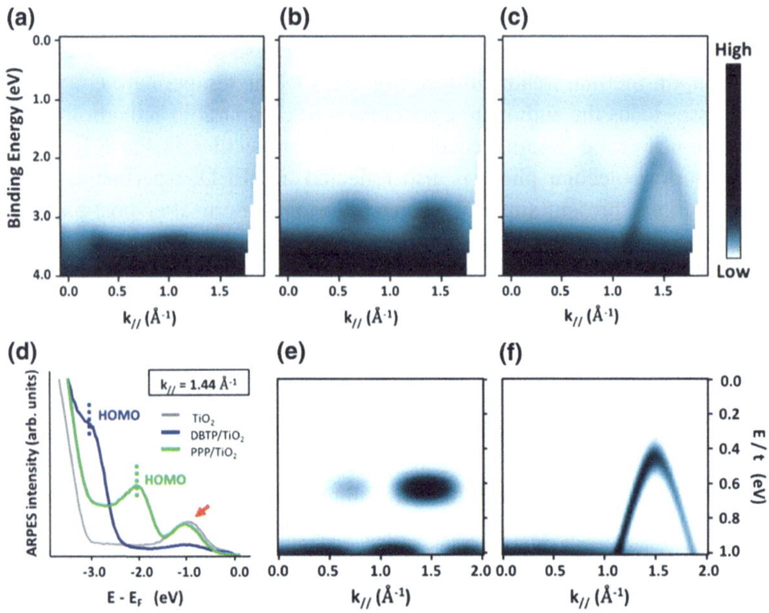

Fig. 6 Angle-resolved photoemission spectroscopy: **a–c** Experimental raw ARPES intensity maps acquired on the pristine surface (**a**), after 1 ML DBTP deposition at room temperature (**b**), and after 1 ML deposition with the sample kept at 575 K (**c**); **d** Energy dispersion curves taken at $k_{\|} = 1.44$ Å^{-1} from the three previous maps; **e**, **f** Theoretical ARPES intensity distributions calculated within a tight-binding model for a terphenyl molecule (**e**) and an infinite poly (p-phenylene) chain (**f**). The energy scale (E) is normalized by the resonance integral (t) between the *pz* orbitals of two neighboring carbon atoms. Reprinted from Ref. [19] with permission from The American Chemical Society

fit to the obtained band Vasseur et al. were able to determine an effective mass of $-0.17\ m_0$, together with the corresponding group velocity reaching about 9.2×10^5 m s^{-1}. These results are the first spectroscopic evidence that atomically precise carbon-based nanostructures can be synthesized on rutile TiO_2. Moreover, they show that the strategy of aryl halide C–C coupling is versatile and can be applied to other faces of rutile titania.

Finally, Vasseur et al. discuss the defect state (DS) behaviour during their ARPES experiment. DS located close to 1 eV below the Fermi energy (red arrow in Fig. 6d) is assigned to mentioned in Sect. 2 excess charge related to surface defects, i.e. oxygen vacancies, hydroxyl groups and interstitial Ti atoms. The authors show that DS disappears after DBTP deposition at RT. In contrast, the very state is clearly seen for bare surface and after DBTP deposition at 300 °C. Accordingly to adduced XPS data there is no signal related to Br 3d levels at the temperature of 300 °C, suggesting there is no Br atoms present on the surface. Basing on this assumption the authors relate the DS disappearing effect to interaction between halogen atoms and surface defects, particularly oxygen vacancies and interpret it as a possible hint for interpretation of the reaction mechanism for on-surface polymerization on $TiO_2(110)$ surface. However, the detailed analysis of this effect is still needed as the resolution of laboratory sources of X-ray for ARPES and XPS experiments do not allow to qualitative verification of possible reaction scenarios and the authors left this issue for future research. Moreover, due to the mentioned difference in the chemical activity between the two surfaces, it is still an open question, whether the reaction on the (110) face of rutile follows the mechanism proposed for the (011) face.

5 Summary and Conclusions

In summary, in this chapter we described the first successful experiments related to aryl halide covalent coupling on two different crystallographic faces of rutile titanium dioxide. Due to complexity of structural and electronic properties of rutile titanium dioxide surfaces detailed description of reaction pathways of aryl–aryl homocoupling is far from being complete. The first experiments suggest, though, future research directions. Particularly, the reduction state of the surface and the bulk seems to be of crucial importance. The reduction state is directly related to surface defects including oxygen vacancies, hydroxyl groups and H atoms, as well as, subsurface titanium interstitials. Detailed studies on the influence of these defects on the aryl halide reaction outcome are needed.

First step in that direction has been made. The experiment relating the surface density of hydroxyl groups on the rutile $TiO_2(011)$-(2×1) with the C–C coupling indeed demonstrated that the efficiency of the reaction could be governed by these defects [18]. Such a result strongly supports the already proposed proton-mediated reaction mechanism [17]. However, presentation of a clear-cut evidence for that reaction pathway is still an open issue. Different physico-chemical properties of other rutile surfaces, including specifically the (110)-(1×1) face, do not allow for direct generalization of the proposed mechanism to all rutile titania faces.

The only available experimental results on the aryl–aryl homocoupling on the (110) face do not address the question on the reaction pathway and role played in it by surface defects and substrate reduction state [19]. Rather, the electronic properties of polymeric chains obtained by aryl-halide coupling are in the focus. By employing angle resolved photoelectron spectroscopic characterisation it is shown that the discussed strategy can deliver carbon based conjugated systems, alike to graphene nanoribbons, directly on top of transition-metal oxide surface. The detailed analysis of the reaction mechanism is in this case still lacking. In the experimental verification of possible reaction pathways one will certainly need to correlate the reaction outcome with a method of substrate preparation. In this respect, the analysis should include variations of typical surface defects, like oxygen vacancies or step edges [82].

One additional, important conclusion can be made. From the experiment on the (110) face follows that judicious choice of molecular precursors that takes into account anisotropy of the substrate may lead to increased reaction yield, in terms of oligomer chains length. Supposedly, a similar effect can be observed on the other face too.

The presented results clearly show that thermally triggered Ullmann-like strategy is a very versatile tool for organic nanostructures formation and can be extended from coinage metal surfaces to other classes of technologically important martials. Given the fact that titanium dioxide is already found in various applications, adding yet another method of well-controlled surface modifications by formation of on-surface organic nanostructures, strengthens position of rutile titania among other materials.

Acknowledgements Support from the Polish Ministry of Science and Higher Education, contract no. 0341/IP3/2016/74 is gratefully acknowledged.

References

1. Grill, L., et al.: Nano-architectures by covalent assembly of molecular building blocks. Nat. Nanotechnol. **2**, 687–691 (2007)
2. Bieri, M., et al.: Porous graphenes: two-dimensional polymer synthesis with atomic precision. Chem. Commun. **45**, 6919–6921 (2009)
3. Lipton-Duffin, J.A., et al.: Synthesis of polyphenylene molecular wires by surface-confined polymerization. Small **5**(5), 592–597 (2009)
4. Perepichka, D.F., Rosei, F.: Extending polymer conjugation into the second dimension. Science **323**(5911), 216–217 (2009)
5. Lafferentz, L., et al.: Conductance of a single conjugated polymer as a continuous function of its length. Science **323**(5918), 1193–1197 (2009)
6. Cai, J., et al.: Atomically precise bottom-up fabrication of graphene nanoribbons. Nature **466**, 470–473 (2010)
7. Lafferentz, L., et al.: Controlling on-surface polymerization by hierarchical and substrate-directed growth. Nat. Chem. **4**, 215–220 (2012)
8. Gourdon, A.: On-surface covalent coupling in ultrahigh vacuum. Angew. Chem. Int. Ed. **47**, 6950–6953 (2008)
9. Franc, G., Gourdon, A.: Covalent networks through on-surface chemistry in ultra-high vacuum: state-of-the-art and recent developments. Phys. Chem. Chem. Phys. **13**, 14283–14292 (2011)
10. Lindner, R., Kuhnle, A.: On-surface reactions. ChemPhysChem **16**, 1582–1592 (2015)
11. Fan, Q., Gottfried, J.M., Zhu, J.: Surface-catalyzed C-C covalent coupling strategies toward the synthesis of low-dimensional carbon-based nanostructures. Acc. Chem. Res. **48**(8), 2484–2494 (2015)
12. Shen, Q., Gao, H.-Y., Fuchs, H.: Frontiers of on-surface synthesis: from principles to applications. Nano Today **13**, 77–96 (2017)
13. Björk, J., Hanke, F., Stafström, S.: Mechanisms of halogen-based covalent self-assembly on metal surfaces. J. Am. Chem. Soc. **135**(15), 5768–5775 (2013)
14. Treier, M., et al.: Surface-assisted cyclodehydrogenation provides a synthetic route towards easily processable and chemically tailored nanographenes. Nat. Chem. **3**(1), 61–67 (2011)

15. Berner, N.C., et al.: Adsorption of 5,10,15,20-tetrakis (4-bromophenyl) porphyrin on germanium(001). Phys. Status Solidi C **9**, 1404–1407 (2012)
16. Berner, N.C.: Towards stable molecular nanostructures on a semiconductor surface. School of Physics, Trinity College, University of Dublin, Dublin, Ireland (2012)
17. Kolmer, M., et al.: Polymerization of polyanthrylene on a titanium dioxide (011)-(2×1) surface. Angew. Chem. Int. Ed. **52**, 10300–10303 (2013)
18. Kolmer, M., et al.: On-surface polymerization on a semiconducting oxide: aryl halide coupling controlled by surface hydroxyl groups on rutile TiO_2(011). Chem. Commun. **51**, 11276–11279 (2015)
19. Vasseur, G., et al.: Pi band dispersion along conjugated organic nanowires synthesized on a metal oxide semiconductor. J. Am. Chem. Soc. **138**(17), 5685–5692 (2016)
20. Olszowski, P., et al.: Aryl halide C–C coupling on Ge(001): H surfaces. J. Phys. Chem. C **119** (49), 27478–27482 (2015)
21. Kittelmann, M., et al.: On-surface covalent linking of organic building blocks on a bulk insulator. ACS Nano **5**, 8420–8425 (2011)
22. Kittelmann, M., et al.: Sequential and site-specific on-surface synthesis on a bulk insulator. ACS Nano **7**(6), 5614–5620 (2013)
23. Bieri, M., et al.: Two-dimensional polymer formation on surfaces: insight into the roles of precursor mobility and reactivity. J. Am. Chem. Soc. **132**, 16669–16676 (2010)
24. Zebari, A.A.A., Kolmer, M., Prauzner-Bechcicki, J.S.: STM tip-assisted engineering of molecular nanostructures: PTCDA islands on Ge(001):H surfaces. Beilstein J. Nanotechnol. **4**, 927–932 (2013)
25. Zebari, A.A.A., Kolmer, M., Prauzner-Bechcicki, J.S.: Characterization of PTCDA nanocrystals on Ge(001):H-(2 × 1) surfaces. Appl. Surf. Sci. **332**, 403–408 (2015)
26. Loske, F., et al.: Growth of ordered C-60 islands on TiO_2(110). Nanotechnology **20**(6) (2009)
27. Prauzner-Bechcicki, J.S., et al.: High-resolution STM studies of terephthalic acid molecules on rutile TiO_2(110)-(1 × 1) surfaces. J. Phys. Chem. C **113**(21), 9309–9315 (2009)
28. Godlewski, S., et al.: Adsorption of organic molecules on the TiO_2(011) surface: STM study. J. Chem. Phys. **134**(22), 224701 (2011)
29. Godlewski, S., et al.: Supramolecular ordering of PTCDA molecules: the key role of dispersion forces in an unusual transition from physisorbed into chemisorbed state. ACS Nano **6**(10), 8536–8545 (2012)
30. Godlewski, S., et al.: [11]Anthrahelicene on TiO_2 surfaces. Surf. Sci. **606**(21–22), 1600–1607 (2012)
31. Grinter, D.C., et al.: Binding of a benzoate dye-molecule analogue to rutile titanium dioxide surfaces. J Phys. Chem. C **116**(1), 1020–1026 (2012)
32. English, C.R., et al.: Formation of self-assembled monolayers of pi-conjugated molecules on TiO_2 surfaces by thermal grafting of aryl and benzyl halides. Langmuir **28**(17), 6866–6876 (2012)
33. Lanzilotto, V., et al.: Commensurate growth of densely packed PTCDI islands on the rutile TiO_2(110) surface. J. Phys. Chem. C **117**(24), 12639–12647 (2013)
34. Zając, Ł., et al.: Ordered heteromolecular overlayers formed by metal phthalocyanines and porphyrins on rutile titanium dioxide surface studied at room temperature. J. Chem. Phys. **143**(22), 224702 (2015)
35. Godlewski, S., et al.: Transformations of PTCDA structures on rutile TiO_2 induced by thermal annealing and intermolecular forces. Beilstein J. Nanotechnol. **6**, 1498–1507 (2015)
36. Zajac, L., et al.: Self-assembling of Zn porphyrins on a (110) face of rutile TiO_2–The anchoring role of carboxyl groups. Appl. Surf. Sci. **379**, 277–281 (2016)
37. Prauzner-Bechcicki, J.S., et al.: Scanning probe microscopy studies on the adsorption of selected molecular dyes on titania. Beilstein J. Nanotechnol. **7**, 1642–1653 (2016)
38. Prauzner-Bechcicki, J., Kolmer, M., Szymonski, M.: Aryl–aryl coupling on semiconductor surfaces. In: Encyclopedia of Interfacial Chemistry: Surface Science and Electrochemistry. Elsevier, Amsterdam. http://doi.org/10.1016/B978-0-12-409547-2.13105-1 (2017)
39. Diebold, U.: The surface science of titanium dioxide. Surf. Sci. Rep. **48**(5–8), 53–229 (2003)

40. Pang, C.L., Lindsay, R., Thornton, G.: Chemical reactions on rutile $TiO_2(110)$. Chem. Soc. Rev. **37**(10), 2328–2353 (2008)
41. Dohnálek, Z., Lyubinetsky, I., Rousseau, R.: Thermally-driven processes on rutile $TiO_2(110)$-(1 × 1): a direct view at the atomic scale. Prog. Surf. Sci. **85**(5–8), 161–205 (2010)
42. Thomas, A.G., Syres, K.L.: Adsorption of organic molecules on rutile TiO_2 and anatase TiO_2 single crystal surfaces. Chem. Soc. Rev. **41**(11), 4207–4217 (2012)
43. Godlewski, S., Szymonski, M.: Adsorption and self-assembly of large polycyclic molecules on the surfaces of TiO_2 single crystals. Int. J. Mol. Sci. **14**(2), 2946–2966 (2013)
44. Pang, C.L., Lindsay, R., Thornton, G.: Structure of clean and adsorbate-covered single-crystal rutile TiO_2 surfaces. Chem. Rev. **113**(6), 3887–3948 (2013)
45. Henderson, M.A.: A surface science perspective on photocatalysis. Surf. Sci. Rep. **66**(6–7), 185–297 (2011)
46. Zhang, Z., Yates, J.T.: Band bending in semiconductors: chemical and physical consequences at surfaces and interfaces. Chem. Rev. **112**(10), 5520–5551 (2012)
47. Thompson, T.L., Yates, J.T.: TiO_2-based photocatalysis: surface defects, oxygen and charge transfer. Top. Catal. **35**(3), 197–210 (2005)
48. Thompson, T.L., Yates, J.T.: Surface science studies of the photoactivation of TiO_2 new photochemical processes. Chem. Rev. **106**(10), 4428–4453 (2006)
49. Guo, Q., et al.: Elementary photocatalytic chemistry on TiO_2 surfaces. Chem. Soc. Rev. **45**(13), 3701–3730 (2016)
50. Cai, Y., Feng, Y.P.: Review on charge transfer and chemical activity of TiO_2: Mechanism and applications. Prog. Surf. Sci. **91**(4), 183–202 (2016)
51. Ramamoorthy, M., Vanderbilt, D., Kingsmith, R.D.: 1st-Principles calculations of the energetics of stoichiometric TiO_2 surfaces. Phys. Rev. B **49**(23), 16721–16727 (1994)
52. Gong, X.-Q., et al.: The 2 × 1 reconstruction of the rutile $TiO_2(011)$ surface: a combined density functional theory, X-ray diffraction, and scanning tunneling microscopy study. Surf. Sci. **603**(1), 138–144 (2009)
53. Torrelles, X., et al.: Geometric structure of $TiO_2(011)$-(2 × 1). Phys. Rev. Lett. **101**(18), 185501 (2008)
54. Yurtsever, A., et al.: Imaging the $TiO_2(011)$-(2 × 1) surface using noncontact atomic force microscopy and scanning tunneling microscopy. J. Phys. Chem. C **120**(6), 3390–3395 (2016)
55. Pang, C.L., et al.: (2n × 1) Reconstructions of $TiO_2(011)$ revealed by noncontact atomic force microscopy and scanning tunneling microscopy. J. Phys. Chem. C **118**(40), 23168–23174 (2014)
56. Kubo, T., Orita, H., Nozoye, H.: Surface structures of rutile $TiO_2(011)$. J. Am. Chem. Soc. **129**(34), 10474–10478 (2007)
57. Dupont, C., et al.: Structure of $TiO_2(011)$ revealed by photoelectron diffraction. Phys. Rev. B **94**(24), 241304 (2016)
58. Woolcot, T., et al.: Scanning tunneling microscopy contrast mechanisms for TiO_2. Phys. Rev. Lett. **109**(15) (2012)
59. Wang, Q., et al.: The unexpectedly rich reconstructions of rutile $TiO_2(011)$-(2 × 1) surface and the driving forces behind their formation: an ab initio evolutionary study. Phys. Chem. Chem. Phys. **18**(29), 19549–19556 (2016)
60. Dulub, O., et al.: Structure, defects, and impurities at the rutile $TiO_2(011)$-(2 × 1) surface: a scanning tunneling microscopy study. Surf. Sci. **600**(19), 4407–4417 (2006)
61. Kolmer, M., et al.: Temperature-dependent orientation of self-organized nanopatterns on ion-irradiated $TiO_2(110)$. Phys. Rev. B **88**(19), 195427 (2013)
62. Tao, J.G., et al.: Diffusion and reaction of hydrogen on rutile $TiO_2(011)$-2 × 1: the role of surface structure. J. Phys. Chem. C **116**(38), 20438–20446 (2012)
63. Li, M., et al.: The influence of the bulk reduction state on the surface structure and morphology of rutile $TiO_2(110)$ single crystals. J. Phys. Chem. B **104**(20), 4944–4950 (2000)
64. Setvin, M., et al.: Direct view at excess electrons in TiO_2 rutile and anatase. Phys. Rev. Lett. **113**(8), 086402 (2014)

65. Deskins, N.A., Rousseau, R., Dupuis, M.: Defining the Role of Excess Electrons in the Surface Chemistry of TiO_2. The Journal of Physical Chemistry C **114**(13), 5891–5897 (2010)
66. Kowalski, P.M., et al.: Charge localization dynamics induced by oxygen vacancies on the TiO_2(110) surface. Phys. Rev. Lett. **105**(14), 146405 (2010)
67. Heckel, W., et al.: The role of hydrogen on the adsorption behavior of carboxylic acid on TiO_2 surfaces. J. Phys. Chem. C **118**(20), 10771–10779 (2014)
68. Yuan, F., et al.: The hydroxylated and reduced rutile TiO_2(011)-2 × 1 surfaces: a first-principles study. Surf. Sci. **628**, 126–131 (2014)
69. Zhang, D., Yang, M., Dong, S.: Hydroxylation of the rutile TiO_2(110) surface enhancing its reducing power for photocatalysis. J. Phys. Chem. C **119**(3), 1451–1456 (2014)
70. Zhang, Z., Yates, J.T.: A new form of chemisorbed photo- and electro-active atomic H species on the TiO_2(110) surface. Surf. Sci. **652**, 195–199 (2016)
71. Mao, X., et al.: Band-gap states of TiO_2(110): major contribution from surface defects. J. Phys. Chem. Lett. **4**(22), 3839–3844 (2013)
72. Yang, W., et al.: Controlled vacancy-assisted C–C couplings of acetaldehyde on rutile TiO_2(110). J. Phys. Chem. C **118**(48), 27920–27924 (2014)
73. Idriss, H., Barteau, M.A.: Selectivity and mechanism shifts in the reactions of acetaldehyde on oxidized and reduced TiO_2(001) surfaces. Catal. Lett. **40**(3), 147–153 (1996)
74. Benz, L., et al.: McMurry chemistry on TiO_2(110): reductive C=C coupling of benzaldehyde driven by titanium interstitials. J. Am. Chem. Soc. **131**(41), 15026–15031 (2009)
75. Benz, L., et al.: Molecular imaging of reductive coupling reactions: interstitial-mediated coupling of benzaldehyde on reduced TiO_2(110). ACS Nano **5**(2), 834–843 (2011)
76. Cremer, T., Jensen, S.C., Friend, C.M.: Enhanced photo-oxidation of formaldehyde on highly reduced o-TiO_2(110). J. Phys. Chem. C **118**(50), 29242–29251 (2014)
77. Koch, M., et al.: Voltage-dependent conductance of a single graphene nanoribbon. Nat. Nanotechnol. **7**(11), 713–717 (2012)
78. Li, S.-C., et al.: Hydrogen bonding controls the dynamics of catechol adsorbed on a TiO_2(110) surface. Science **328**(5980), 882–884 (2010)
79. Saywell, A., et al.: Polymerization on stepped surfaces: alignment of polymers and identification of catalytic sites. Angew. Chem. Int. Ed. **51**(21), 5096–5100 (2012)
80. Kawai, S., et al.: Quantifying the atomic-level mechanics of single long physisorbed molecular chains. Proc. Natl. Acad. Sci. U S A **111**(11), 3968–3972 (2014)
81. Di Giovannantonio, M., et al.: Insight into organometallic intermediate and its evolution to covalent bonding in surface-confined Ullmann polymerization. ACS Nano **7**(9), 8190–8198 (2013)
82. Miccio, L.A., et al.: Interplay between steps and oxygen vacancies on curved TiO_2(110). Nano Lett. **16**(3), 2017–2022 (2016)

On-Surface Synthesis of Two-Dimensional Polymers: Rational Design and Electronic Properties

Sabine Maier

Abstract The fabrication of long-range ordered 2D polymers directly on metal surfaces still presents a great challenge in ultra-high vacuum surface science. The structure of the polymer networks is in general predetermined by the coupling chemistry and the symmetry of the molecular precursors. However, the irreversible nature of the C–C coupling reaction readily leads to the formation of defects. Over the last decade, several strategies in the on-surface synthesis have been suggested to improve the structural order. Among them, the programmed hierarchical synthesis through a sequential polymerization proved to be suitable to reduce defects in the formation of 2D polymers. This chapter provides a review on the state-of-the-art structural characterization of surface-supported 2D polymers by established surface science techniques. Further, the chapter focuses on the electronic structure of 2D polymers, which remained experimentally widely unexplored until now.

1 Introduction

Two-dimensional polymers are sheet-like, covalently bonded molecular networks that extend in exactly two dimensions [1, 2]. The ultimate goal in fabricating novel 2D polymeric materials is motivated by a potentially limitless range of structures with unprecedented properties, which can be created. In these organic analogs of graphene, the band gap and the charge carrier mobility may be tuned by using specifically designed molecular building blocks. On-surface polymerization proved successful for the fabrication of one-dimensional structures with atomically defined width, edge termination, and dopants on the surface [3–11]. In contrast, the bottom-up fabrication of crystalline 2D polymers presents a far greater challenge owing to the limited control during their synthesis on the surface despite the wide range of possible surface reactions [12]. Some of the main issues are the bond

---

S. Maier (✉)
Department of Physics, Friedrich-Alexander University Erlangen-Nürnberg, Erwin-Rommel-Strasse 1, 91058 Erlangen, Germany
e-mail: sabine.maier@fau.de

D. G. de Oteyza and C. Rogero (eds.), *On-Surface Synthesis II*, Advances in Atom and Single Molecule Machines,
https://doi.org/10.1007/978-3-319-75810-7_8

flexibility in the molecular building blocks that allows the formation of a range of polygons and the irreversible nature of the created covalent C–C bonds that prohibits an error correction during the network formation in ultra-high vacuum (UHV). These difficulties lead readily to equilibrium disordered structures of limited size [13]. Hence, the on-surface synthesis of two-dimensional periodic structures and the characterization of their structural and physical properties is a growing field of research.

The 2D covalent organic polymers hold great potential in particular for applications in future electronic and optoelectronic devices. The strong coupling between the molecular building blocks by covalent bonds promotes, apart from a great mechanical and thermal stability, the ability to conduct electrons in the extended π-conjugated system. The inherent porous structure induces an electronic band gap that can be controlled by the pore size, density, and geometry [14]. The topology of the network may lead in some cases to unique electronic band structures with linear electronic band dispersion and associated Dirac physics as predicted by density-functional theory (DFT) [15–17]. Hence, some of the 2D polymers promise to combine the outstanding electronic properties of graphene with a tunable band gap, which makes them more suitable than pristine graphene for many electronic and optoelectronic applications that require gapped semiconductor materials [4]. In addition, the use of functionalized building blocks allows for chemically tuning the electronic properties through doping. However, requirement for these exciting electronic properties are long-range ordered structures. The electronic structure of bottom-up fabricated 2D networks remained experimentally so far widely unexplored due to a lack of structural order. Moreover, the pores of the 2D polymers offer fascinating options in host-guest chemistry. Hence, 2D polymer networks may be suitable for future applications in molecular electronics and as membranes for gas sensing, separation, and storage owing to their tunable porous structure and their improved thermal stability compared to non-covalently linked self-assemblies [2, 18, 19].

Despite the numerous potential applications, the synthesis of large-scale, periodic 2D polymers *via* wet chemistry or by on-surface synthesis is still challenging. In organic chemistry, mainly boronic acid-based condensations and Schiff-base reactions have been applied as major reversible reactions so far to successfully fabricate covalent organic frameworks [20]. Numerous graphene-like, 2D carbon materials have been successfully synthesized in flask [21]. However, the synthesis of a whole range of novel 2D carbon allotropes such as graphyne and graphdiyne could so far not be accomplished [22]. In this respect, on-surface synthesis may be a powerful method, since it has been shown that atomically defined carbon-based nanostructures can be fabricated in a bottom-up approach directly on the surface [23–26]. Apart from acting as catalyst, the surface provides a templating effect for the fabrication of novel 2D materials by confining the reacting species in two dimensions. Thereby, the molecular precursors are commonly activated by thermal annealing, but also irradiation with electrons or light [27, 28] have been successfully applied. For their characterization, various surface science techniques are used. However, scanning probe microscopy (SPM) is currently one of the favorite characterization tools for the surface supported 2D polymers, since molecular resolution is readily accomplished, and imaging in real space facilitates a direct assessment of the structural quality. While scanning tunneling microscopy (STM) and spectroscopy provide direct insights into the local topographic and electronic structure, non-contact atomic force (nc-AFM) microscopy has proven important in the unambiguous identification of reaction mechanisms and intermediates, since functionalized tips allow to resolve the molecular structure with sub-molecular resolution [29, 30]. Complementary, X-ray photoelectron spectroscopy (XPS) is commonly used to determine the chemical state of the reaction products, near edge X-ray absorption fine structure (NEXAFS) to obtain information on the adsorption geometry, and angle-resolved photoemission spectroscopy (ARPES) to determine the band dispersion.

In this chapter, recent progress towards the synthesis of ordered 2D surface-supported polymers is reviewed. Thereby, their geometric structure as consequence of various synthesis strategies as well as their electronic structure are discussed.

2 Bottom-Up Fabrication of 2D Networks

2.1 Limited Structural Control

One of the key issues preventing structural control in the bottom-up fabrication of periodic 2D polymers under UHV condition is the irreversible nature of the newly formed intermolecular covalent bonds. In contrast to self-assembled supramolecular structures, which are held together by weak, reversible interactions—e.g. hydrogen bonds, van-der-Waals, or metal-ligand interactions—the covalent C-C bond formation prohibits an error correction in the network formation. Therefore, the experimentally observed 2D network are usually kinetically trapped far from equilibrium. Coupling reactions used in dynamic covalent chemistry, a common concept in supramolecular chemistry that is based on the reversible bond formation, were so far only applied with limited success in on-surface polymerization experiments in UHV. One of the reasons is the irreversible loss of smaller products, e.g. water, which are essential to realize the back-reaction. For instance, Abel and co-workers [31] investigated aryl boronic acids as building blocks, but could not achieve periodic networks on an Ag(111) surface as a consequence of the irreversible bond formation in UHV. In contrast, Yaghi et al. [32] employed successfully the same coupling reaction in a solution approach to obtain the first covalent organic frameworks (COF). Similarly, branched and irregular network structures formed by condensation polymerization of trialdehydes and diamines co-adsorbed on Au(111) in UHV, which is a reversible reaction in solution [33].

As a consequence of the irreversible bond formation, polygonal defects are frequently observed in 2D networks fabricated in UHV. The defect density is strongly related to the bond flexibility in the molecular building blocks. The energy relaxation for arranging threefold symmetric molecules into isolated, regular *n*-gons favors the hexagon, whose geometry is commensurate with the symmetry of the molecule [13, 34]. However, molecules may form other polygons, at an energy cost of around an eV per ring, which leads readily to equilibrium disordered structures. In conclusion, 2D covalent networks are inherently limited with respect to their structural perfection [13]. The structural quality of covalent networks can however be improved by an optimization of the reaction parameters, as it was shown in the kinetically controlled Ullmann-type coupling on Au(111) [35].

In addition, we also have to consider effects of the molecule-surface interaction on the structural control in the bottom-up fabrication of 2D networks. The substrate does not only act as support but is actively involved in all reaction steps and therefore significantly influences the morphology of the self-organized covalent nanostructures. Bieri et al. showed in an Ullmann-type coupling reaction that polyphenylene networks adapt morphologies from branched, fractal-like structures on Cu(111) to extended, regular 2D networks on Ag(111) [36]. Their DFT analysis of the diffusion and coupling pathways on Cu and Ag revealed that the balance between the diffusion and coupling steps is significantly different on the two substrates and plays an important role in the structural control. While the radicals after debromination spontaneously form covalent intermolecular bonds on Cu, on the Ag surface, diffusion prevails over intermolecular coupling, which results in an overall increased mobility of the molecules on the surface and in the formation of regular 2D networks. Hence, a high mobility (or low coupling affinity of the reactants) is a prerequisite for the growth of dense 2D polymer networks. In addition, also adsorbed side products from the on-surfaces synthesis can influence the diffusion of the networks and prohibit the fabrication of extended two-dimensional networks.

2.2 Rational Design of 2D Polymers

A crucial step in the rational design of 2D polymers in on-surface synthesis experiments is to identify reaction protocols that irrespective of the reaction kinetics lead to periodic networks. The molecular design of the building blocks is hereby a crucial step to obtain 2D polymers with a well-defined structure. Key parameters for the structural control are the symmetry and rigidity of the molecular scaffolds, as well as the functional groups.

In general, the networks symmetry affects the susceptibility to forming defects; the angular strain to form irregular polygons increases by about a factor two for fourfold-symmetric networks compared to threefold-symmetric networks. Therefore, the likeliness for the irregular polygon formation is drastically reduced in a squared network, while nearly inevitable for honeycomb networks built from larger compounds without using a rational design. The geometrical stability of squared networks was already demonstrated in the pioneering work by Grill et al. in the first thermally induced on-surface Ullmann-type coupling. They could obtain defect-free patches of covalently-linked tetra(4-bromophenyl)porphyrins [37]. For honeycomb networks, a rigid molecular scaffold of the precursors is crucial to improve the structural quality, because weak and rigid bonds yield a highly ordered network, while strong and flexible bonds give rise to disordered structures [13]. The prevalence of hexagonal pores has been successfully shown for example for the rigid hexaiodo-substituted macrocycle cyclohexa-*m*-phenylene precursor, that has been applied to synthesize porous graphene (see Fig. 1a) [38].

Ullmann-type coupling is the best studied and most applied polymerization reaction on surfaces. However, organic chemistry is extremely diverse; thus, a good

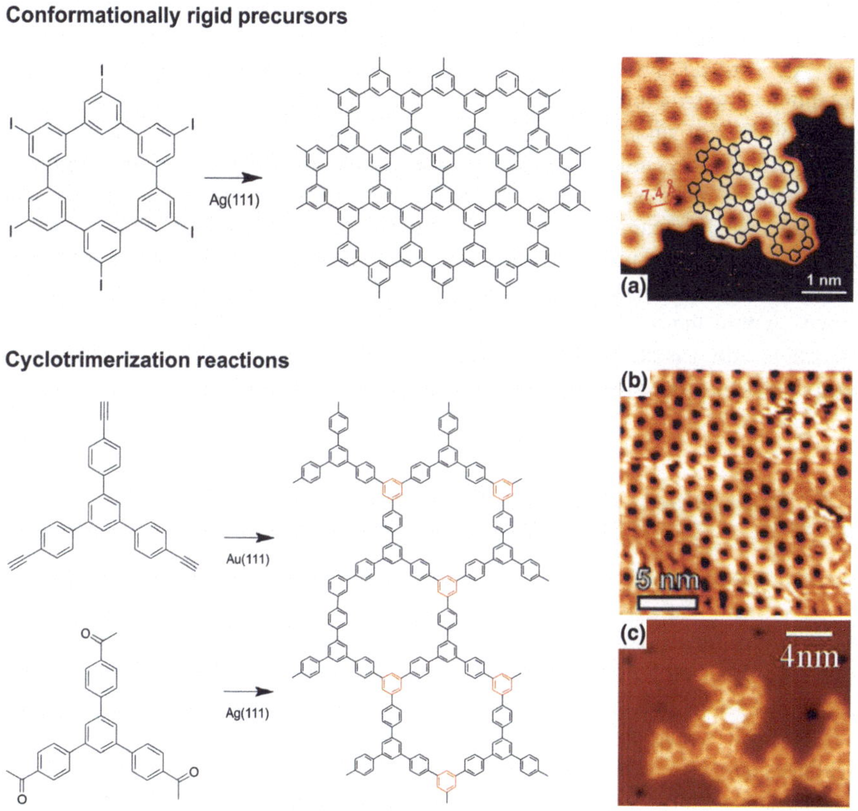

Fig. 1 Selected design strategies in the on-surface synthesis of ordered 2D polymers. **a** Formation of porous graphene *via* an Ullmann-type reaction of the hexaiodo-substituted macrocycle cyclohexa-*m*-phenylene on Ag(111). Adapted with permission from Ref. [38], copyright 2015, RCS. **b–c** Cyclotrimerization of (**b**) 1,3,5-tris(4-ethynylphenyl)benzene and (**c**) 1,3,5-tris (4-acetylphenyl)benzene into 2D conjugated covalent networks. (**b**) Adapted with permission from Ref. [39], copyright 2014, RCS and (**c**) adapted with permission from Ref. [40], copyright 2015, ACS

strategy is to explore different coupling reactions on surfaces that provide more geometrical stability. In contrast to linear reaction schemes, cycloadditions and cyclization reactions have great potential for the fabrication of defect-free 2D polymers in on-surface chemistry, since byproducts in this "node design" are less frequently generated. The high selectivity of the cyclotrimerization reaction was convincingly demonstrated in the on-surface synthesis of well-ordered, two-dimensional networks on Au(111) using 1,3,5-tris(4-ethynylphenyl)benzene (TEB) by Liu et al. (see Fig. 1b), while homocoupling of the same precursor on Ag (111) led to unordered networks [41]. Also, the cyclotrimerization of acetyls to aromatics has shown to be a promising approach to form 2D conjugated covalent networks on surfaces under UHV (see Fig. 1c) [40].

The underlying substrate also provides for templating effects to promote long-range order. For instance, intrinsic adatoms form intermediate organometallic structures with carbon-metal-carbon bonds in the Ullmann-type reaction on the Cu and Ag surfaces instead of directly establishing covalent C-C bonds. These organometallic structures can serve as a template considering the advantage of the reversibility of organometallic networks to equilibrate into highly ordered structures in a self-assembly process. However, the covalent-linking is far from being a topochemical reaction, because the conversion from organometallic to covalent structures is accompanied by a considerable shrinkage of intermolecular distances, which makes it difficult to obtain large domains [42]. In topochemical polymerization reactions the monomers polymerize without the movement of their center of gravity within the molecular lattice, only a minimal movement around the polymerizable group is allowed while the rest remains static. Diacetylene compounds are popular to proceed in topochemical polymerizations [43].

2.3 *Hierarchical Structure Formation*

Hierarchical structure formation is a major step towards increasing the complexity of on-surface synthesis experiments. In a hierarchical synthesis, sequential polymerization is applied using the same or dissimilar reaction types to improve the structural quality of the reaction products. An advantage of the sequential polymerization is the fabrication of geometrically stable and larger intermediate structures, which would be too large for thermal deposition. Sequential on-surface synthesis strategies comprising two different reaction types are commonly applied in one-dimensional structures. The fabrication of graphene nanoribbons often combines an Ullmann-type coupling reaction with a subsequent intramolecular Scholl reaction (cyclodehydrogenation) to planarize the structures [3, 4, 44]. Also, intermolecular dehydrogenation reactions combined with Ullmann-type coupling are used in 1D structures, for instance, to increase the width of graphene nanoribbons [9, 45].

In the hierarchical synthesis of two-dimensional networks mainly the sequential synthesis *via* the same reaction type has been employed. In Ullmann-type coupling reactions, one takes advantage of the bond energy differences between C–I, C–Br, and C–Cl bonds, which leads to a lower activation temperature for aryl iodides than aryl bromides and aryl chlorides. On Au(111), the reaction hierarchy is preserved because stable organometallic intermediates are only rarely observed [46], which makes it an ideal substrate. In the pioneering work done by Lafferentz et al. [47], the successful interlinking of porphyrin building blocks was demonstrated in a hierarchical manner through an Ullmann-type coupling on Au(111) (see Fig. 2a). Thereby, selective activation of the weaker C-I bonds was used to form 1D polymer chains in a first step, which were laterally interconnected by subsequent thermal activation of C–Br bonds at higher temperatures to obtain 2D network structures with high regularity.

For the hierarchical fabrication of open-porous honeycomb networks, Eichhorn et al. used 1,3-bis(*p*-bromophenyl)-5-(*p*-iodophenyl)benzene to create in a first step dimers *via* C-I cleavage and subsequently 2D networks *via* C–Br cleavage (see Fig. 2b) [35]. Similarly, Shi et al. used 1,3-bis(*p*-bromophenyl)-5-(*p*-chlorophenyl)benzene to form dimers in a C–Br cleavage reaction and subsequently networks in a C–Cl cleavage reaction. However, in both studies, no substantial difference could be found in the network quality for the direct and hierarchical polymerization protocols. This suggests that dimers as intermediate products due their large structural flexibility do not provide a

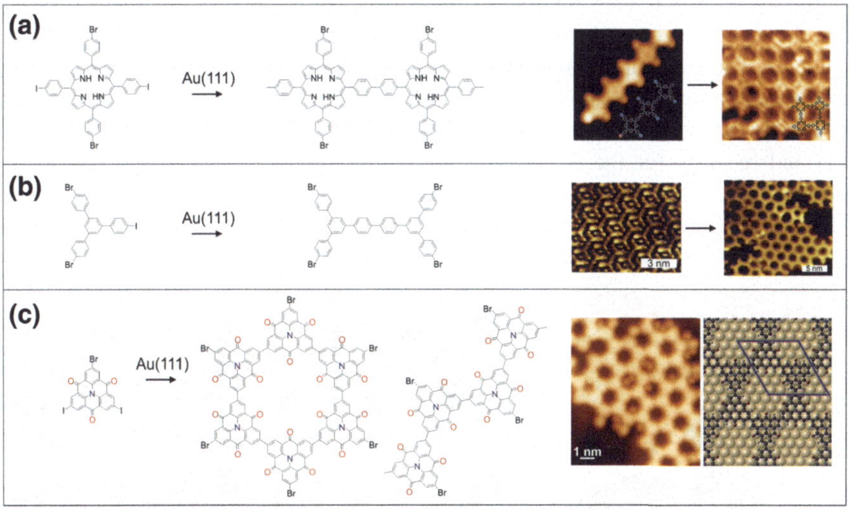

Fig. 2 Hierarchical synthesis of two-dimensional polymers *via* Ullmann-type reactions on Au (111). **a** Formation of square porphyrin networks with chains as intermediate structures. Adapted with permission from Ref. [47], copyright 2012, Nature Chemistry. **b** On-surface synthesis of porous phenylene networks using dimers as intermediate structures. Adapted with permission from Ref. [35], copyright 2014, ACS. **c** Hierarchical synthesis of carbonyl-bridged triphenylamine networks using macrocycles and chains as intermediate structures. The STM image of the 2D triphenylamine-based polymer with the corresponding DFT model is shown on the right. Adapted with permission from Ref. [49], copyright 2017, Nature Communications

suitable template to prevent the formation of pentagonal and heptagonal pores in hexagonal networks [48].

A benefit of the hierarchical synthesis for the fabrication of well-ordered honeycomb networks was shown by Steiner et al. who used a more rigid precursor molecule and larger intermediate structures (see Fig. 2c) [49]. In the hierarchical synthesis, the diiodinated/brominated carbonyl-bridged triphenylamines (CTPA) formed covalently-linked six-membered macrocycles and 1D chains as intermediate structures, which were subsequently connected to a porous 2D polymer as final product. Using brominated macrocycles and chains as reaction intermediates constrains the defects to the rim of the formed 2D networks, leaving the properties of the covalent polymer largely preserved. The covalent linkage of the well-ordered six-membered macrocycles leads by design to the formation of honeycomb networks. The only possible defect would be a combination of 4–8 membered ring pairs, which were, however, not observed likely due to the large strain resulting from the inherent rigidity of the CTPA moieties. Pentagonal and heptagonal defects at the periphery of the porous graphene islands may be created by interlinking short chains before the second reaction occurs. Hence, covalent linking of macrocycles form defect-free 2D networks after the second reaction step, while chains are still susceptible to build pentagonal and heptagonal defects. Unfortunately, there is currently no method to tune the selectivity towards the macrocycle formation. The key of this hierarchical approach is that macrocycles can be fabricated *via* a bottom-up approach with high structural quality and rarely show other polygons than hexagons, whose geometry is commensurate with the symmetry of the molecule. The weak, non-covalent interactions among the macrocycles makes them structurally controllable and leads to equilibrated structures. This was also observed previously in the on-surface synthesis of hyperbenzene [50] and honeycombene [51] and recently for the diiodo-substituted triphenylamines [49] (see Fig. 3). In addition, the self-assembly of intermediate macrocycles and chains is essential to

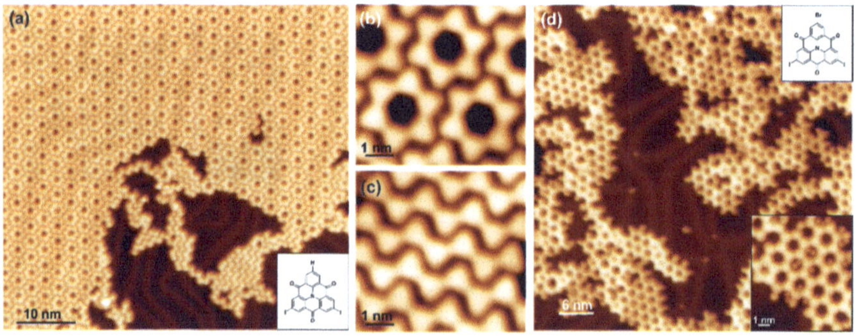

Fig. 3 **a–c** Self-assembly of macrocycles (**b**) and chains (**c**) fabricated by on-surface synthesis from diiodo-substituted carbonyl-bridged triphenylamine precursor molecules on Au(111). **d** Hierarchically synthesized 2D polymer with carbonyl-functionalized pores on Au(111). Adapted with permission from Ref. [49], copyright 2017, Nature Communications

achieve structurally well-ordered covalent 2D networks, because undesired structural elements can segregate to the periphery of the network.

3 Electronic Structure of 2D Polymers

Much of the research of organic 2D polymers is driven by their unique electronic properties predicted by DFT. This includes the expectation to correlate the geometric structure with electronic properties [15]. While the electronic properties of 1D structures fabricated by on-surface synthesis have thoroughly been characterized with several surface science techniques including scanning tunneling spectroscopy (STS) and ARPES [5, 7, 8, 52], similar approaches to study the band structure in two-dimensional π-conjugated systems remain largely unexplored. The concept of band gap engineering in 2D is especially interesting as the number of intermolecular connections scales linearly with the oligomer length in 1D but superlinearly in 2D, which leads to a different HOMO-LUMO gap (HLG) for otherwise similar polymers (see Fig. 4) [53]. More specifically, the number of connections (conjugated links k) in 1D structures with the oligomer length n is $k = n - 1$ and in 2D structures $k = 2\sqrt{n}(\sqrt{n} - 1)$.

A number of DFT studies reported the dependence of the HOMO-LUMO gap on the 1D oligomer length and compared it also to the corresponding 2D structures [53–56]. Gutzler et al. [53] systematically studied by DFT the HLG of polymer structures previously fabricated by on-surface synthesis experiments as a function

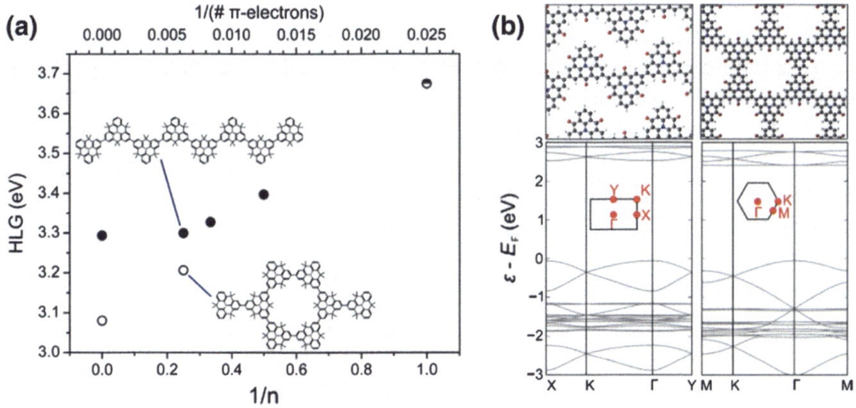

Fig. 4 **a** Evolution of the HOMO-LUMO gap (HLG) calculated by DFT (B3LYP) as a function of oligomer size for 1D and 2D dimethyl-bridged triphenylamine structures. Reprinted with permission from Ref. [53], copyright 2013, ACS. **b** DFT band structures (HSE) for 1D (left) and 2D (right) carbonyl-bridged triphenylamine structures revealing a decrease of the band gap of about 120 meV in the 2D case. Adapted with permission from Ref. [49], copyright 2017, Nature Communications

of oligomer size and revealed the interplay between the dimensionality and the effective conjugation. One of the main conclusions of this DFT study was that the HLG of 2D conjugated polymers is always smaller than that of the 1D counterpart, but the difference critically depends on the connectivity between the repeat units and varied between 0.1 and 1 eV. The electron delocalization in most of the studied 2D polymers is limited by cross-conjugation (as in meta-substituted benzene) or by a large twist angle between the π-orbitals, although several of the reported 2D polymers possess a continuous network of sp^2-carbons. In contrast, only a few experimental surface science studies have addressed the changes in the electronic structure from the precursor molecule to oligomers and 2D networks on metals [55, 57, 58]. Wiengarten et al. confirmed the reduction of the HLG in the transition from molecular precursors to 1D structures fabricated by surface-assisted dehydrogenative homocoupling of porphine molecules [57]. Similarly, Cardenas et al. used a surface-confined Ullmann-type polymerization to fabricate 2D networks of polythiophene and showed in ultraviolet photoelectron spectra (UPS) experiments that the HOMO of the molecular overlayer shifts towards the Fermi level by 0.6 eV upon polymerization [55]. However, the functional groups of the precursor molecules may play a decisive role in such experiments. Morchutt et al. [58] showed in a combined STS/UPS/DFT study that in the polymerization reaction of 1,3,5-tris (4-carboxyphenyl)-benzene *via* decarboxylation on Cu(111) the carboxyl groups in the precursor molecule give rise to empty states that lie closer to the Fermi level than the empty states of polymerized structures. This is in contrast to the LUMO stabilization and shift toward the Fermi level as a result of the extending π-conjugation.

Structural control with atomic precision is prerequisite for studying the electronic structure of extended 2D polymer networks, because their band structure sensitively depends on the specific atomic configuration and size. The hierarchical approach presented by Steiner et al. [49] yields 2D and 1D polymers built from identical building blocks with satisfactory structural quality (see Fig. 5). It provides, therefore, a direct comparison between the electronic structure of bottom-up fabricated 2D polymers and their analogous 1D counterparts, which has so far only been explored theoretically for extended π-systems [53]. STS experiments showed that the triphenylamine-based 2D networks are organic semiconductors with a bandgap of 2.45 ± 0.09 eV, which is compared to the HOMO-LUMO gap of the monomer of 3.91 ± 0.03 eV significantly reduced. Further, it was concluded that the band gap in the CTPA compounds changes most from the monomer to the 1D chains, while it only decreases around 170 meV from 1D to 2D structures. The decrease of the band gap from the monomer to the 2D polymer is a result of the increased effective π-conjugation length, which indicates an efficient conjugation among the triphenylamine building blocks within the nanostructure. The small difference between the band gaps of 1D and 2D compared to 0D and 1D shows that the π-electron delocalization length is only little increased between 1D and 2D CTPA polymers. In addition, constant height dI/dV maps at the energies of the observed valence band edge (VBE) and conduction band (CB) states provided further insights into the localization of the states. While the VBE shows an almost homogeneous electron density across the monomer backbone, the first unfilled states are localized near and at the newly formed C-C bond, respectively.

The band gap of the triphenylamine-based networks obtained by DFT calculations at the HSE level are in good agreement with the experimentally measured values [49]. In addition, band structure calculations provided further insights into the reduction of the band gap from 1D to 2D structures. In the 1D case, the band dispersion of the interacting π-bands around the Fermi level is decreased along $\overline{KY}$ compared with the 2D case (Fig. 4b). This is observed in the valence band and (to a smaller extent) in the conduction band, leading to an overall decrease of the band gap by 0.12 eV. Moreover, the dispersionless band of the conduction band edge is missing in the 1D case.

The inherent porous structure of 2D polymers leads by itself already to promising electronic properties. The pores induce a band gap, which can be controlled for instance by the pore size, density, and geometry [14]. In addition, the electronic properties can be further tuned by either including dopant atoms in the carbon

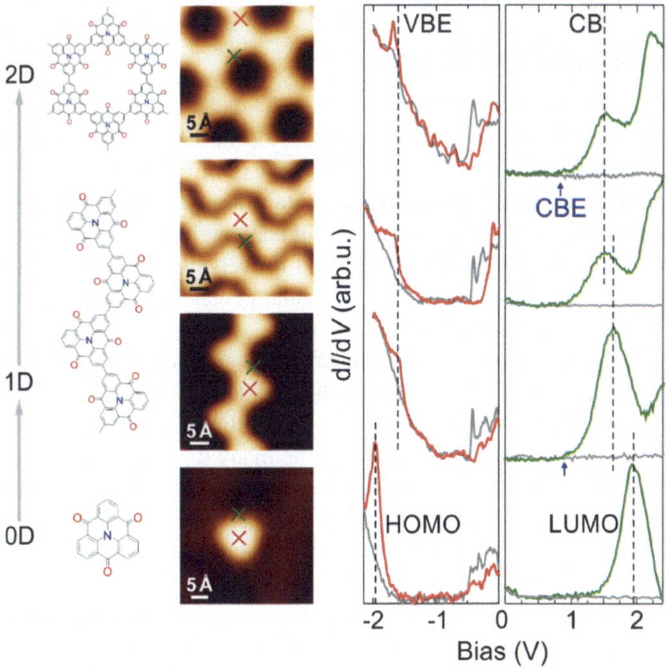

Fig. 5 d*I*/d*V* curves measured on the porous 2D polymer, 1D chains, and CTPA monomer. The STM images indicate the location of the recorded STS spectra of the valence band edge (VBE, red) and the conduction band (CB, green) of the 2D network (top), self-assembled 1D chains (2nd row), and isolated 1D chains (3rd row), and the corresponding HOMO and LUMO of the CTPA monomer (bottom). The grey spectra represent the reference on the bare Au(111) surface. The blue arrows indicate the onset of the conduction band edge (CBE). Adapted with permission from Ref. [49], copyright 2017, Nature Communications

backbone [59] or by trapping dopants in the pores *via* host-guest chemistry. Mostly carbon- and nitrogen-containing [59, 60] 2D porous polymers have been studied so far to avoid competing interactions among the functional groups during the reaction. With suitable precursors, there is no inherent limitation, however, to extend the surface-assisted synthesis to more complex 2D polymers that feature functionalized pores [49, 61, 62]. The functional groups facing the pores can be used to selectively bind molecules through host-guest chemistry. Most of the host-guest chemistry experiments in physisorbed networks on solid surfaces have focused to date on supramolecular networks [63]. The reversible nature of the non-covalent interaction frequently leads to alterations in structure and pore sizes by the inclusion of guest molecules [64, 65]. In contrast, the geometry and pore sizes of the covalently linked porous frameworks can hardly be impacted by the accommodation of guest species independent of their concentration and therefore provide more stability. Two-dimensional nanoporous covalent organic frameworks have been used to control the organization of various host molecules at the solid-liquid interface [66, 67] and in UHV [68].

4 Summary and Outlook

On-surface polymerization is a successful method in the bottom-up fabrication of novel covalent nanostructures that are to some extent otherwise synthetically not accessible. While the structural control to obtain long-range ordered structures is commonly achieved in 1D structures, it is still challenging in 2D structures. One of the main reasons is the kinetically driven and irreversible formation of topological defects in form of various polygons, e.g. irregular pentagonal or heptagonal pores in hexagonal networks. Some preprogrammed synthesis strategies such as cyclotrimerization reactions or a hierarchical synthesis showed promise to suppress the defect formation, however so far with limited scalability. The long-range order is a requirement for any application, since crystallinity is necessary for the promotion of a defined band structure. Despite the many exciting electronic properties predicted by DFT, only a few experimental surface science studies have so far reported on the electronic structure of 2D polymers. Recently, the change in the HOMO-LUMO gap from the precursor molecule to oligomers has been shown for a few systems. STS experiments on hierarchically synthesized structures unveiled a narrowing of the band gap with increasing dimensionality. For applications of 2D polymers in devices, the 2D polymers need to be electrically decoupled. This requires the use of insulating surfaces either by in situ synthesis or subsequent transfer. First experiments on calcite surfaces have shown that on-surface synthesis on bulk insulator surfaces is feasible, however strategies to form well ordered 2D networks remain to be explored [69, 70]. In the other hand, the surface supported networks can also be decoupled *via* intercalation of halogens [71–73]. In conclusion, the research on the surface synthesis of 2D polymer networks is a growing field, which is promising in particular for the synthesis of novel 2D carbon allotropes beyond graphene, such as graphyne or graphdiyne.

Acknowledgements The fruitful collaboration with Prof. Dr. Milan Kivala and Prof. Dr. Andreas Görling on the on-surface synthesis of 2D polymers at the Friedrich-Alexander University Erlangen-Nürnberg is gratefully acknowledged. Financial support by the German Research Foundation (DFG) through the Collaborative Research Center SFB 953 "Synthetic Carbon Allotropes" and the Cluster of Excellence EXC 315 "Engineering of Advanced Materials" at the Friedrich-Alexander University Erlangen-Nürnberg and the ERC Starting Grant SURFLINK (contract No. 637831) is acknowledged.

References

1. Sakamoto, J., van Heijst, J., Lukin, O., Schlüter, A.D.: Two-dimensional polymers: just a dream of synthetic chemists? Angew. Chem. Int. Ed. **48**, 1030–1069 (2009)
2. Colson, J.W., Dichtel, W.R.: Rationally synthesized two-dimensional polymers. Nat. Chem. **5**, 453–465 (2013)
3. Cai, J., Ruffieux, P., Jaafar, R., Bieri, M., Braun, T., Blankenburg, S., Muoth, M., Seitsonen, A.P., Saleh, M., Feng, X., et al.: Atomically precise bottom-up fabrication of graphene nanoribbons. Nature **466**, 470–473 (2010)
4. Talirz, L., Ruffieux, P., Fasel, R.: On-surface synthesis of atomically precise graphene nanoribbons. Adv. Mater. **28**, 6222–6231 (2016)
5. Söde, H., Talirz, L., Gröning, O., Pignedoli, C.A., Berger, R., Feng, X., Müllen, K., Fasel, R., Ruffieux, P.: Electronic band dispersion of graphene nanoribbons via Fourier-transformed scanning tunneling spectroscopy. Phys. Rev. B **91**, 045429 (2015)
6. Cai, J., Pignedoli, C.A., Talirz, L., Ruffieux, P., Söde, H., Liang, L., Meunier, V., Berger, R., Li, R., Feng, X., et al.: Graphene nanoribbon heterojunctions. Nat. Nanotechnol. **9**, 896–900 (2014)
7. Chen, Y.-C., de Oteyza, D.G., Pedramrazi, Z., Chen, C., Fischer, F.R., Crommie, M.F.: Tuning the band gap of graphene nanoribbons synthesized from molecular precursors. ACS Nano **7**, 6123–6128 (2013)
8. Chen, Y.C., Cao, T., Chen, C., Pedramrazi, Z., Haberer, D., de Oteyza, D.G., Fischer, F.R., Louie, S.G., Crommie, M.F.: Molecular bandgap engineering of bottom-up synthesized graphene nanoribbon heterojunctions. Nat. Nanotechnol. **10**, 156–160 (2015)
9. Kawai, S., Saito, S., Osumi, S., Yamaguchi, S., Foster, A.S., Spijker, P., Meyer, E.: Atomically controlled substitutional boron-doping of graphene nanoribbons. Nat. Commun. **6**, 8098 (2015)
10. Cloke, R.R., Marangoni, T., Nguyen, G.D., Joshi, T., Rizzo, D.J., Bronner, C., Cao, T., Louie, S.G., Crommie, M.F., Fischer, F.R.: Site-specific substitutional Boron doping of semiconducting armchair graphene nanoribbons. J. Am. Chem. Soc. **137**, 8872–8875 (2015)
11. van der Lit, J., Boneschanscher, M.P., Vanmaekelbergh, D., Ijäs, M., Uppstu, A., Ervasti, M., Harju, A., Liljeroth, P., Swart, I.: Suppression of electron-vibron coupling in graphene nanoribbons contacted via a single atom. Nat. Commun. **4**, 2023 (2013)
12. Held, P.A., Fuchs, H., Studer, A.: Covalent-bond formation via on-surface chemistry. Chem. Eur. J. **23**, 5874–5892 (2017)
13. Whitelam, S., Tamblyn, I., Haxton, T.K., Wieland, M.B., Champness, N.R., Garrahan, J.P., Beton, P.H.: Common physical framework explains phase behavior and dynamics of atomic, molecular, and polymeric network formers. Phys. Rev. X **4**, 011044 (2014)

14. Pedersen, T.G., Flindt, C., Pedersen, J., Mortensen, N.A., Jauho, A.-P., Pedersen, K.: Graphene antidot lattices: designed defects and spin qubits. Phys. Rev. Lett. **100**, 136804 (2008)
15. Adjizian, J.-J., Briddon, P., Humbert, B., Duvail, J.-L., Wagner, P., Adda, C., Ewels, C.: Dirac cones in two-dimensional conjugated polymer networks. Nat. Commun. **5**, 5842 (2014)
16. Adjizian, J.-J., Lherbier, A., M.-M., Dubois, S., Botello-Mendez, A.R., Charlier, J.-C.: The electronic and transport properties of two-dimensional conjugated polymer networks including disorder. Nanoscale, **8**, 1642–1651 (2016)
17. Malko, D., Neiss, C., Viñes, F., Görling, A.: Competition for graphene: graphynes with direction-dependent Dirac cones. Phys. Rev. Lett. **108**, 086804 (2012)
18. Perepichka, D.F., Rosei, F.: Extending polymer conjugation into the second dimension. Science **323**, 216 (2009)
19. Blankenburg, S., Bieri, M., Fasel, R., Müllen, K., Pignedoli, C.A., Passerone, D.: Porous graphene as an atmospheric nanofilter. Small **6**, 2266–2271 (2010)
20. Xiang, Z., Cao, D., Dai, L.: Well-defined two dimensional covalent organic polymers: rational design, controlled syntheses, and potential applications. Polym. Chem. **6**, 1896–1911 (2015)
21. Narita, A., Wang, X.-Y., Feng, X., Mullen, K.: New advances in nanographene chemistry. Chem. Soc. Rev. **44**, 6616–6643 (2015)
22. Li, Y., Xu, L., Liu, H., Li, Y.: Graphdiyne and graphyne: from theoretical predictions to practical construction. Chem. Soc. Rev. **43**, 2572–2586 (2014)
23. Gourdon, A.: On-surface covalent coupling in ultrahigh vacuum. Angew. Chem. Int. Ed. **47**, 6950–6953 (2008)
24. Fan, Q., Gottfried, J.M., Zhu, J.: Surface-catalyzed C-C covalent coupling strategies toward the synthesis of low-dimensional carbon-based nanostructures. Acc. Chem. Res. **48**, 2484–2494 (2015)
25. Dong, L., Liu, P.N., Lin, N.: Surface-activated coupling reactions confined on a surface. Acc. Chem. Res. **48**, 2765–2774 (2015)
26. Björk, J., Hanke, F.: Towards design rules for covalent nanostructures on metal surfaces. Chem. Eur. J. **20**, 928–934 (2014)
27. Lindner, R., Rahe, P., Kittelmann, M., Gourdon, A., Bechstein, R., Kühnle, A.: Substrate templating guides the photoinduced reaction of C_{60} on calcite. Angew. Chem. Int. Ed. **53**, 7952–7955 (2014)
28. Palma, C.-A., Diller, K., Berger, R., Welle, A., Björk, J., Cabellos, J.L., Mowbray, D.J., Papageorgiou, A.C., Ivleva, N.P., Matich, S., et al.: Photoinduced C-C reactions on insulators toward photolithography of graphene nanoarchitectures. J. Am. Chem. Soc. **136**, 4651–4658 (2014)
29. Kawai, S.: Revealing mechanical and structural properties of molecules on surface by high-resolution atomic force microscopy. Polym. J. **49**, 3–11 (2017)
30. Gross, L.: Recent advances in submolecular resolution with scanning probe microscopy. Nat. Chem. **3**, 273–278 (2011)
31. Zwaneveld, N.A.A., Pawlak, R., Abel, M., Catalin, D., Gigmes, D., Bertin, D., Porte, L.: Organized formation of 2D extended covalent organic frameworks at surfaces. J. Am. Chem. Soc. **130**, 6678–6679 (2008)
32. Côté, A.P., Benin, A.I., Ockwig, N.W., O'Keeffe, M., Matzger, A.J., Yaghi, O.M.: Porous, crystalline. Covalent organic frameworks. Science **310**, 1166–1170 (2005)
33. Weigelt, S., Busse, C., Bombis, C., Knudsen, M.M., Gothelf, K.V., Lægsgaard, E., Besenbacher, F., Linderoth, T.R.: Surface synthesis of 2D branched polymer nanostructures. Angew. Chem. Int. Ed. **47**, 4406–4410 (2008)
34. Schlögl, S., Heckl, W.M., Lackinger, M.: On-surface radical addition of triply iodinated monomers on Au(111)—the influence of monomer size and thermal post-processing. Surf. Sci. **606**, 999–1004 (2012)
35. Eichhorn, J., Nieckarz, D., Ochs, O., Samanta, D., Schmittel, M., Szabelski, P.J., Lackinger, M.: On-surface Ullmann coupling: the influence of kinetic reaction parameters on the morphology and quality of covalent networks. ACS Nano **8**, 7880–7889 (2014)

36. Bieri, M., Nguyen, M.-T., Gröning, O., Cai, J., Treier, M., Aït-Mansour, K., Ruffieux, P., Pignedoli, C.A., Passerone, D., Kastler, M., et al.: Two-dimensional polymer formation on surfaces: insight into the roles of precursor mobility and reactivity. J. Am. Chem. Soc. **132**, 16669–16676 (2010)
37. Grill, L., Dyer, M., Lafferentz, L., Persson, M., Peters, M.V., Hecht, S.: Nano-architectures by covalent assembly of molecular building blocks. Nat. Nanotechnol. **2**, 687–691 (2007)
38. Bieri, M., Treier, M., Cai, J., Ait-Mansour, K., Ruffieux, P., Groning, O., Groning, P., Kastler, M., Rieger, R., Feng, X., et al.: Porous graphenes: two-dimensional polymer synthesis with atomic precision. Chem. Commun. 6919–6921 (2009)
39. Liu, J., Ruffieux, P., Feng, X., Müllen, K., Fasel, R.: Cyclotrimerization of arylalkynes on Au (111). Chem. Commun. **50**, 11200–11203 (2014)
40. Yang, B., Björk, J., Lin, H., Zhang, X., Zhang, H., Li, Y., Fan, J., Li, Q., Chi, L.: Synthesis of surface covalent organic frameworks via dimerization and cyclotrimerization of acetyls. J. Am. Chem. Soc. **137**, 4904–4907 (2015)
41. Liu, J., Ruffieux, P., Feng, X., Müllen, K., Fasel, R.: Cyclotrimerization of arylalkynes on Au (111). Chem. Commun. **50**, 11200–11203 (2014)
42. Eichhorn, J., Strunskus, T., Rastgoo-Lahrood, A., Samanta, D., Schmittel, M., Lackinger, M.: On-surface Ullmann polymerization via intermediate organometallic networks on Ag(111). Chem. Commun. **50**, 7680–7682 (2014)
43. Biradha, K., Santra, R.: Crystal engineering of topochemical solid state reactions. Chem. Soc. Rev. **42**, 950–967 (2013)
44. Ruffieux, P., Wang, S., Yang, B., Sánchez-Sánchez, C., Liu, J., Dienel, T., Talirz, L., Shinde, P., Pignedoli, C.A., Passerone, D., et al.: On-surface synthesis of graphene nanoribbons with zigzag edge topology. Nature **531**, 489–492 (2016)
45. Basagni, A., Sedona, F., Pignedoli, C.A., Cattelan, M., Nicolas, L., Casarin, M., Sambi, M.: Molecules–oligomers–nanowires–graphene nanoribbons: a bottom-up stepwise on-surface covalent synthesis preserving long-range order. J. Am. Chem. Soc. **137**, 1802–1808 (2015)
46. Zhang, H., Franke, J.-H., Zhong, D., Li, Y., Timmer, A., Arado, O.D., Mönig, H., Wang, H., Chi, L., Wang, Z., et al.: Surface supported gold-organic hybrids: on-surface synthesis and surface directed orientation. Small **10**, 1361–1368 (2014)
47. Lafferentz, L., Eberhardt, V., Dri, C., Africh, C., Comelli, G., Esch, F., Hecht, S., Grill, L.: Controlling on-surface polymerization by hierarchical and substrate-directed growth. Nat. Chem. **4**, 215–220 (2012)
48. Shi, K.J., Zhang, X., Shu, C.H., Li, D.Y., Wu, X.Y., Liu, P.N.: Ullmann coupling reaction of aryl chlorides on Au(111) using dosed Cu as a catalyst and the programmed growth of 2D covalent organic frameworks. Chem. Commun. **52**, 8726–8729 (2016)
49. Steiner, C., Gebhardt, J., Ammon, M., Yang, Z., Heidenreich, A., Hammer, N., Görling, A., Kivala, M., Maier, S.: Hierarchical on-surface synthesis and electronic structure of carbonyl-functionalized one- and two-dimensional covalent nanoarchitectures. Nat. Commun. **8**, 14765 (2017)
50. Fan, Q., Wang, C., Han, Y., Zhu, J., Hieringer, W., Kuttner, J., Hilt, G., Gottfried, J.M.: Surface-assisted organic synthesis of hyperbenzene nanotroughs. Angew. Chem. Int. Ed. **52**, 4668–4672 (2013)
51. Chen, M., Shang, J., Wang, Y., Wu, K., Kuttner, J., Hilt, G., Hieringer, W., Gottfried, J.M.: On-Surface synthesis and characterization of honeycombene oligophenylene macrocycles. ACS Nano **11**, 134–143 (2017)
52. Vasseur, G., Fagot-Revurat, Y., Sicot, M., Kierren, B., Moreau, L., Malterre, D., Cardenas, L., Galeotti, G., Lipton-Duffin, J., Rosei, F., et al.: Quasi one-dimensional band dispersion and surface metallization in long-range ordered polymeric wires. Nat. Commun. **7**, 10235 (2016)
53. Gutzler, R., Perepichka, D.F.: π-electron conjugation in two dimensions. J. Am. Chem. Soc. **135**, 16585–16594 (2013)
54. Wen, J., Luo, D., Cheng, L., Zhao, K., Ma, H.: Electronic structure properties of two-dimensional π-conjugated polymers. Macromolecules **49**, 1305–1312 (2016)

55. Cardenas, L., Gutzler, R., Lipton-Duffin, J., Fu, C., Brusso, J.L., Dinca, L.E., Vondracek, M., Fagot-Revurat, Y., Malterre, D., Rosei, F., et al.: Synthesis and electronic structure of a two dimensional π-conjugated polythiophene. Chem. Sci. **4**, 3263–3268 (2013)
56. Baryshnikov, G.V., Minaev, B.F., Karaush, N.N., Minaeva, V.A.: The art of the possible: computational design of the 1D and 2D materials based on the tetraoxa[8]circulene monomer. RSC Adv. **4**, 25843–25851 (2014)
57. Wiengarten, A., Seufert, K., Auwärter, W., Ecija, D., Diller, K., Allegretti, F., Bischoff, F., Fischer, S., Duncan, D.A., Papageorgiou, A.C., et al.: Surface-assisted dehydrogenative homocoupling of porphine molecules. J. Am. Chem. Soc. **136**, 9346–9354 (2014)
58. Morchutt, C., Björk, J., Straßer, C., Starke, U., Gutzler, R., Kern, K.: Interplay of chemical and electronic structure on the single-molecule level in 2D polymerization. ACS Nano **10**, 11511–11518 (2016)
59. Sánchez-Sánchez, C., Brüller, S., Sachdev, H., Müllen, K., Krieg, M., Bettinger, H.F., Nicolaï, A., Meunier, V., Talirz, L., Fasel, R., et al.: On-surface synthesis of BN-substituted heteroaromatic networks. ACS Nano **9**, 9228–9235 (2015)
60. Bieri, M., Blankenburg, S., Kivala, M., Pignedoli, C.A., Ruffieux, P., Müllen, K., Fasel, R.: Surface-supported 2D heterotriangulene polymers. Chem. Comm. **47**, 10239–10241 (2011)
61. Faury, T., Dumur, F., Clair, S., Abel, M., Porte, L., Gigmes, D.: Side functionalization of diboronic acid precursors for covalent organic frameworks. CrystEngComm **15**, 2067–2075 (2013)
62. Xu, L., Cao, L., Guo, Z., Zha, Z., Lei, S.: Side-functionalized two-dimensional polymers synthesized via on-surface Schiff-base coupling. Chem. Comm. **51**, 8664–8667 (2015)
63. Teyssandier, J., Feyter, S.D., Mali, K.S.: Host-guest chemistry in two-dimensional supramolecular networks. Chem. Comm. **52**, 11465–11487 (2016)
64. Kong, X.-H., Deng, K., Yang, Y.-L., Zeng, Q.-D., Wang, C.: H-bond switching mediated multiple flexibility in supramolecular host-guest architectures. J. Phys. Chem. C **111**, 17382–17387 (2007)
65. Banerjee, K., Kumar, A., Canova, F.F., Kezilebieke, S., Foster, A.S., Liljeroth, P.: Flexible self-assembled molecular templates on graphene. J. Phys. Chem. C **120**, 8772–8780 (2016)
66. Sun, J., Zhou, X., Lei, S.: Host-guest architectures with a surface confined imine covalent organic framework as two-dimensional host networks. Chem. Comm. **52**, 8691–8694 (2016)
67. Cui, D., MacLeod, J.M., Ebrahimi, M., Rosei, F.: Selective binding in different adsorption sites of a 2D covalent organic framework. CrystEngComm (2017)
68. Blunt, M.O., Russell, J.C., Champness, N.R., Beton, P.H.: Templating molecular adsorption using a covalent organic framework. Chem. Comm. **46**, 7157–7159 (2010)
69. Kittelmann, M., Nimmrich, M., Lindner, R., Gourdon, A., Kühnle, A.: Sequential and site-specific on-surface synthesis on a bulk insulator. ACS Nano **7**, 5614–5620 (2013)
70. Lindner, R., Kühnle, A.: On-surface reactions. ChemPhysChem **16**, 1582–1592 (2015)
71. Rastgoo-Lahrood, A., Björk, J., Lischka, M., Eichhorn, J., Kloft, S., Fritton, M., Strunskus, T., Samanta, D., Schmittel, M., Heckl, W.M., et al.: Post-synthetic decoupling of on-surface-synthesized covalent nanostructures from Ag(111). Angew. Chem. Int. Ed. **55**, 7650–7654 (2016)
72. Rastgoo-Lahrood, A., Lischka, M., Eichhorn, J., Samanta, D., Schmittel, M., Heckl, W.M., Lackinger, M.: Reversible intercalation of iodine monolayers between on-surface synthesised covalent polyphenylene networks and Au(111). Nanoscale **9**, 4995–5001 (2017)
73. Peyrot, D., Silly, F.: On-surface synthesis of two-dimensional covalent organic structures versus halogen-bonded self-assembly: competing formation of organic nanoarchitectures. ACS Nano **10**, 5490–5498 (2016)

On-Surface Coupling Reactions with Extrinsic Catalysts

Wei Zhao, Lei Dong, Ran Zhang and Nian Lin

Abstract On-surface coupling reactions have opened a novel route toward synthesizing various organic nanostructures on surfaces. In this chapter, we discuss the catalytic effects of Cu and Pd deposits in the on-surface Ullmann coupling and Sonogashira cross-coupling reactions. The stepwise reaction paths, intermediates, and activation energies are deliberated at a single molecular level using scanning tunneling microscopy (STM). Such studies offer mechanistic insights into the on-surface coupling reactions catalyzed by extrinsic metals.

1 Introduction

In the last decades, molecular self-assembly on surface had been widely used in the bottom-up fabrication of nanostructures. Various intermolecular interactions have been employed to construct molecular networks, such as van der Waals interactions, hydrogen bonds, and metal-organic coordination bonds [1–4]. These non-covalent intermolecular interactions normally result in good long-range ordering, however, give rise to low mechanical stability and poor charge-transport properties. In contrast, covalent interactions lead to robust molecular architectures with appreciable electronic characteristics. Hence, on-surface synthesis, a process involving formation of covalent bonds on surfaces, renders a promising direction for the fabrication of molecular nanostructures. The first and most studied on-surface reaction is Ullmann coupling, whose typical motif is to create reactive species by dehalogenation of aryl halide precursors for subsequent aryl-aryl coupling [5–7]. Most on-surface Ullmann coupling reactions were conducted on metal substrates, taking advantages of the inherent reactivity of the

W. Zhao · L. Dong · R. Zhang · N. Lin (✉)
Department of Physics, The Hong Kong University of Science and Technology, Hong Kong, China
e-mail: phnlin@ust.hk

W. Zhao
Department of Chemistry, University of Washington, Seattle, WA 98195, USA

D. G. de Oteyza and C. Rogero (eds.), *On-Surface Synthesis II*,
Advances in Atom and Single Molecule Machines,
https://doi.org/10.1007/978-3-319-75810-7_9

substrates. In Sect. 2.1, we discuss the on-surface Ullmann reactions assisted by Cu and Pd. We found that both metals significantly reduce the reaction temperature, but delineate different mechanism in terms of reaction intermediate and reaction path. Considering the metal substrates often strongly alter the electronic properties of the on-surface molecular structures, to achieve coupling reactions on a non-metallic substrate is highly desirable. Because non-metallic substrates are normally chemically inert, the catalytic activation using extrinsic metals becomes imperative for the Ullmann reaction taking place on such substrates. In Sect. 2.2, we show that Cu and Pd deposits effectively catalyze Ullmann coupling on an atomically thin hexagonal boron nitride (h-BN) layer [8–15]. It is found that both Cu and Pd show notable catalytic effect, leading to much lower threshold reaction temperatures and higher efficiency of oligomeric production compared to the previous studies without external catalysts. In Sect. 3, we present a study of on-surface Sonogashira cross-coupling reactions. Sonogashira cross-coupling, which form a carbon-carbon bond between a terminal alkyne and an aryl or vinyl halide [16, 17], has been widely used in organic synthesis. The major challenge to apply this reaction on surfaces is to suppress the homo-coupling among the precursors and enhance the yield of the cross-coupling products [18–20]. We demonstrate using coordination templates and metal catalysts can remarkably promote the cross-coupling reaction [21].

2 Ullmann Reaction with Extrinsic Catalysts

2.1 On Metal Substrate

Cu is known to be a prototypical catalyst for Ullmann coupling [5]. In the on-surface Ullmann coupling on a Cu surface, Cu acts not only as a platform but also as a promoter for the reaction [22]. Organo-copper intermediates in the reaction have been discovered in experiments and calculations [6, 23]. Pd is also a versatile catalyst widely used in organic synthesis [24, 25]. However, Pd catalysis has rarely been studied in the on-surface reactions. Here, we use 5,15-bis-(4-bromophenyl)-10,20-diphenyl porphyrin (**1**, see the inset of Fig. 1a) to study the catalytic activity of Cu and Pd in the Ullmann coupling on an Au(111) substrate. The molecules of **1** were thermally deposited onto a Au(111) surface that was pre-dosed with Cu or Pd deposits. Reaction yields, rate constants and activation energy were deduced by monitoring isothermal reaction products in situ using STM at a single-molecule level. Kinetic Monte Carlo (KMC) simulations were performed to understand the reaction pathways.

In the STM images shown in Fig. 1a–c, polymeric chains can be found all over the Pd-deposited surface with post annealing at 447 K. The adjacent monomers in a polymeric chain (inset of Fig. 1c) are spaced with a center-to-center distance of 1.74 ± 0.04 nm, which is consistent with the distance between covalently linked

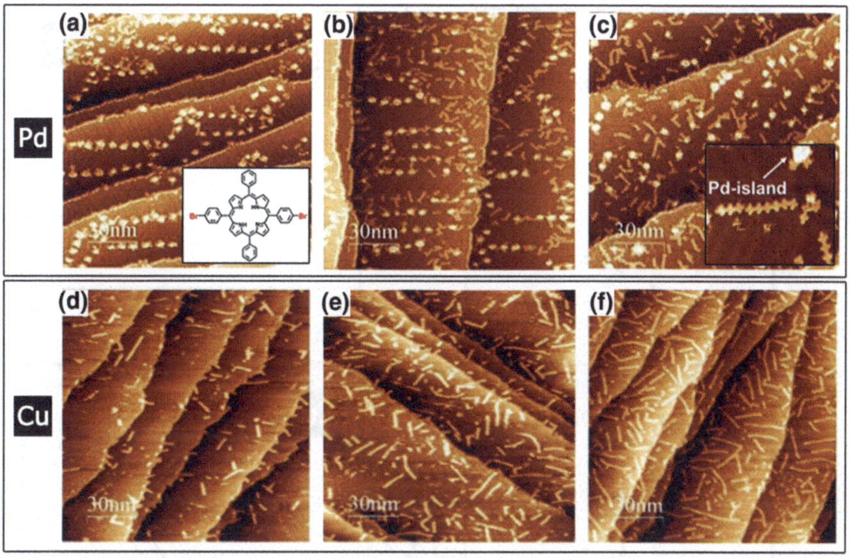

Fig. 1 Representative STM images show the Pd or Cu catalyzed polymerization of 1. **a–c** With Pd annealed at 447 K for 5, 45 and 105 min, respectively. Inset in (**a**): chemical structure of 1. Inset in (**c**): Polymeric chains and Pd island. **d–e** With Cu annealed at 453 K for 5, 75 and 160 min, respectively. Reproduced with permission from Ref. [22]. Copyright Wiley-VCH

porphyrin molecules [26, 27]. It is hence believed that the molecules of **1** were debrominated and coupled together via C–C bonds. Similar chains were obtained when we conducted the experiment on Cu-deposited surface (Fig. 1d–f). In the absence of Pd or Cu, annealing up to 453 K resulted in a very low yield of the covalently-linked species [27]. Therefore, the polymeric chains formed at lower temperatures in the presence of Pd or Cu can be attributed to the Ullmann coupling reaction catalyzed by the extrinsically dosed Pd or Cu deposits [28, 29].

In the case of Pd-catalyzed reaction, the molecules were deposited on the Au (111) surface precovered with Pd at room temperature (RT) and the sample was annealed at certain temperatures for 8–10 min then cooled down to RT for STM measurement. Dimers and longer chains were counted so that the number of bonds formed as a function of reaction time and temperature could be derived. The STM images in Fig. 1a–c clearly show an increase of the polymeric chains with longer reaction time. This time-dependent increment in C–C bond concentration was observed at different annealing temperatures (Fig. 2). At 429 K, bond concentration first rises rapidly and then gradually reaches saturation; At 411 K, however, the initial rapid increase is not so apparent; At 393 K, a two-phase behavior appears: a relatively slow increase (0–60 min, defined as phase I) is followed by a rapid one (60–140 min, phase II). The two-phase character suggests that the coupling reaction involves multiple steps, which resembles the Pd-catalyzed homo-coupling of aryl halides in solution [24]. We propose a two-step reaction pathway: phase I is

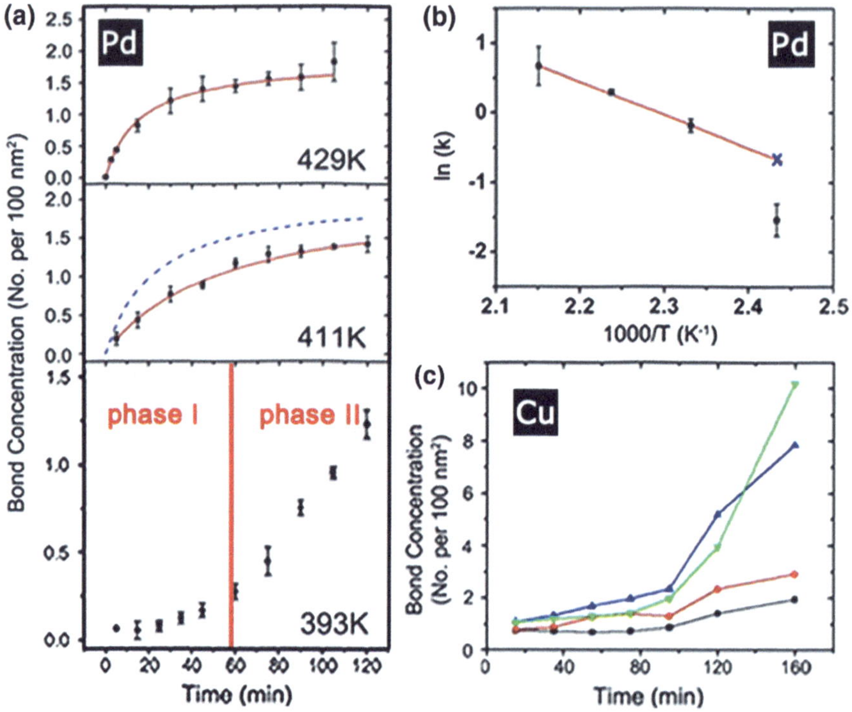

Fig. 2 **a** Bond concentration in Pd-catalyzed coupling of **1** on Au(111) as a function of reaction time at various temperatures. The solid lines are fitting curves using Eq. (2). The dashed line is a curve calculated according to point "x" in (**b**). **b** Arrhenius plot of rate constant k obtained from experimental data. **c** Bond concentration in Cu-catalyzed coupling of **1** on Au(111) as a function of reaction time at 399 K (black), 417 K (red), 435 K (blue) and 453 K (green). The 453 K data (blue) are scaled down by a factor of five. Reproduced with permission from Ref. [22]. Copyright Wiley-VCH

associated with an initial activation process, which forms an intermediate organo-palladium complex; and phase II involves C–C bond formation. Such a conclusion is also derived from the Pd-catalyzed coupling reaction on h-BN layer, which will be discussed later.

Due to the high mobility, monomers of **1** were not detected at RT by STM and the short lifetime of phase I even indiscernible above 411 K, the rate-limiting step lies in phase II, that is, the C–C bond formation. We define [bond] as the concentration of the C–C bonds formed and [phenyl-Br] as the concentration of the unreacted phenyl bromide, the rate equation can be written as Eq. (1):

$$d[\text{bond}]/dt = k[\text{phenyl-Br}]^2 \tag{1}$$

In which k is the rate constant. Then the bond concentration could be presented as Eq. (2):

$$[\text{bond}] = \frac{kt[\text{phenyl-Br}]_0^2}{1 + 2kt[\text{phenyl-Br}]_0} + [\text{bond}]_0 \tag{2}$$

where $[\text{phenyl-Br}]_0$ refers to the initial concentration of the phenyl bromide, while $[\text{bond}]_0$ the bond concentration prior to annealing. By fitting the experimental data at different annealing temperatures using Eq. (2), we got values of k which depend on temperature T according to the Arrhenius equation, so the plot of ln(k) versus T^{-1} gave a linear relationship in Fig. 2b and an activation energy of 0.41 ± 0.03 eV for the overall reaction was deduced by the fitting. Actually the data point of 411 K deviates obviously from the fitting line, because at that temperature phase I is not negligible and then it cannot be well described by Eq. (2). A higher rate constant would be expected if phase I is not counted in, as labeled with "x" in Fig. 2b. Accordingly a hypothesized bond concentration profile is displayed as dashed curve in the middle panel of Fig. 2a, which owns a similar shape as 429 K with a higher yield than the experimental one. On the other hand, a much faster kinetics of phase I implies a lower activation energy than phase II. In comparison to the activation energy of 1.12 eV reported for the Ullmann coupling of iodobenzene [30] and a weaker reactivity of Br than I, we believe that as a catalyst Pd is more efficient than Cu.

Similar experiments were carried out using Cu as catalyst, whose representative STM images are shown in Fig. 1d–e. From the corresponding bond concentration plotted in Fig. 2c, we found that all of them looked like the curve of Pd at 393 K in Fig. 2a with a slow starting phase. Taking account of the multiple reaction steps suggested for Cu-mediated Ullmann coupling, we assume that the initial activation step, that is, the cleavage of halogen atoms to generate radicals, has a higher energy value in the reaction catalyzed by Cu than Pd, though we could not identify each single step in our experiments.

It is also notable that Cu-catalyzed reaction always has a much higher C–C bond formation yield after the whole annealing process than the Pd case. The former shows a rapid increase at higher temperatures rather than saturating like the latter. Furthermore, in Fig. 1, it could be seen that longer polymeric chains were generated by the Cu-catalyzed reactions than those catalyzed by Pd. As the length distribution exhibited in Fig. 3, Pd-catalyzed reactions resulted in short chains, from dimers to hexamers, with dimers predominating (Fig. 3a). On the contrary, longer chains consisting of up to 17 monomers were found in the Cu-catalyzed reactions and it presents a much broader distribution without noticeable length preference (Fig. 3b).

To interpret the different behaviors between Pd and Cu catalysts, we simulated the coupling reaction using KMC algorithm [27, 31]. The entire process is divided into two steps: first, the phenyl bromide of monomer is activated with energy barrier E_1; Second, C–C bond is formed with energy barrier E_2 when one activated phenyl bromide is encountered by an inactivated end of another molecule. According to our experiments, phase I was undetectable for Pd-catalyzed reactions at higher temperatures while for Cu-catalyzed reaction it always existed, so we set a lower E_1 for Pd (0.35 eV) than for Cu (0.50 eV). E_2 was set to be 0.40 eV for both.

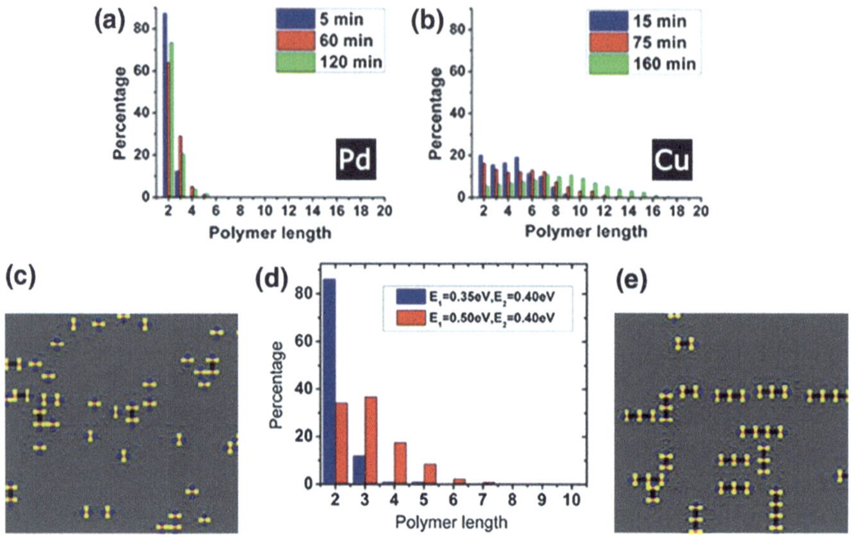

Fig. 3 **a**, **b** Length distribution of the polymeric chains in the reactions catalyzed by Pd annealed at 465 K (**a**) and Cu annealing at 453 K (**b**) for various durations. **c**–**e** KMC-simulated results with different E_1: (**c**) 0.35 eV; (**e**) 0.50 eV. Yellow tips represent the activated phenyl bromide, and black bars are C–C bonds. Both of the length distributions are shown in (**d**). Reproduced with permission from Ref. [22]. Copyright Wiley-VCH

As shown in Fig. 3c–e, lower E_1 produces short chains with a narrow length distribution. In addition, a lower E_1 for Pd also gives rise to lower yield. This simulated result implies that a sufficient activation of the molecules reduces the chance of encounter between an activated end and an inactivated one, then the formation of C–C bond is suppressed and longer chain growth is unfavorable.

2.2 *On BN Surface*

Direct fabrication of molecular nanostructures on inert or insulating substrates is thought to be an effective way to decouple the on-surface synthesized products from the metallic substrates. The h-BN layer grown on metal substrates was used as the substrate for Ullmann coupling. However, the reactions were found to happen at much higher temperature as compared with those on the metal surfaces [13, 14]. To lower down the reaction temperature, we employed Cu and Pd as catalysts for the Ullmann reaction on a h-BN layer [15]. A single layer of h-BN was first synthesized on a clean Ni(111) surface using CVD method by exposing the substrate to borazine ($B_3N_3H_6$) vapor at 780 °C [32], and then the polyphenylene precursor, 1,3,5-tris-(4-bromophenyl)benzene (TBB, named **2**), was thermally deposited onto the h-BN/Ni(111) sample held at RT. Two kinds of long-range ordered pattern were revealed by STM as the self-assembly of molecules **2** which were believed to be intact on the BN layer. When we annealed the sample up to 400 °C, the molecules **2** coupled into disordered units with various sizes and shapes where dimeric structures predominated. A center-to-center distance of ~1.3 nm between the linked monomers indicates the formation of covalent bonds between molecules [33]. Such an activation of dehalogenation and subsequent coupling started at ~220 °C, being consistent with a previous study [14].

We used Cu to catalyze the coupling reaction on h-BN/Ni(111). Two recipes were adopted. Recipe 1 (**R1**): deposition of Cu on **2**/h-BN/Ni(111) at RT, followed with stepped annealing (up to 400 °C); Recipe 2 (**R2**): direct deposition of Cu onto **2**/h-BN/Ni(111) kept at elevated temperatures (50–185 °C). Surprisingly, for **R1**, coupling of **2** was found to start from 220 °C, showing a quite similar scenario to direct annealing of **2** on h-BN/Ni(111) without catalyst. On the contrary, **R2** did considerably lower down the threshold temperature of the coupling to about 70 °C. To understand the distinct catalytic efficiency of Cu reflected in the two recipes, reaction yields as a function of annealing temperatures are statistically plotted in Fig. 4c. Here the yield is defined as the ratio between the number of molecules **2** taking part in the covalent coupling reaction and the number of the molecules **2** in total on the surface. For all the annealing steps, we can see that the yield with Cu along **R1**(red) almost coincides the value of direct annealing of molecules **2** on the substrate (black). It implies that Cu plays an ignorable role in the coupling reaction with **R1**. However, the curve of **R2** (green) shows that the coupling is initiated already to 70 °C and rapidly accelerated above 150 °C. It indicates that Cu efficiently catalyzes the coupling reaction only when directly deposited onto a 'hot' surface. In **R1**, Cu clusters (diameter: 1–4 nm) were observed after the deposition at RT and molecules **2** kept the same self-assembled motif, implying that the agglomerated Cu atoms do not activate the coupling reaction even in the following annealing process. However, in **R2**, Cu atoms were directly dosed on the sample held at elevated temperatures, where the coupling reactions can occur with more mobile Cu atoms available as catalyst. Hence, it is concluded that individual

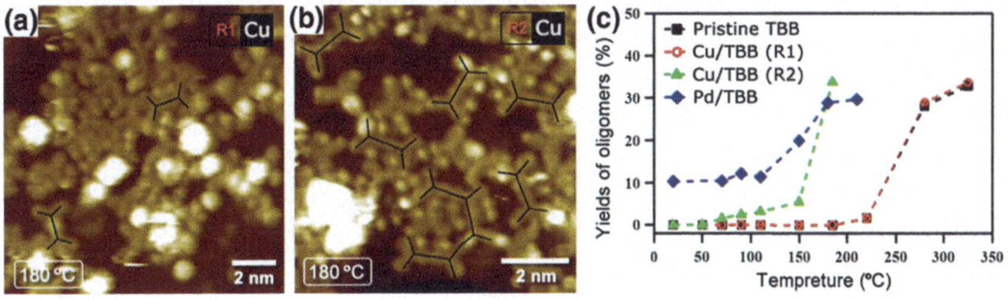

Fig. 4 Ullmann coupling reaction of molecules **2** on h-BN/Ni(111) catalyzed by dosed metal atoms. **a**, **b** Representative STM images of the sample after annealing at 180 °C with Cu catalyst along **R1** and **R2**, respectively. Sketches in black highlight the produced oligomers. **c** Statistical plots of the yields of oligomers versus temperatures for different experimental methods. Reproduced with permission from Ref. [15]. Copyright Royal Society of Chemistry

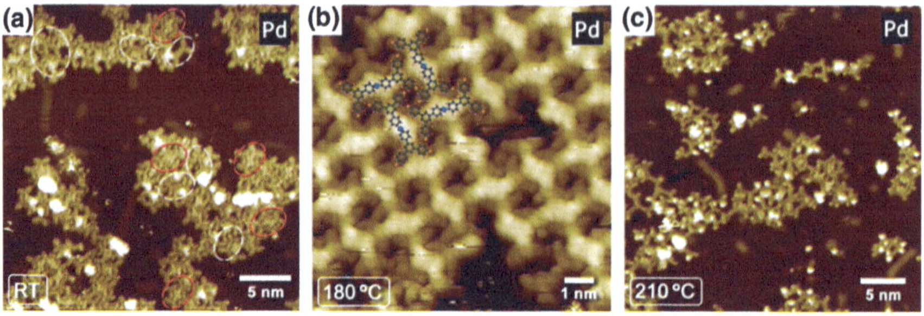

Fig. 5 Ullmann coupling of molecules **2** catalyzed by Pd on h-BN/Ni(111). **a** Dosage of Pd atoms to **2**/h-BN/Ni(111) triggered the formation of organometallic intermediates (red dashed oval) and oligomers (white dashed oval). **b** Ordered array of kinked dimeric intermediates after annealing at 180 °C. Tentative models are superimposed. **c** Most of the molecules have been coupled in oligomers after being heated to 210 °C

diffusing Cu adatoms are the key agent for catalyzing the Ullmann coupling on h-BN/Ni(111). In addition, unlike the Ullmann coupling on Cu(111) surface [6, 23, 33, 34], on h-BN no organometallic intermediates were observed.

In contrast to the Cu-catalyzed reaction, deposition of Pd atoms onto **2**/h-BN/Ni (111) surface at RT leads to both covalently linked structures and organometallic intermediate, as shown in Fig. 5a. Gentle annealing would generate islands made up of these intermediates regularly-arranged as the pattern displayed in Fig. 5b. The shape and size of these kinked dimers hint that they are composed of two head-to-head linked molecules **2** with a central joint. A tentative model is overlaid on the assembly where two monomers are linked via C–Pd–C bridge. After being heated up to 210 °C, the ordered islands were dissolved and turned into covalently-linked oligomers dispersed on the surface, which could be seen in Fig. 5c.

Apparently, Pd exhibits higher catalytic activity than Cu especially at lower temperatures (<180 °C), as plotted in blue in Fig. 4c. It could be easily distinguished by the comparison of Fig. 4a, b. Above 180 °C, the yield of Cu-catalyzed reaction exceeds the Pd-catalyzed one. Besides the catalytic activity, the organometallic intermediates are observed in the Pd-catalyzed reaction, but which is absent in Cu-catalyzed coupling, implying that Cu and Pd render distinctive reaction paths. We propose the following reaction mechanism: The Pd-catalyzed reaction overcomes two reaction barriers: one from brominated molecules to the organometallic intermediate and the other from the intermediate to the covalently-linked final products. In contrast, the Cu-catalyzed reaction does not involve a stable organometallic intermediate, and thus only have one energy barrier to go over. Presumably the two-step Pd-catalyzed reaction involves moderate reaction barriers, while the single-step Cu-catalyzed reaction involves a higher reaction barrier. Such a conclusion agrees well with the result on Au(111) in Sect. 2.1 and supports the previous interpretation.

3 Sonogashira Cross-Coupling

On-surface Sonogashira cross-coupling (SCC) is of great potential for designing molecular nanostructures. The challenge is to suppress the unwanted homo—coupling of the precursor molecules. Herein we employed a metal-organic template to control the SCC reaction process, as illustrated in Fig. 6a [27, 35–37]. The template (***T***) was formed through the coordination of 5,10,15-tri-(4-pyridyl)-20-bromophenyl porphyrin (**3**) and Cu atoms on Au(111) surface [38]. ***T*** features a double-row ladder structure involving py-Cu-py coordination in two orthogonal directions. Figure 6b reveals clearly that the bromides are present at the two sides of the ladders, which are fully accessible for SCC reaction between molecules **3** and another molecular precursor, 4-ethynylbiphenyl (***A***). (see the scheme in Fig. 6a) After molecules of ***A*** were dosed onto the sample (Fig. 6c), the areas between the ladders appeared blurry due to the presence of diffusive 2D gas-phase ***A*** molecules on the surface. No SCC products were observed. Annealing this sample to 170 °C resulted in many rod-like and Y-shaped features on both sides of ***T***, marked in Fig. 6d with short and long arrows, respectively. The distance from the center of molecule **3** to the end of the rod is 1.86 ± 0.03 nm, matching the expected value for a SCC product. We propose that the Y-shaped structures (denoted as CP2) also are products of cross-coupling. Figure 6e presents a higher-resolution STM topograph of the SCC and CP2 products with the proposed molecular models overlaid. We postulate that a CP2 product is formed in a two-step process: first a SCC

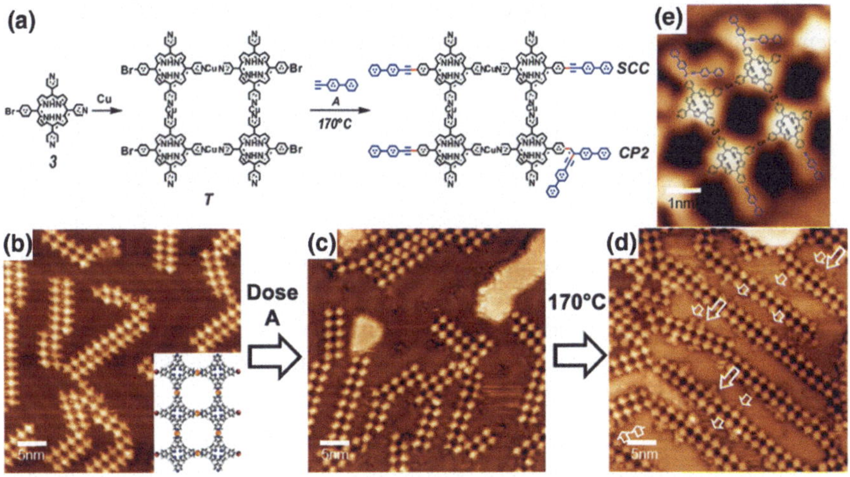

Fig. 6 **a** SCC between **3** and ***A*** with template ***T***. STM topographs showing templated reactions at different steps: **b** Cu-coordinated template ***T***, with the inset showing a structural model (white, H; grey, C; blue, N; red, Br; orange, Cu); **c** after dosing ***A*** molecules; **d** after annealing to 170 °C, with SCC products indicated by short arrows and CP2 products by long arrows; **e** a magnified image showing SCC products and CP2 products overlaid with chemical structures. Reproduced with permission from Ref. [21]. Copyright Royal Society of Chemistry

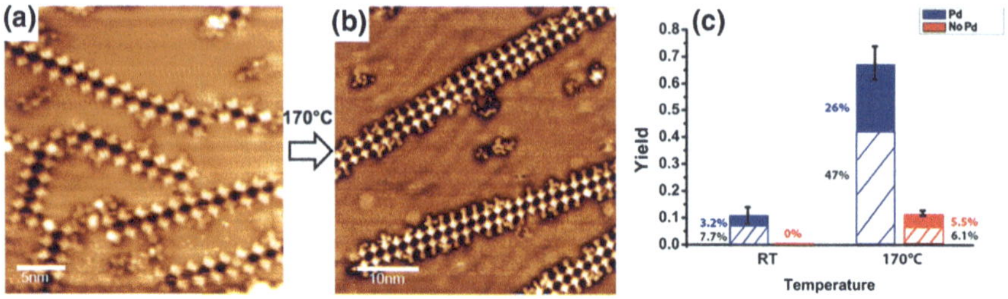

Fig. 7 STM images of Pd-catalyzed SCC on ***T*** at **a** room temperature and **b** with annealing to 170 °C. **c** Yields of SCC (shaded) and CP2 products (solid) in Pd-free and Pd-catalyzed reactions

product forms; in the second step, another ***A*** molecule undergoes an addition reaction with the triple bond in the SCC product, forming a butene moiety. Such addition processes were reported in homo-coupling of aryl acetylenes [39–43].

To explore the Pd catalysis, we loaded Pd to a sample with ***T*** and ***A*** at room temperature. Figure 7a clearly shows the SCC and CP2 products on the sides of the ladder structures. The amount of these products increased substantially after annealing the sample to 170 °C. As shown in Fig. 7b, most sites along the two sides of the ladders are occupied with either SCC or CP2 products. Figure 7c compares the yields of Pd-free reactions with those of Pd-catalyzed ones: The yield of SCC products increased from 0 to 7.7% at room temperature, and from 6 to 41% at 170 °C. The yield of CP2 products increased from 0 to 3.2% at room temperature, and from 5.5 to 26% at 170 °C. Apparently, Pd is an effective catalyst for on-surface SCC. The ratio of SCC to CP2 products increased from 6:5.5 to 41:26 when Pd catalyst was added, suggesting that Pd promotes SCC more efficiently than it does for CP2.

4 Summary

On-surface organic synthesis is developing rapidly and being studied with tremendous efforts. Various prototypal covalent reactions have been found to work on the surfaces to generate robust molecular architectures. In this chapter, we focus on the catalytic effects of using the extrinsic metal deposits to promote the on-surface reactions. Our findings are summarized as follows:

(1) Cu or Pd-catalyzed Ullmann coupling on Au(111) show a two-phase reaction pathway that involves an initial activation followed by C–C bond formation. Furthermore, the two catalysts account for different reaction mechanisms: the initial activation phase is the rate limiting step for the Cu-catalyzed reaction at all temperatures tested, whereas the later phase of C–C formation is the rate-limiting one for the Pd-catalyzed reaction;

(2) Cu or Pd-catalyzed Ullmann reactions on h-BN/Ni(111) show that both Cu and Pd are capable of catalyzing the Ullmann coupling on the inert h-BN surface, but delineate different reaction mechanisms in terms of reaction intermediate and reaction path;
(3) Sonogashira cross-coupling can be controlled using a metal-organic template. Higher production yields could be achieved by using Pd as catalyst.

The introduction of extrinsic catalysts is proved to be a feasible approach for activating the Ullmann coupling and Sonogashira cross-coupling reactions. We anticipate that various extrinsic metal catalysts will be also useful for other on-surface reactions.

Acknowledgements This work was supported by Hong Kong RGC 16303514.

References

1. Barth, J.V., Costantini, G., Kern, K.: Engineering atomic and molecular nanostructures at surfaces. Nature **437**, 671–679 (2005). https://doi.org/10.1038/nature04166
2. Slater, A.G., Perdigão, L.M.A., Beton, P.H., Champness, N.R.: Surface-based supramolecular chemistry using hydrogen bonds. Acc. Chem. Res. **47**, 3417–3427 (2014). https://doi.org/10.1021/ar5001378
3. Otero, R., Gallego, J.M., Vázquez de Parga, A.L., Martín, N., Miranda, R.: Molecular self-assembly at solid surfaces. Adv. Mater. **23**, 5148–5176 (2011). https://doi.org/10.1002/adma.201102022
4. Dong, L., Gao, Z., Lin, N.: Self-assembly of metal-organic coordination structures on surfaces. Prog. Surf. Sci. **91**, 101–135 (2016). https://doi.org/10.1016/j.progsurf.2016.08.001
5. Ullmann, F., Bielecki, J.: Synthesis in the biphenyl series. Ber. Dtsch. Chem. Ges. **34**, 2174–2185 (1901)
6. Dong, L., Liu, P.N., Lin, N.: Surface-activated coupling reactions confined on a surface. Acc. Chem. Res. **48**, 2765–2774 (2015). https://doi.org/10.1021/acs.accounts.5b00160
7. Fan, Q., Gottfried, J.M., Zhu, J.: Surface-catalyzed C-C covalent coupling strategies toward the synthesis of low-dimensional carbon-based nanostructures. Acc. Chem. Res. **48**, 2484–2494 (2015). https://doi.org/10.1021/acs.accounts.5b00168
8. Dil, H., Lobo-Checa, J., Laskowski, R., Blaha, P., Berner, S., Osterwalder, J., et al.: Surface trapping of atoms and molecules with dipole rings. Science **319**, 1824–1826 (2008). https://doi.org/10.1126/science.1154179
9. Schulz, F., Drost, R., Hämäläinen, S.K., Liljeroth, P.: Templated self-assembly and local doping of molecules on epitaxial hexagonal boron nitride. ACS Nano **7**, 11121–11128 (2013). https://doi.org/10.1021/nn404840h
10. Joshi, S., Bischoff, F., Koitz, R., Écija, D., Seufert, K., Seitsonen, A.P., et al.: Control of molecular organization and energy level alignment by an electronically nanopatterned boron nitride template. ACS Nano **8**, 430–442 (2013). https://doi.org/10.1021/nn406024m
11. Erler, P., Schmitt, P., Barth, N., Irmler, A., Bouvron, S., Huhn, T., et al.: Highly ordered surface self-assembly of Fe_4 single molecule magnets. Nano Lett. **15**, 4546–4552 (2015). https://doi.org/10.1021/acs.nanolett.5b01120
12. Liu, L., Dienel, T., Widmer, R., Gröning, O.: Interplay between energy-level position and charging effect of manganese phthalocyanines on an atomically thin insulator. ACS Nano **9**, 10125–10132 (2015). https://doi.org/10.1021/acsnano.5b03741

13. Dienel, T., Gomez-Diaz, J., Seitsonen, A.P., Widmer, R., Iannuzzi, M., Radican, K., et al.: Dehalogenation and coupling of a polycyclic hydrocarbon on an atomically thin insulator. ACS Nano **8**, 6571–6579 (2014). https://doi.org/10.1021/nn501906w
14. Morchutt, C., Björk, J., Krotzky, S., Gutzler, R., Kern, K.: Covalent coupling via dehalogenation on Ni(111) supported boron nitride and graphene. Chem. Commun. **51**, 2440–2443 (2015). https://doi.org/10.1039/c4cc07107g
15. Zhao, W., Dong, L., Huang, C., Win, Z.M., Lin, N.: Cu- and Pd-catalyzed Ullmann reaction on a hexagonal boron nitride layer. Chem. Commun. **52**, 13225–13228 (2016). https://doi.org/10.1039/C6CC05029H
16. Sonogashira, K., Tohda, Y., Hagihara, N.: Convenient synthesis of acetylenes: catalytic substitutions of acetylenic hydrogen with bromoalkenes, iodoarenes and bromopyridines. Tetrahedron Lett. **16**, 4467–4470 (1975). https://doi.org/10.1016/S0040-4039(00)91094-3
17. Sonogashira, K.: Development of Pd–Cu catalyzed cross-coupling of terminal acetylenes with sp^2-carbon halides. J. Organomet. Chem. **653**, 46–49 (2002). https://doi.org/10.1016/S0022-328X(02)01158-0
18. Kanuru, V.K., Kyriakou, G., Beaumont, S.K., Papageorgiou, A.C., Watson, D.J., Lambert, R. M.: Sonogashira coupling on an extended gold surface in vacuo: reaction of phenylacetylene with iodobenzene on Au(111). J. Am. Chem. Soc. **132**, 8081–8086 (2010). https://doi.org/10.1021/ja1011542
19. Sánchez-Sánchez, C., Yubero, F., González-Elipe, A.R., Feria, L., Sanz, J.F., Lambert, R.M.: The Flexible Surface Revisited: Adsorbate-Induced Reconstruction, Homocoupling, and Sonogashira Cross-Coupling on the Au(100) Surface. J. Phys. Chem. C **118**, 11677–11684 (2014). https://doi.org/10.1021/jp501321u
20. Sánchez-Sánchez, C., Orozco, N., Holgado, J.P., Beaumont, S.K., Kyriakou, G., Watson, D. J., et al.: Sonogashira cross-coupling and homocoupling on a silver surface: chlorobenzene and phenylacetylene on Ag(100). J. Am. Chem. Soc. **137**, 940–947 (2015). https://doi.org/10.1021/ja5115584
21. Zhang, R., Lyu, G., Li, D.Y., Liu, P.N., Lin, N.: Template-controlled sonogashira cross-coupling reactions on a Au(111) surface. Chem. Commun. **53**, 1731–1734 (2017). https://doi.org/10.1039/c6cc10091k
22. Adisoejoso, J., Lin, T., Shang, X.S., Shi, K.J., Gupta, A., Liu, P.N., et al.: A single-molecule-level mechanistic study of Pd-catalyzed and Cu-catalyzed homocoupling of aryl bromide on an Au(111) surface. Chem. Eur. J. **20**, 4111–4116 (2014). https://doi.org/10.1002/chem.201304443
23. Wang, W., Shi, X., Wang, S., Van Hove, M.A., Lin, N.: Single-molecule resolution of an organometallic intermediate in a surface-supported Ullmann coupling reaction. J. Am. Chem. Soc. **133**, 13264–13267 (2011). https://doi.org/10.1021/ja204956b
24. Yin, L., Liebscher, J.: carbon–carbon coupling reactions catalyzed by heterogeneous palladium catalysts. Chem. Rev. **107**, 133–173 (2007). https://doi.org/10.1021/cr0505674
25. Molnar, A.: Efficient, selective, and recyclable palladium catalysts in carbon-carbon coupling reactions. Chem. Rev. **111**, 2251–2320 (2011). https://doi.org/10.1021/cr100355b
26. Grill, L., Dyer, M., Lafferentz, L., Persson, M., Peters, M.V., Hecht, S.: Nano-architectures by covalent assembly of molecular building blocks. Nat. Nanotech. **2**, 687–691 (2007). https://doi.org/10.1038/nnano.2007.346
27. Lin, T., Shang, X.S., Adisoejoso, J., Liu, P.N., Lin, N.: Steering on-surface polymerization with metal-directed template. J. Am. Chem. Soc. **135**, 3576–3582 (2013). https://doi.org/10.1021/ja311890n
28. Cai, J., Ruffieux, P., Jaafar, R., Bieri, M., Braun, T., Blankenburg, S., et al.: Atomically precise bottom-up fabrication of graphene nanoribbons. Nature **466**, 470–473 (2010). https://doi.org/10.1038/nature09211
29. Lafferentz, L., Eberhardt, V., Dri, C., Africh, C., Comelli, G., Esch, F., et al.: Controlling on-surface polymerization by hierarchical and substrate-directed growth. Nat. Chem. **4**, 215–220 (2012). https://doi.org/10.1038/NCHEM.1242

30. Meyers, J.M., Gellman, A.J.: Effect of substituents on the phenyl coupling reaction on Cu (111). Surf. Sci. **337**, 40–50 (1995). https://doi.org/10.1016/0039-6028(95)00528-5
31. Li, Y., Lin, N.: Combined scanning tunneling microscopy and kinetic Monte Carlo study on kinetics of Cu-coordinated pyridyl-porphyrin supramolecular self-assembly on a Au(111) surface. Phys. Rev. B **84**, 125418 (2011). https://doi.org/10.1103/PhysRevB.84.125418
32. Auwärter, W., Kreutz, T.J., Greber, T., Osterwalder, J.: XPD and STM investigation of hexagonal boron nitride on Ni(111). Surf. Sci. **429**, 229–236 (1999). https://doi.org/10.1016/S0039-6028(99)00381-7
33. Walch, H., Gutzler, R., Sirtl, T., Eder, G., Lackinger, M.: Material- and orientation-dependent reactivity for heterogeneously catalyzed carbon–bromine bond homolysis. J. Phys. Chem. C **114**, 12604–12609 (2010). https://doi.org/10.1021/jp102704q
34. Gutzler, R., Walch, H., Eder, G., Kloft, S., Heckl, W.M., Lackinger, M.: Surface mediated synthesis of 2D covalent organic frameworks: 1,3,5-tris(4-bromophenyl)benzene on graphite (001), Cu (111), and Ag (110). Chem. Commun. **0**, 4456–4458(2009). https://doi.org/10.1039/b906836h
35. Hoffmann, M., Wilson, C.J., Odell, B., Anderson, H.L.: Template-directed synthesis of a π-conjugated porphyrin nanoring. Angew. Chem. Int. Ed. Engl. **46**, 3122–3125 (2007). https://doi.org/10.1002/anie.200604601
36. O'Sullivan, M.C., Sprafke, J.K., Kondratuk, D.V., Rinfray, C., Claridge, T.D.W., Saywell, A., et al.: Vernier templating and synthesis of a 12-porphyrin nano-ring. Nature **469**, 72–75 (2011). https://doi.org/10.1038/nature09683
37. Favereau, L., Cnossen, A., Kelber, J.B., Gong, J.Q., Oetterli, R.M., Cremers, J., et al.: Six-coordinate zinc porphyrins for template-directed synthesis of spiro-fused nanorings. J. Am. Chem. Soc. **137**, 14256–14259 (2015). https://doi.org/10.1021/jacs.5b10126
38. Adisoejoso, J., Li, Y., Liu, J., Liu, P.N., Lin, N.: Two-dimensional metallo-supramolecular polymerization: toward size-controlled multi-strand polymers. J. Am. Chem. Soc. **134**, 18526–18529 (2012). https://doi.org/10.1021/ja308480x
39. Gao, H.-Y., Wagner, H., Zhong, D., Franke, J.-H., Studer, A., Fuchs, H.: Glaser coupling at metal surfaces. Angew. Chem. Int. Ed. **52**, 4024–4028 (2013). https://doi.org/10.1002/anie.201208597
40. Klappenberger, F., Zhang, Y.-Q., Björk, J., Klyatskay, S., Ruben, M., Barth, J.V.: On-surface synthesis of carbon-based scaffolds and nanomaterials using terminal alkynes. Acc. Chem. Res. **48**, 2140–2150 (2015). https://doi.org/10.1021/acs.accounts.5b00174
41. Zhou, H., Liu, J., Du, S., Zhang, L., Li, G., Zhang, Y., Tang, B.Z., Gao, H.-J.: Direct visualization of surface-assisted two-dimensional diyne polycyclotrimerization. J. Am. Chem. Soc. **136**, 5567–5570 (2014). https://doi.org/10.1021/ja501308s
42. Cirera, B., Zhang, Y.-Q., Björk, J., Klyatskaya, S., Chen, Z., Ruben, M., Barth, J.V., Klappenberger, F.: Synthesis of extended graphdiyne wires by vicinal surface templating. Nano Lett. **14**, 1891–1897 (2014). https://doi.org/10.1021/nl4046747
43. Cirera, B., Zhang, Y.-Q., Klyatskaya, S., Ruben, M., Klappenberger, F., Barth, J.V.: 2D self-assembly and catalytic homo-coupling of the terminal alkyne 1,4-Bis (3,5-diethynyl-phenyl)butadiyne-1,3 on Ag(111). ChemCatChem **5**, 3281–3288 (2013). https://doi.org/10.1002/cctc.201300299

Addressing Long-Standing Chemical Challenges by AFM with Functionalized Tips

Diego Peña, Niko Pavliček, Bruno Schuler, Nikolaj Moll, Dolores Pérez, Enrique Guitián, Gerhard Meyer and Leo Gross

Abstract In this chapter, we illustrate the great potential of combining organic synthesis with atomic resolution AFM and STM to address relevant and classic issues in chemistry, by summarizing selected examples in which we were involved in recent years. As case studies, the long-standing chemical challenges covered here include the experimental discrimination of bond orders in single molecules, the characterization of extremely insoluble compounds such as nanographenes, the analysis of the individual components of complex mixtures, and the on-surface generation and identification of highly reactive molecules and intermediates.

1 Introduction

Undoubtedly, scanning probe microscopy (SPM) is one of the most relevant tools to study surfaces at the nanoscale. First, the introduction of scanning tunneling microscopy (STM) [1], followed by the discovery of atomic force microscopy (AFM) [2], have contributed extensively to the development of diverse scientific fields. In the last decade, SPM has demonstrated to be the crucial technique in the

D. Peña (✉) · D. Pérez · E. Guitián
Centro Singular de Investigación en Química Biolóxica e Materiais Moleculares (CiQUS), Departamento de Química Orgánica, Universidade de Santiago de Compostela, 15782 Santiago de Compostela, Spain
e-mail: diego.pena@usc.es

N. Pavliček · B. Schuler · N. Moll · G. Meyer · L. Gross
IBM Research–Zurich, 8803 Rüschlikon, Switzerland
e-mail: LGR@zurich.ibm.com

Present Address:
B. Schuler
Molecular Foundry, Lawrence Berkeley National Laboratory, California 94720, USA

Present Address:
N. Pavliček
ABB Corporate Research, 5405 Baden-Dättwil, Switzerland

D. G. de Oteyza and C. Rogero (eds.), *On-Surface Synthesis II*,
Advances in Atom and Single Molecule Machines,
https://doi.org/10.1007/978-3-319-75810-7_10

emerging field of on-surface synthesis, to identify and to characterize products resulting from the thermally-induced polymerization of selected monomers on different surfaces [3–8]. On-surface synthesis is an ideal arena for fruitful scientific collaborations between organic chemists and surface physicists who are specialists in SPM: the former design and synthesize suitable molecules in solution while the latter use these molecules as building blocks for on-surface polymerizations (Fig. 1). In a seminal experiment [9], 10,10′-dibromo-9,9′-bianthracene was heated on a Au(111) surface to obtain an armchair graphene nanoribbon (7-AGNR) by a sequence of Ullmann couplings followed by cyclodehydrogenation reactions (Fig. 2). Since then, new dibromo-substituted polycyclic aromatic compounds have been designed and provided as "à la carte" building blocks to produce GNRs with well-defined widths, edges, and substitution to control the properties and functionality of the final nanostructure. This successful approach is extensively discussed in other chapters of this volume, as well as in some recent reviews [10, 11].

Another interesting perspective in this collaborative scenario, which is frequently performed in parallel to the previously mentioned approach, is the topic of this chapter: to take advantage of the clues captured by SPM to answer long-standing chemical questions which are difficult to address with more conventional techniques. In this respect, it has been crucial to achieve sub-molecular resolution in SPM of compounds studied on-surface, a recent breakthrough enabled by the introduction of functionalized tips in AFM under ultrahigh vacuum (UHV) conditions at cryogenic temperatures (5 K) [12]. This finding has provided a privileged methodology to image single molecules with unprecedented resolution and has stimulated a plethora of exciting results in the following years, including the characterization of natural products [13, 14], the conformational analysis of diverse molecules [15–17], the identification of intermediates and products of on-surface reactions [18–26], the study of GNR derivatives [27–31] and the characterization of metal-organic complexes on-surface [32, 33], among other examples [34, 35]. In this chapter we discuss the powerful combination of organic synthesis and atomic resolution AFM to address relevant issues in chemistry, by summarizing selected projects in which the authors, chemists from CIQUS and physicists from IBM Research–Zurich, have been involved in recent years.

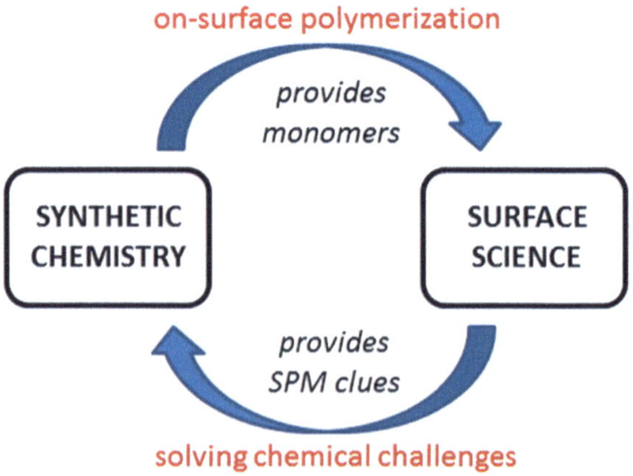

Fig. 1 Interrelation between synthetic chemistry and surface science

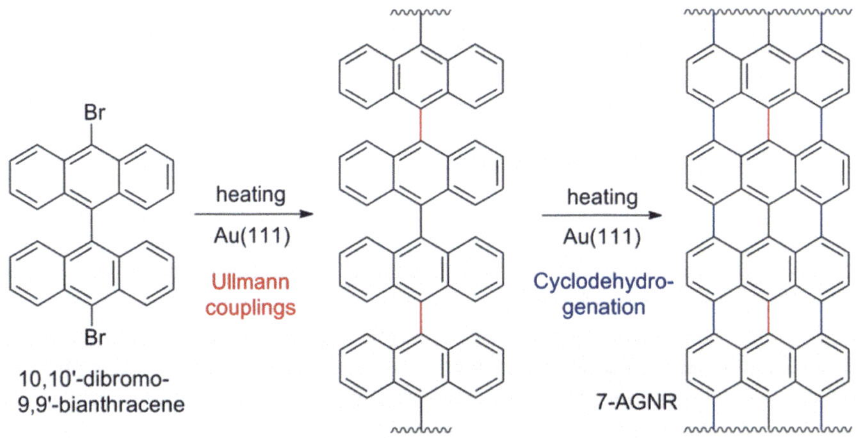

Fig. 2 On-surface preparation of 7-AGNR [9]

2 Experimental Details

The contrast above bonds and atoms in AFM images of molecules arises from repulsive force contributions due to Pauli repulsion [12, 36]. To reach this regime, in which the electron distributions of the molecule and the tip apex start to overlap, the tip is scanned parallel to the surface at a vertical distance of about 4 Å to the molecule. Metallic tip apices are too reactive to reach such small tip-molecule separations. Therefore, it is crucial to terminate (functionalize) the tip apex with an inert front atom or molecule [12]. Most frequently, CO is used, but a wide variety of different terminations has been explored, ranging from single atoms like Cl [12], Br [37], Xe, [37], or O [38] to larger naphthalenetetracarboxylic diimide (NTCDI) molecules [39]. CO-tips are most popular because of their stability, ease of preparation, the localized probing electron density that protrudes from the oxygen atom and the flexibility of the CO at the tip.

Other prerequisites are small oscillation amplitudes of a few tens of picometers to increase the sensitivity to the short-range contributions responsible for the contrast on bonds and atoms [40]. To this end, piezoelectric quartz sensors are operated in frequency-modulation mode [41]. That is, the sensor oscillates with a fixed amplitude at its resonance frequency. The measurement signal is the shift in that resonance frequency, Δf, due to the tip-sample interaction. In the limit of small oscillation amplitudes, Δf is proportional to the vertical gradient of the interaction force. When the frequency shift is recorded as a function of distance, the force can then be calculated numerically [42].

While all experiments presented in this chapter were carried out at temperatures of about T = 5 K using qPlus sensors oscillating at about 30 kHz [43], other works use length extension sensors oscillating at 1 MHz [44, 45]. However, resolution on molecules has also been demonstrated at higher temperatures, at 77 K [39], and even at room temperature using a Si cantilever [46].

3 Pauling Bond-Order Analysis

In 1935, Linus Pauling introduced the concept of bond order to explain the double-bond character found in molecules with single-double bond resonances [47]. Since then, the bond-order analysis has been widely used in chemistry to predict bond lengths and reactivity in a simple manner by drawing the different resonance structures of a molecule. For example, naphthalene has 11 C–C bonds, which can be classified in four kinds of bonds (Fig. 3) and three different resonance structures (**1a, 1b, 1c**). Focusing on the C1–C2 bond (in red, which is equivalent to C3–C4, C5–C6 or C7–C8), this bond can be drawn as a double bond in structures **1a** and **1b**, but as a single bond in structure **1c**, leading to a 2/3 of double-bond character by the Pauling bond-order analysis. By contrast, the C2–C3 bond (in green, equivalent to C6–C7) can only be drawn as a double bond in the structure **1c**, that means 1/3 of double-bond character. Since greater double-bond character implies shorter C–C bond distance, this simple analysis suggests C1–C2 to be shorter than C2–C3. This prediction was experimentally proved by X-ray diffraction analysis, which shows that the C1–C2 is shorter than C2–C3 (1.36 Å vs. 1.40 Å) [48]. However, diffraction techniques are based on averaged values over a large set of molecules. A tool to analyze bonds within individual molecules with the necessary precision was lacking.

In 2012, it was shown that AFM with CO-terminated tips can discriminate bond orders of organic compounds with high precision [49], based on two effects: the frequency shift of the bonds and the apparent length in constant-height AFM images. The first one is directly related to the different electron densities: the electron density increases with the bond order and greater electron densities give rise to greater repulsive force contributions, which lead to increased frequency shifts. The second effect is related to the tilting of the CO tip which magnifies the

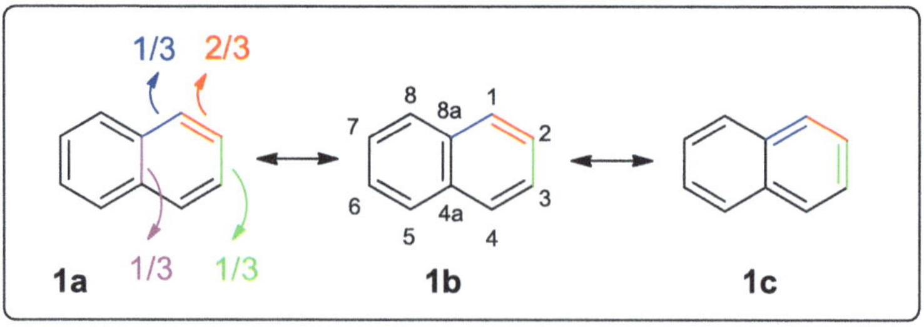

Fig. 3 Double-bond character in the four kinds of C–C bonds of naphthalene. The three main resonance structures for naphthalene are **1a**, **1b** and **1c**

apparent bond length differences and provides an additional mechanism to discriminate bond orders. This analysis can be difficult at the edge of the molecules or with non-planar adsorption geometries. On the contrary, bond-order discrimination by AFM is particularly accurate in internal C–C bonds of flat molecules. For example, Fig. 4 shows this analysis for dibenzo[*cd,n*]naphtho[3,2,1,8-*pqra*]perylene (DBNP, Fig. 4a), a polycyclic aromatic hydrocarbon (PAH) which was prepared in solution by a sequence of domino Diels-Alder cycloadditions, followed by a reduction and a cyclodehydrogenation reaction [50]. Interestingly, DBNP has five different C–C bonds in the central part of the molecule (*q*, *r*, *s*, *t* and *u*) with calculated bond orders varying from 0.163 (bond *t*) to 0.490 (bond *r*). Indeed, AFM analysis confirmed that bond *r* has the greatest bond order out of these five bonds, by showing the largest frequency shift (Fig. 4b, d) and the smallest apparent length (Fig. 4c, e) [49]. As we will discuss later, bond-order analysis was found to be particularly useful in the characterization of organic intermediates such as *ortho*-arynes [51] as well as the study of oligoacetylene chains [19].

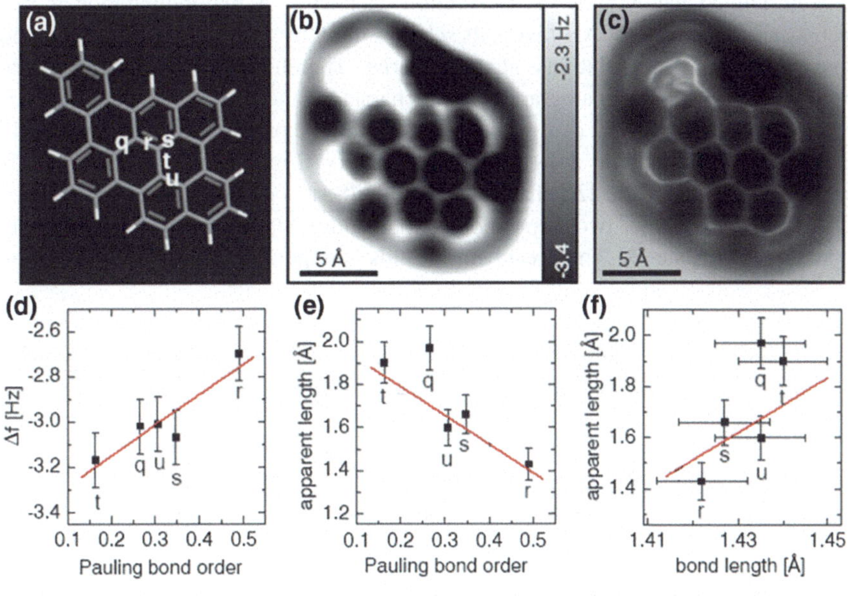

Fig. 4 Bond-order analysis for dibenzo [*cd,n*] naphtho [3,2,1,8-*pqra*]perylene (DBNP). **a** Structure of DBNP. **b** Constant height AFM measurements on bilayer NaCl on Cu(111) at z = 3.6 Å (oscillation amplitude = 0.5 Å). **c** Pseudo-3D representation of (**b**). **d** Measured values of the frequency shift as a function of the Pauling bond order. **e** Measured values of the apparent bond length as a function of the Pauling bond order. **f** Apparent bond length as a function of the bond length obtained by DFT calculations. Linear regressions (in red) are drawn as a guide to the eye. Adapted from [49]. Reprinted with permission from AAAS

4 Characterization of Insoluble Compounds: Nanographenes

Organic chemists have a deep experience in the manipulation and identification of soluble molecules. Even most nano-sized compounds can be fully characterized by modern spectroscopic techniques, in particular by NMR, if the molecules are soluble enough [52]. On the contrary, the characterization of insoluble compounds is a great challenge for researchers working in an organic chemistry laboratory. This is often the case when preparing large unsubstituted polycyclic aromatic hydrocarbons (PAHs). Interestingly, these molecules are not only ubiquitous environmental pollutants [53, 54] and the largest molecules ever detected in space [55], but PAHs are also promising compounds for organic electronics which have attracted a great deal of attention in material science [56]. Moreover, since a piece of graphene is nothing but a large PAH, the synthesis of large PAHs has undergone an important revival in recent years [57–61]. Synthetic methodologies that were traditionally used for decades for the preparation of PAHs are now employed in the preparation of well-defined nanographenes by solution chemistry. A major limitation is that increasing the size of a PAH favors its self-assembly by means of π-interaction and thereby reducing the solubility of the material. Therefore, the purification of flat nanographenes by chromatography or its characterization by NMR in solution is not possible. This was the case when we attempted the preparation of the three-fold symmetric nanographene **2** by solution chemistry (Fig. 5) [62]. This is a PAH formed by the fusion of 22 benzene rings with 78 sp^2 carbon atoms. We designed an extremely simple two-steps/one-pot procedure starting with commercially available bistriflate **3**. The treatment of this compound with CsF in solution

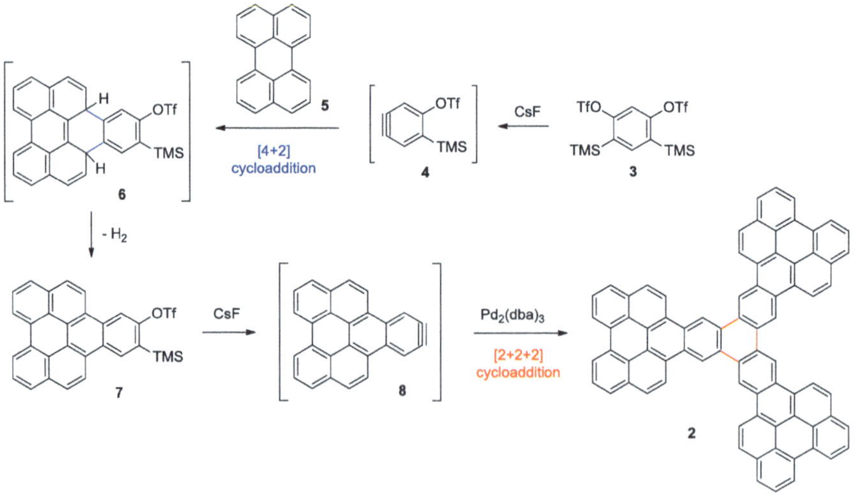

Fig. 5 Synthetic route to obtain three-fold symmetric nanographene 2 by solution chemistry [62]

generated the intermediate aryne **4**. This aryne reacted with one of the bay regions of perylene (**5**) by means of a [4+2] cycloaddition (Diels-Alder reaction) to afford adduct **6**, which evolved in the reaction mixture to form triflate **7**. In a second step, which was performed in the same reaction flask, the treatment of compound **7** in solution with CsF in the presence of a catalytic amount of a palladium complex led to the formation of a highly insoluble grayish solid. Based on our experience on aryne chemistry [61], we expected this solid to be nanographene 2, resulting from the Pd-catalyzed [2+2+2] cycloaddition of the in-situ formed aryne **8**. In fact, the mass spectrum of the solid showed a molecular ion at m/z = 972.2 consistent with the molecular formula expected for this compound ($C_{78}H_{36}$). However, the extreme insolubility of this compound precluded structural confirmation by NMR spectroscopy.

Inspired by the successful use of AFM for the identification of natural products [13], we attempted to confirm the structure of nanographene **2** analogously. Compound **2** was expected to be planar, an important advantage for the accurate characterization by atomic resolution AFM. However, since purification by chromatography was not possible, the solid isolated from the reaction was far from being a pure compound. In addition, the large size of the molecule could complicate its sublimation in order to deposit the molecule on the surface. Despite these limitations, the solid was sublimated by rapid heating from a Si wafer onto a Cu(111) sample partially covered with islands of NaCl held at $T = 10$ K. Satisfactorily, the analysis by AFM with CO-functionalized tip confirmed the structure of nanographene **2** (Fig. 6a). STM was also carried out to obtain molecular orbital images. The image of the negative ion resonance (Fig. 6b) corresponds to the lowest unoccupied molecular orbital (LUMO) as seen by comparison with the orbital density obtained by DFT calculation (Fig. 6c). This agreement is an additional confirmation of the molecular structure, therefore the combination of AFM and STM proved the successful synthesis of compound **2**. Interestingly, careful inspection of other adsorbates by AFM discovered the presence of a tetranaphthoheptacene, an unexpected compound probably formed by the head-to-tail dimerization of aryne **8** [62]. This finding gave

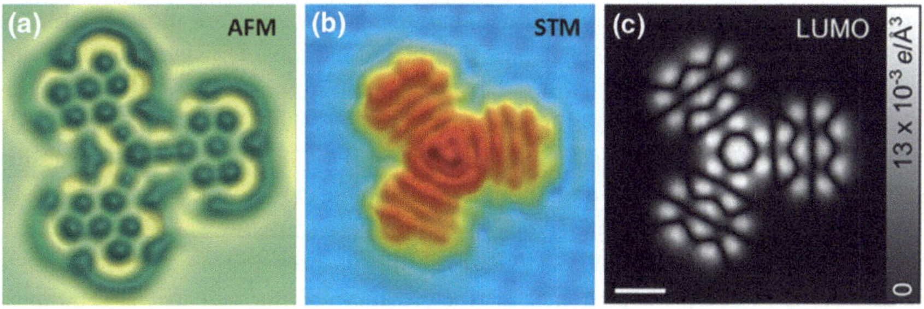

Fig. 6 On-surface identification of nanographene **2**. **a** AFM image. **b** STM orbital image on bilayer NaCl on Cu(111) using a CO tip. **c** Calculated orbital density of the LUMO at a distance of 1.2 Å above the molecular plane. The scale bar is 5 Å. Adapted from [62]. Reprinted with permission from Wiley

additional value to the analysis of reaction mixtures by AFM, where the on-surface identification of byproducts provided key clues to understand the reaction of aryne intermediates in solution.

5 Characterization of Complex Mixtures: Asphaltenes

Usually, SPM studies are performed with ultrapure compounds, since the presence of small amounts of impurities can complicate both the preparation and the study of the sample. However, high resolution SPM can also be extremely useful to analyze complex mixtures by providing the molecular characterization of individual constituents, as it was demonstrated in 2015 by means of asphaltene characterization with atomic resolution AFM [63]. Asphaltenes are the solid components of crude oil, which consist of an extremely complex mixture of polyaromatic molecules with high economic relevance in oil processing [64]. This chemical complexity has limited the structural analysis of asphaltenes by conventional techniques and consequently their molecular architecture has been subject to debates. For example, some studies suggest that the asphaltene molecular structure is based on a single PAH core (*island model*) while other studies point to a structure with multiple PAH cores (*archipelago model*). Notably, the structure elucidation of single molecules of asphaltenes by atomic resolution AFM indicate that the *island*-type architecture is clearly predominant (Fig. 7) [63]. As expected, AFM analysis showed the enormous molecular diversity of this fraction. Orbital imaging with STM contributed to the structural assignment of the polyaromatic molecules, which in some cases have sophisticated molecular architectures (e.g. 18-ring nanographene CA6). These findings suggest the exciting possibility of using asphaltenes as an unlimited source of new specimens for single-molecule electronics. Many different molecules can be screened by SPM in a single preparation without going through a lengthy chemical synthesis, purification, and characterization for each one of them beforehand.

Obviously, this AFM-based method is not limited to analyze asphaltenes but also related carbon-based aromatic mixtures, such as petroleum and coal vacuum residues, heavy fractions, tar, pitch, bitumen, etc. SPM techniques are most suitable for flat molecules composed mainly by sp^2 hybridized carbons. However, some of the compounds constituting these carbon-based aromatic mixtures present alkyl chains with sp^3 hybridized carbons and non-planar fragments, which are difficult to study by AFM and complicate the structure elucidation. For this reason, the preparation and AFM characterization of model compounds featuring aliphatic moieties can be extremely useful to identify molecules in carbon-based mixtures [17].

For example, compound **9** (CHNP, Fig. 8) was prepared as a cycloaliphatic model compound for asphaltene characterization, with a partially saturated six-membered ring with four sp^3 hybridized carbons. AFM analysis on Cu and NaCl surfaces showed that this cyclohexene moiety adopted the expected chiral half-chair conformation (Fig. 9). In fact it was possible to identify both enantiomers (**9a** and **9b**), which were configurationally stable at cryogenic temperatures.

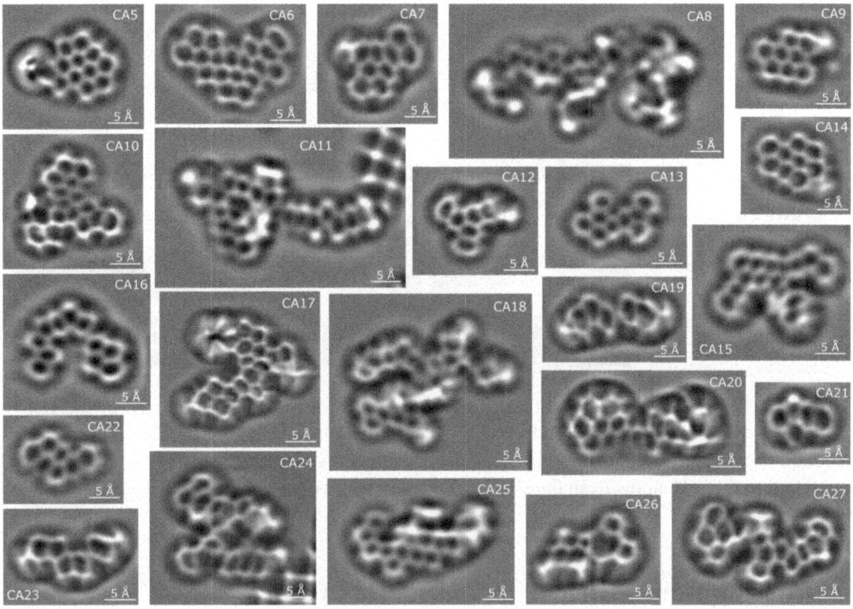

Fig. 7 CO-tip AFM images of individual molecules of a coal-derived asphaltene. Reprinted with permission from Ref. [63]. Copyright 2015 American Chemical Society

Demonstrating the ultimate control of SPM manipulation over on-surface deposited single molecules, reversible enantiomerization (chiral switching) between **9a** and **9b** was induced by applying a STM voltage pulse of 0.4 V (Fig. 9f, g). In addition, in a remarkable example of an on-surface oxidation reaction, it was possible to abstract four hydrogen atoms from CHNP by voltage pulses between 3.1 and 3.8 V, to obtain PAH **10** (Figs. 8 and 9h) [17]. To sum up, these experiments illustrate the powerful of on-surface atomic manipulation by SPM, not only to provide key fingerprints of both aromatic and aliphatic moieties for structural elucidation of natural samples, but also to study basic chemical transformations and even to perform conformational analysis of organic molecules on surface.

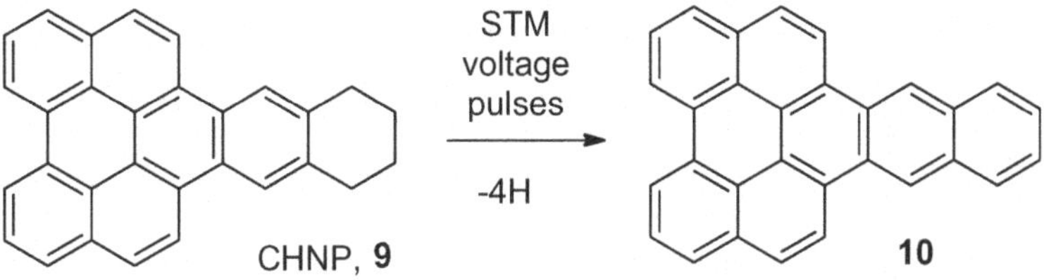

Fig. 8 Structure of 10,11,12,13-tetrahydroanthra[1,2,3,4-*ghi*]perylene (CHNP, **9**) used as cycloaliphatic model compound and product **10** resulting from the on-surface STM tip-induced dehydrogenation [17]

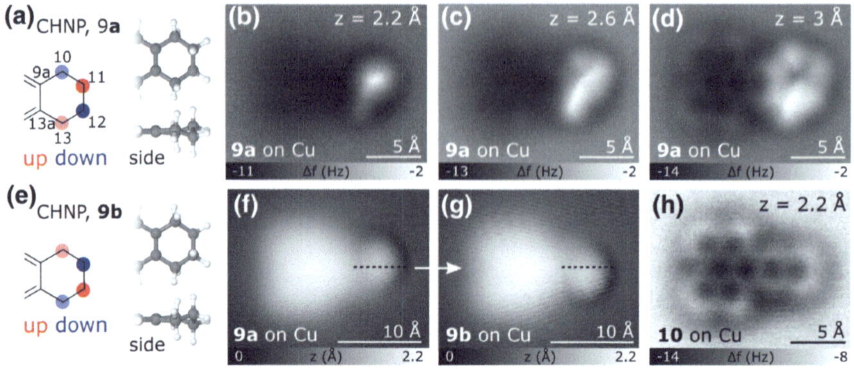

Fig. 9 AFM and STM characterization of CHNP, **9**. **a**, **e** Enantiomers **9a** and **9b** of the CHNP structure. **b–d** CO tip AFM images of **9a** on Cu(111) at different tip heights. **f**, **g** CO tip STM image of **9a** before (**f**) and **9b** after (**g**) enantiomerization by means of a voltage pulse of 0.4 V. **h** Molecule **10** after removing four H atoms from the aliphatic ring of **9** with voltage pulses between 3.1 and 3.8 V. Adapted from [17]

6 Investigation of Elusive Molecules: Acenes

Acenes are PAHs formed by the linear fusion of several benzene rings, which have been extensively used in organic electronics due to their outstanding semiconducting properties [65]. However, the synthesis and manipulation of acenes under ambient conditions is demanding due to low solubility and high reactivity, in particular with increasing acene length. Qualitative estimation on the stability of PAHs can be deduced by application of the Clar's sextet rule, grouping the π electrons of the PAH into sextets (i.e. three conjugated double bonds within a six-membered ring) [66]. These sextets provide an aromatic stabilization to the molecule, such that PAH with only few Clar sextets are particularly reactive under ambient conditions. Acenes have only one sextet, regardless of the length of the molecule, which explains their instability. Therefore, the synthesis of large acenes has fascinated organic chemists for decades and remains a long-standing synthetic challenge [56]. Up to now, the largest unsubstituted acene ever prepared was nonacene, which was photogenerated and detected in an Ar matrix at cryogenic temperatures [67]. On-surface generation of large acenes under UHV conditions from soluble and stable precursors offers the possibility to circumvent the major difficulties of solution-based acene preparation. This concept has been recently demonstrated with the on-surface reduction of diepoxytetracenes **11** to form tetracene (**13**) on Cu(111), through the formation of intermediate **12** (Fig. 10) [68]. It was shown that this transformation can be achieved either by thermal activation or by STM tip-induced manipulation, where the Cu–O interaction is crucial to achieve the deoxygenation of the starting epoxyacenes. This method was latter employed for the on-surface generation of larger acenes, such as the preparation of hexacene

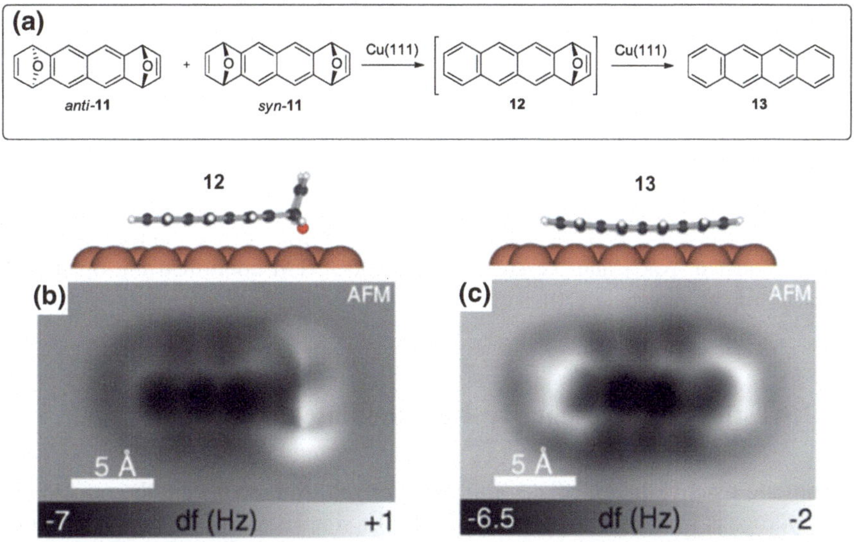

Fig. 10 On-surface generation of tetracene (**13**). **a** Reaction route to obtain tetracene (**13**) from diepoxytetracenes *syn/anti*-**11**. **b** Constant-height AFM image of the epoxytetracene **12** intermediate after the first oxygen detachment. **c** Constant-height AFM image of the resulting tetracene (**13**) after complete deoxygenation. Adapted with permission from [68]. Copyright 2016 American Chemical Society

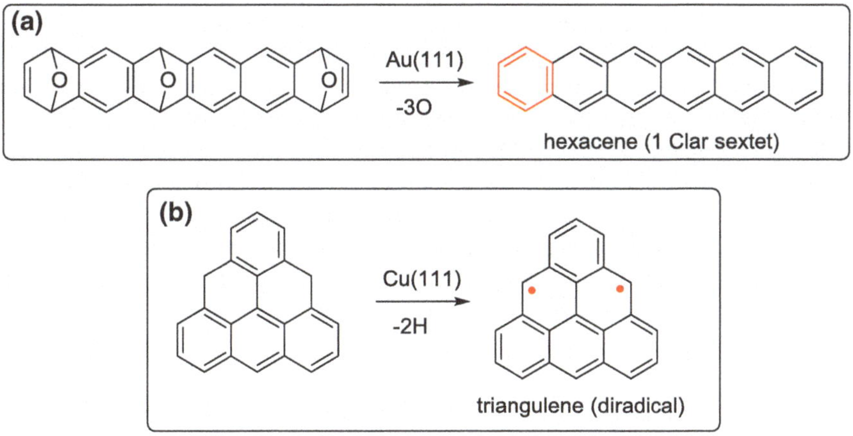

Fig. 11 On-surface generation of hexacene (**a**) [69] and triangulene (**b**) [70]

on Au(111) by deoxygenation of triepoxyhexacenes (Fig. 11a) [69]. As a result, it was possible to study the electronic structure of single hexacene molecules with intramolecular resolution.

A recent breakthrough on the investigation of elusive PAHs was the generation and characterization of triangulene by combined STM/AFM [70]. Triangulene is formed by six benzene rings fused in a triangle shape, so it is not possible to draw a Kekulé resonance structure and triangulene presents two unpaired electrons (Fig. 11b). Therefore, the molecule is a π-diradical which is unstable under ambient conditions, so attempts to generate triangulene by solution chemistry were unsuccessful [71]. Finally, atom manipulation has allowed the preparation of this three-fold symmetric PAH for the first time by means of tip-induced dehydrogenation on Cu, NaCl and Xe surfaces. Notably, STM confirmed the open-shell character of the molecule [70].

7 Investigation of Reactive Intermediates: Arynes

Arynes are reactive intermediates formally derived from arenes by the removal of two hydrogens from one benzene ring [72]. The simplest arynes are benzynes (one single benzene ring), which are classified depending on the relative position of the two unpaired electrons: *ortho* (1,2), *meta* (1,3) or *para* (1,4) (Fig. 12). Although arynes were first suggested in 1902 [73], the first evidence of the intermediacy of benzyne in a chemical reaction was established in 1953 by Roberts and coworkers in a seminal isotopic labeling experiment [74]. Arynes are so reactive that they only exist for several milliseconds in solution, making them extremely difficult to characterize. However, structural data of arynes were obtained from UV, IR, microwave and NMR spectroscopies, either in gas phase, using matrix-isolation techniques or in solution after encapsulation inside a hemicarcerand [75], but the interpretation of these data have been the matter of debate. In particular, there is a long-standing discussion about the contribution of diradical **14a**, alkyne **14b** and cumulene **14c** resonance forms to the real structure of *ortho*-arynes (Fig. 12a).

Recently, combined STM/AFM with functionalized tips allowed the on-surface generation and visualization of an aryne (10,11-didehydronaphtho[1,2,3,4-*g,h,i*]perylene) [51]. The aryne was generated by tip induced cleavage of the two vicinal C–I bonds of 10,11-diiodonaphtho[1,2,3,4-*g,h,i*]perylene (DINP, Fig. 13). To this aim, after positioning the tip above DINP, a voltage pulse (V > 1.6 V) was applied. The analysis by atomically resolved AFM using a CO terminated tip, showed the

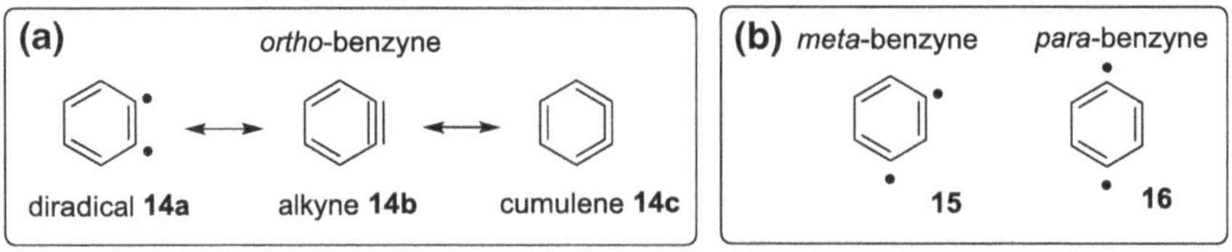

Fig. 12 Structure of benzynes, the simplest arynes: *ortho*-benzyne (**a**), *meta*- and *para*-benzyne (**b**)

Fig. 13 On-surface generation of an *ortho*-aryne (10,11-didehydronaphtho[1,2,3,4-*g,h,i*]perylene) from 10,11-diiodonaphtho[1,2,3,4-*g,h,i*]perylene (DINP). Images with atomic resolution were obtained by non-contact AFM using a CO tip. Adapted from [51] with permission from Nature Publishing Group

dissociation of both C–I bonds and the formation of the corresponding aryne. To proof that the formed specie retains its arynic nature and reactivity at cryogenic temperature, an additional experiment was performed. After lateral manipulation to place two iodine atoms close to the aryne ring, a voltage pulse was applied causing the reconstruction of the intact DINP molecule, in contrast to the reactivity of arynes in solution. In addition, the extended planar polyaromatic core facilitated the bond-order analysis of this aryne, following the methodology described in detail in Sect. 2, with the results shown in Fig. 14. The bond order evaluation, focused on the central bonds marked on Fig. 14a, pointed out a dominant contribution of the cumulene resonance structure under these conditions, in good agreement with DFT calculations [51], contributing to clarifying a classic debate in aryne chemistry [72].

Another type of closely related intermediates are *para*-arynes (Fig. 12) [72], σ-diradical species which can be involved in relevant rearrangement reactions such as the Bergman cyclization. This reaction was first proposed in 1972 to explain the thermal isomerization of enediynes [76], and it became an important transformation in synthetic chemistry when enediyne anticancer antibiotics were introduced in the 80s [77]. Recently, a reversible Bergman cyclization on NaCl was induced by atomic manipulation and studied by combined AFM/STM [78]. Starting with dibromoantharacene (DBA) a sequence of STM voltage pulses induced the cleavage of the two C-Br bonds to generate a *para*-diradical, which evolved to a

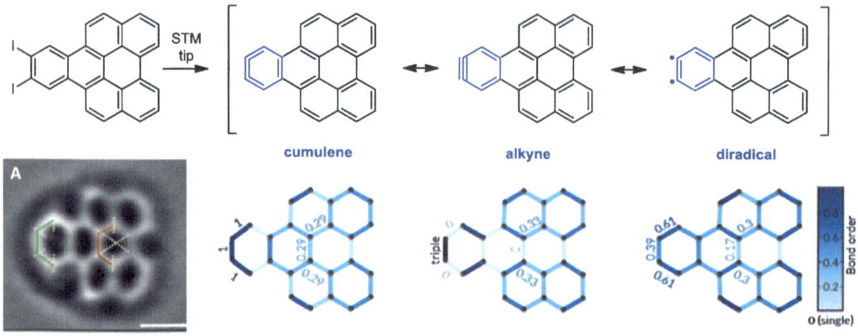

Fig. 14 Bond-order analysis of the three major resonance structures of an *ortho*-aryne. (A) Laplace-filtered representation of the AFM image of an individual aryne. Adapted from [51] with permission from Nature Publishing Group

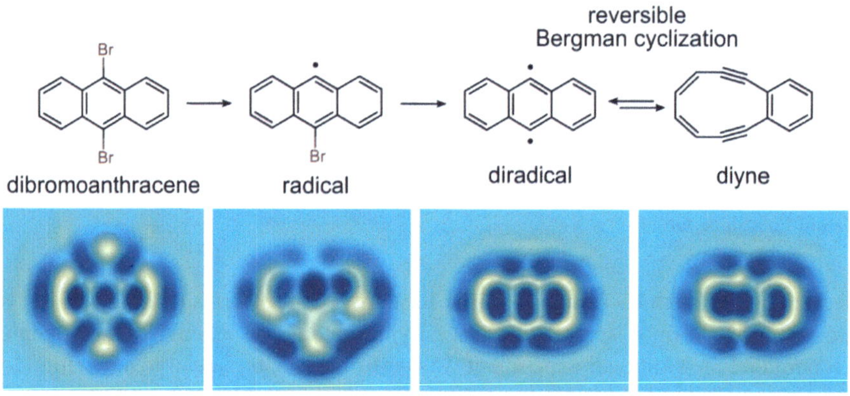

Fig. 15 Reversible Bergman cyclization. STM voltage pulses induce the sequential dissociation of two bromine atoms from dibromoanthracene (DBA) to generate a *para*-diradical, which evolved to a diyne by a retro-Bergman cyclization. The latter transformation can be reversibly triggered by tunneling electrons with a minimum energy of 1.6 V. All measurements were performed on bilayer NaCl on Cu(111). Adapted from [78] with permission from Nature Publishing Group

strained 10-membered ring diyne by means of a retro-Bergman reaction (Fig. 15). Notably, the strained diyne could be transformed back into the diradical, demonstrating switching on demand between two intermediates with different reactivity, electronic and magnetic properties.

8 Conclusions

Atomic resolution AFM and STM by means of functionalized tips is in its infancy, but has already demonstrated an impressive potential regarding to the visualization and manipulation of single atoms and molecules. We are convinced that classic issues in organic chemistry will be revisited using the combination of AFM/STM, mostly to confirm the thesis of chemistry pioneers such as Pauling [47], Clar [66, 71], or Roberts [74] as shown in this chapter, but also to discover new molecules, intriguing molecular behavior or unexpected reactions. In fact, the combination of organic chemistry and AFM/STM with sub-molecular resolution is an ideal serendipity playground which is expected to provide exiting findings in the near future. More than ever, we are closer to the inspiring prediction by Richard P. Feynman, masterfully outlined in his famous talk *Plenty of Room at the Bottom* in 1959:

> …it would be, in principle, possible (I think) for a physicist to synthesize any chemical substance that the chemist writes down. Give the orders and the physicist synthesizes it. How? Put the atoms down where the chemist says, and so you make the substance. The problems of chemistry and biology can be greatly helped if our ability to see what we are doing, and to do things on an atomic level, is ultimately developed— a development which I think cannot be avoided.

Acknowledgements We are deeply grateful to all our coworkers. We acknowledge financial support from the ERC Grants CEMAS (agreement no. 291194) and AMSEL (682144), the EU project PAMS (610446), the Agencia Estatal de Investigación (MAT2016-78293-C6-3-R and CTQ2016-78157-R), the Xunta de Galicia (Centro singular de investigación de Galicia accreditation 2016–2019, ED431G/09) and the European Regional Development Fund (ERDF).

References

1. Binnig, G., Rohrer, H., Gerber, C., Weibel, E.: Tunneling through a controllable vacuum gap. Appl. Phys. Lett. **40**, 178–180 (1982)
2. Binnig, G., Quate, C.F., Gerber, C.: Atomic force microscope. Phys. Rev. Lett. **56**, 930–933 (1986)
3. Grill, L., Dyer, M., Lafferentz, L., Persson, M., Peters, M.V., Hecht, S.: Nano-architectures by covalent assembly of molecular building blocks. Nat. Nanotechnol. **2**, 687–691 (2007)
4. Franc, G., Gourdon, A.: Covalent networks through on-surface chemistry in ultra-high vacuum: state-of-the-art and recent developments. Phys. Chem. Chem. Phys. **13**, 14283–14292 (2011)
5. Klappenberger, F., Zhang, Y.-Q., Björk, J., Klyatskaya, S., Ruben, M., Barth, J.V.: On-surface synthesis of carbon-based scaffolds and nanomaterials using terminal alkynes. Acc. Chem. Res. **48**, 2140–2150 (2015)
6. Lindner, R., Kühnle, A.: On-surface reactions. Chem. Phys. Chem. **16**, 1582–1592 (2015)
7. Méndez, J., Francisca López, M., Martín-Gago, J.A.: On-surface synthesis of cyclic organic molecules. Chem. Soc. Rev. **40**, 4578–4590 (2011)
8. Nacci, C., Hecht, S., Grill, L.: The emergence of covalent on-surface polymerization. In: On-Surface Synthesis, S. 1–21. Springer (2016)

9. Cai, J., Ruffieux, P., Jaafar, R., Bieri, M., Braun, T., Blankenburg, S., Muoth, M., Seitsonen, A.P., Saleh, M., Feng, X., Müllen, K., Fasel, R.: Atomically precise bottom-up fabrication of graphene nanoribbons. Nature **466**, 470–473 (2010)
10. Narita, A., Feng, X., Müllen, K.: Bottom-up synthesis of chemically precise graphene nanoribbons. Chem. Rec. **15**, 295–309 (2015)
11. Talirz, L., Ruffieux, P., Fasel, R.: On-surface synthesis of atomically precise graphene nanoribbons. Adv. Mater. **28**, 6222–6231 (2016)
12. Gross, L., Mohn, F., Moll, N., Liljeroth, P., Meyer, G.: The chemical structure of a molecule resolved by atomic force microscopy. Science **325**, 1110–1114 (2009)
13. Gross, L., Mohn, F., Moll, N., Meyer, G., Ebel, R., Abdel-Mageed, W.M., Jaspars, M.: Organic structure determination using atomic resolution scanning probe microscopy. Nat. Chem. **2**, 821–825 (2010)
14. Hanssen, K.O., Schuler, B., Williams, A., Demissie, T.B., Hansen, E., Andersen, J.H., Svenson, J., Blinov, K., Repisky, M., Mohn, F., Meyer, G., Svendsen, J.-S., Ruud, R., Elyashberg, M., Gross, L., Jaspars, M., Isaksson, J.: A combined atomic force microscopy and computational approach for structural elucidation of breitfussin A and B, highly modified halogenated dipeptides from the Arctic hydrozoan *Thuiaria breitfussi*. Angew. Chem. Int. Ed. **51**, 12238–12241 (2012)
15. Pavliček, N., Fleury, B., Neu, M., Niedenführ, J., Herranz-Lancho, C., Ruben, M., Repp, J.: Atomic force microscopy reveals bistable configurations of dibenzo[a, h]thianthrene and their interconversion pathway. Phys. Rev. Lett. **108**, 086101 (2012)
16. Albrecht, F., Bischoff, F., Auwärter, W., Barth, J.V., Repp, J.: Direct identification and determination of conformational response in adsorbed individual nonplanar molecular species using noncontact atomic force microscopy. Nano Lett. **16**, 7703–7709 (2016)
17. Schuler, B., Zhang, Y., Collazos, S., Fatayer, S., Meyer, G., Pérez, D., Guitián, E., Harper, M. R., Kushnerick, J.D., Peña, D., Gross, L.: Characterizing aliphatic moieties in hydrocarbons with atomic force microscopy. Chem. Sci. **8**, 2315–2320 (2017)
18. de Oteyza, D.G., Gorman, P., Chen, Y.-C., Wickenburg, S., Riss, A., Mowbray, D.J., Etkin, G., Pedramrazi, Z., Tsai, H.-Z., Rubio, A., Crommie, M.F., Fischer, F.R.: Direct imaging of covalent bond structure in single-molecule chemical reactions. Science **340**, 1434–1437 (2013)
19. Riss, A., Wickenburg, S., Gorman, P., Tan, L.Z., Tsai, H.-Z., de Oteyza, D.G., Chen, Y.-C., Bradley, A.J., Ugeda, M.M., Etkin, G., Louie, S.G., Fischer, F.R., Crommie, M.F.: Local electronic and chemical structure of oligo-acetylene derivatives formed through radical cyclizations at a surface. Nano Lett. **14**, 2251–2255 (2014)
20. Rogers, C., Chen, C., Pedramrazi, Z., Omrani, A.A., Tsai, H.-Z., Jung, H.S., Lin, S., Crommie, M.F., Fischer, F.R.: Closing the nanographene gap: surface-assisted synthesis of peripentacene from 6,6′–bipentacene precursors. Angew. Chem. Int. Ed. **54**, 15143–15146 (2015)
21. Albrecht, F., Pavliček, N., Herranz-Lancho, C., Ruben, M., Repp, J.: Characterization of a surface reaction by means of atomic force microscopy. J. Am. Chem. Soc. **137**, 7424–7428 (2015)
22. Kawai, S., Haapasilta, V., Lindner, B.D., Tahara, K., Spijker, P., Buitendijk, J.A., Pawlak, R., Meier, T., Tobe, Y., Foster, A.S., Meyer, E.: Thermal control of sequential on-surface transformation of a hydrocarbon molecule on a copper surface. Nat. Commun. **7**, 12711 (2016)
23. He, Y., Garnica, M., Bischoff, F., Ducke, J., Bocquet, M.-L., Batzill, M., Auwärter, W., Barth, J.V.: Fusing tetrapyrroles to graphene edges by surface-assisted covalent coupling. Nat. Chem. **9**, 33–38 (2017)
24. Kocic, N., Liu, X., Chen, S., Decurtins, S., Krejcí, O., Jelínek, P., Repp, J., Liu, S.-X.: Control of reactivity and regioselectivity for on-surface dehydrogenative aryl-aryl bond formation. J. Am. Chem. Soc. **138**, 5585–5593 (2016)
25. Riss, A., Paz, A.P., Wickenburg, S., Tsai, H.-Z., de Oteyza, D.G., Bradley, A.J., Ugeda, M. M., Gorman, P., Jung, H.S., Crommie, M.F., Rubio, A., Fischer, F.R.: Imaging

single-molecule reaction intermediates stabilized by surface dissipation and entropy. Nat. Chem. **8**, 678–683 (2016)
26. Stetsovych, O., Švec, M., Vacek, J., Chocholoušová, J.V., Jancark, A., Rybácek, J., Kosmider, K., Stará, I.G., Jelnek, P., Stary, I.: From helical to planar chirality by on-surface chemistry. Nat. Chem. **9**, 213–218 (2017)
27. van der Lit, J., Boneschanscher, M.P., Vanmaekelbergh, D., Ijäs, M., Uppstu, A., Ervasti, M., Harju, A., Liljeroth, P., Swart, I.: Suppression of electron–vibron coupling in graphene nanoribbons contacted via a single atom. Nat. Commun. **4**, 2023 (2013)
28. Ruffieux, P., Wang, S., Yang, B., Sánchez-Sánchez, C., Liu, J., Dienel, T., Talirz, L., Shinde, P., Pignedoli, C.A., Passerone, D., Dumslaff, T., Feng, X., Müllen, K., Fasel, R.: On-surface synthesis of graphene nanoribbons with zigzag edge topology. Nature **531**, 489–492 (2016)
29. Schulz, F., Jacobse, P.H., Canova, F.F., van der Lit, J., Gao, D.Z., van den Hoogenband, A., Han, P., Klein Gebbink, R.J.M., Moret, M.-E., Joensuu, P.M., Swart, I., Liljeroth, P.: Precursor geometry determines the growth mechanism in graphene nanoribbons. J. Phys. Chem. C **121**, 2896–2904 (2017)
30. Kawai, S., Saito, S., Osumi, S., Yamaguchi, S., Foster, A.S., Spijker, P., Meyer, E.: Atomically controlled substitutional boron-doping of graphene nanoribbons. Nat. Commun. **6**, 8098 (2015)
31. Dienel, T., Kawai, S., Söde, H., Feng, X., Müllen, K., Ruffieux, P., Fasel, R., Gröning, O.: Resolving atomic connectivity in graphene nanostructure junctions. Nano Lett. **15**, 5185–5190 (2015)
32. Mohn, F., Repp, J., Gross, L., Meyer, G., Dyer, M.S., Persson, M.: Reversible bond formation in a gold-atom–organic-molecule complex as a molecular switch. Phys. Rev. Lett. **105**, 266102 (2010)
33. Albrecht, F., Neu, M., Quest, C., Swart, I., Repp, J.: Formation and characterization of a molecule–metal–molecule bridge in real space. J. Am. Chem. Soc. **135**, 9200–9203 (2013)
34. Gross, L., Schuler, B., Mohn, F., Moll, N., Repp, J., Meyer, G.: Atomic resolution on molecules with functionalized tips. In: Morita, S., Giessibl, F.J., Meyer, E., Wiesendanger, R. (eds.) Noncontact Atomic Force Microscopy, S. 223–246. Springer (2015)
35. Pavliček, N., Gross, L.: Generation, manipulation and characterization of molecules by atomic force microscopy. Nat. Rev. Chem. **1**, 0005 (2017)
36. Moll, N., Gross, L., Mohn, F., Curioni, A., Meyer, G.: A simple model of molecular imaging with noncontact atomic force microscopy. New J. Phys. **14**, 083023 (2012)
37. Mohn, F., Schuler, B., Gross, L., Meyer, G.: Different tips for high-resolution atomic force microscopy and scanning tunneling microscopy of single molecules. Appl. Phys. Lett. **102**, 073109 (2013)
38. Mönig, H., Hermoso, D.R., Arado, O.D., Todorović, M., Timmer, A., Schüer, S., Langewisch, G., Pérez, R., Fuchs, H.: Submolecular imaging by noncontact atomic force microscopy with an oxygen atom rigidly connected to a metallic probe. ACS Nano **10**, 1201–1209 (2015)
39. Sweetman, A.M., Jarvis, S.P., Sang, H., Lekkas, I., Rahe, P., Wang, Y., Wang, J., Champness, N.R., Kantorovich, L.N., Moriarty, P.: Mapping the force field of a hydrogen-bonded assembly. Nat. Commun. **5**, 3931 (2014)
40. Giessibl, F.J.: Advances in atomic force microscopy. Rev. Mod. Phys. **75**, 949–983 (2003)
41. Albrecht, T.R., Grütter, P., Horne, D., Rugar, D.: Frequency modulation detection using high-Q cantilevers for enhanced force microscope sensitivity. J. Appl. Phys. **69**, 668–673 (1991)
42. Sader, J.E., Jarvis, S.P.: Accurate formulas for interaction force and energy in frequency modulation force spectroscopy. Appl. Phys. Lett. **84**, 1801–1803 (2004)
43. Giessibl, F.J.: High-speed force sensor for force microscopy and profilometry utilizing a quartz tuning fork. Appl. Phys. Lett. **73**, 3956–3958 (1999)
44. Hapala, P., Švec, M., Stetsovych, O., van der Heijden, N.J., Ondráček, M., van der Lit, J., Mutombo, P., Swart, I., Jelínek, P.: Mapping the electrostatic force field of single molecules from high-resolution scanning probe images. Nat. Commun. **7**, 11560 (2016)

45. Stetsovych, O., Švec, M., Vacek, J., Chocholoušová, J.V., Jančařík, A., Rybáček, J., Kosmider, K., Stará, I.G., Jelínek, P., Starý, I.: From helical to planar chirality by on-surface chemistry. Nat. Chem. **9**, 213–218 (2017)
46. Iwata, K., Yamazaki, S., Mutombo, P., Hapala, P., Ondráček, M., Jelínek, P., Sugimoto, Y.: Chemical structure imaging of a single molecule by atomic force microscopy at room temperature. Nat. Commun. **6**, 7766 (2015)
47. Pauling, L., Brockway, L.O., Beach, J.Y.: The dependence of interatomic distance on single bond-double bond resonance. J. Am. Chem. Soc. **57**, 2705–2709 (1935)
48. Abrahams, S.C., Robertson, J.M., White, J.G.: The crystal and molecular structure of naphthalene. Acta Cryst. **2**, 238–244 (1949)
49. Gross, L., Mohn, F., Moll, N., Schuler, B., Criado, A., Guitián, E., Peña, D., Gourdon, A., Meyer, G.: Bond-order discrimination by atomic force microscopy. Science **337**, 1326–1329 (2012)
50. Criado, A., Peña, D., Cobas, A., Guitián, E.: Domino Diels-Alder cycloadditions of arynes: new approach to elusive perylene derivatives. Chem. Eur. J. **16**, 9736–9740 (2010)
51. Pavliček, N., Schuler, B., Collazos, S., Moll, N., Pérez, D., Guitián, E., Meyer, G., Peña, D., Gross, L.: On-surface generation and imaging of arynes by atomic force microscopy. Nat. Chem. **7**, 623–628 (2015)
52. Silverstein, R.M., Webster, F.X., Kiemle, D.J. Bryce, D. L.: Spectrometric Identification of Organic Compounds, 8th edn. Wiley (2014)
53. Harvey, R.G.: Polycyclic Aromatic Hydrocarbons. Wiley, New York (1997)
54. Lunch, A.: The Carcinogenic Effects of Polycyclic Aromatic Hydrocarbons. Imperial College Press, London (2005)
55. Tielens, A.G.M.: Interstellar polycyclic aromatic hydrocarbon molecules. Ann. Rev. Astron. Astrophys. **46**, 289–337 (2008)
56. Anthony, J.E.: The larger acenes: versatile organic semiconductors. Angew. Chem. Int. Ed. **47**, 452–483 (2008)
57. Peña, D.: Ideas in Chemistry and Molecular Sciences: Advances in Synthetic Chemistry, pp. 237–261. Wiley, Weinheim (2010)
58. Chen, L., Hernandez, Y., Feng, X., Müllen, K.: From nanographene and graphene nanoribbons to graphene sheets: chemical synthesis. Angew. Chem. Int. Ed. **51**, 7640–7654 (2012)
59. Narita, A., Wang, X.-Y., Feng, X., Müllen, K.: New advances in nanographene chemistry. Chem. Soc. Rev. **44**, 6616–6643 (2015)
60. Itami, K.: Toward controlled synthesis of carbon nanotubes and graphenes. Pure Appl. Chem. **84**, 907–916 (2012)
61. Pérez, D., Peña, D., Guitián, E.: Aryne cycloaddition reactions in the synthesis of large polycyclic aromatic compounds. Eur. J. Org. Chem. 5981–6013 (2013)
62. Schuler, B., Collazos, S., Gross, L., Meyer, G., Pérez, D., Guitián, E., Peña, D.: From Perylene to a 22-ring aromatic hydrocarbon in one-pot. Angew. Chem. Int. Ed. **53**, 9004–9006 (2014)
63. Schuler, B., Meyer, G., Peña, D., Mullins, O.C., Gross, L.: Unraveling the molecular structures of asphaltenes by atomic force microscopy. J. Am. Chem. Soc. **137**, 9870 (2015)
64. Mullins, O.C., Sheu, E.Y., Hammami, A., Marshall, A.G.: Asphaltenes, Heavy Oils, and Petroleomics, vol. 1. Springer, New York (2007)
65. Ye, Q., Chi, C.: Recent highlights and perspectives on acene based molecules and materials. Chem. Mater. **26**, 4046–4056 (2014)
66. Clar, E.: The Aromatic Sextet. Wiley, New York (1972)
67. Tönshoff, C., Bettinger, H.F.: Photogeneration of octacene and nonacene. Angew. Chem. Int. Ed. **49**, 4125–4128 (2010)
68. Krüger, J., Pavliček, N., Alonso, J.M., Pérez, D., Guitián, E., Lehmann, T., Cuniberti, G., Gourdon, A., Meyer, G., Gross, L., Moresco, F., Peña, D.: Tetracene formation by on-surface reduction. ACS Nano **10**, 4538–4542 (2016)

69. Krüger, J., Eisenhut, F., Alonso, J.M., Lehmann, T., Guitián, E., Pérez, D., Skidin, D., Gamaleia, F., Ryndyk, D.A., Joachim, C., Peña, D., Moresco, F., Cuniberti, G.: Imaging the electronic structure of on-surface generated hexacene. Chem. Commun. **53**, 1583–1586 (2017)
70. Pavliček, N., Mistry, A., Majzik, Z., Moll, N., Meyer, G., Fox, D.J., Gross, L.: Synthesis and characterization of triangulene. Nat. Nanotechnol. **12**, 308–312 (2017)
71. Clar, E., Stewart, D.: Aromatic hydrocarbons. LXV. Triangulene derivatives 1. J. Am. Chem. Soc. **75**, 2667–2672 (1953)
72. Wenk, H.H., Winkler, M., Sander, W.: One century of aryne chemistry. Angew. Chem. Int. Ed. **42**, 502–528 (2003)
73. Störmer, R., Kahlert, B.: Über das 1- und 2-Brom-cumaron. Ber. Dtsch. Chem. Ges. **35**, 1633–1640 (1902)
74. Roberts, J.D., Simmons, H.E., Carlsmith, L.A., Vaughan, C.W.: Rearrangement in the reaction of chlorobenzene-1-C^{14} with potassium amide. J. Am. Chem. Soc. **75**, 3290–3291 (1953)
75. Warmuth, R.: *o*-Benzyne: strained alkyne or cumulene?—NMR characterization in a molecular container. Angew. Chem. Int. Ed. Engl. **36**, 1347–1350 (1997)
76. Jones, R.R., Bergman, R.G.: *p*-Benzyne. Generation as an intermediate in a thermal isomerization reaction and trapping evidence for the 1,4-benzenediyl structure. J. Am. Chem. Soc. **94**, 660–661 (1972)
77. Nicolaou, K.C., Dai, W.-M., Tsay, S.-C., Estevez, V.A., Wrasidlo, W.: Designed enediynes: a new class of DNA-cleaving molecules with potent and selective anticancer activity. Science **256**, 1172–1178 (1992)
78. Schuler, B., Fatayer, S., Mohn, F., Moll, N., Pavliček, N., Meyer, G., Peña, D., Gross, L.: Reversible Bergman cyclization by atomic manipulation. Nat. Chem. **8**, 220–224 (2016)

The manufacturer's authorised representative in the EU is Springer Nature Customer Service Centre GmbH, Europaplatz 3, 69115 Heidelberg, Germany. If you have any concerns regarding our products, please contact ProductSafety@springernature.com

Printed and bound by CPI Group (UK) Ltd, Croydon, CR0 4YY
05/10/2026
02228141-0002